PROTEIN RESEARCH PROGRESS

ALAN B. BOSCOE
AND
CHARLES R. LISTOW
EDITORS

Nova Science Publishers, Inc.
New York

For permission to use material from this book please contact us:
Telephone 631-231-7269; Fax 631-231-8175
Web Site: http://www.novapublishers.com

LIBRARY OF CONGRESS CATALOGING-IN-PUBLICATION DATA
Available upon request

ISBN 13: 978-1-60021-663-3
ISBN 10: 1-60021-663-3

Published by Nova Science Publishers, Inc. ≈ New York

CONTENTS

PREFACE

Proteins are large organic compounds made of amino acids arranged in a linear chain and joined together by peptide bonds between the carboxyl and amino groups of adjacent amino acid residues. The sequence of amino acids in a protein is defined by a gene and encoded in the genetic code. Although this genetic code specifies 20 "standard" amino acids, the residues in a protein are often chemically altered in post-translational modification: either before the protein can function in the cell, or as part of control mechanisms. Proteins can also work together to achieve a particular function, and they often associate to form stable complexes.

Like other biological macromolecules such as polysaccharides and nucleic acids, proteins are essential parts of organisms and participate in every process within cells. Many proteins are enzymes that catalyze biochemical reactions, and are vital to metabolism. Proteins also have structural or mechanical functions, such as actin and myosin in muscle, and the proteins in the cytoskeleton, which forms a system of scaffolding that maintains cell shape. Other proteins are important in cell signaling, immune responses, cell adhesion, and the cell cycle. Protein is also a necessary part of animals' diets, since they cannot synthesise all the amino acids and must obtain essential amino acids from food. Through the process of digestion, animals break down ingested protein into free amino acids that can be used for protein synthesis. This new book presents the latest research in the field.

Expert Commentary A – The influence of deamidation processes in lactoglobulin preparation on its antioxidative properties was studied in a model system of yolk lipoprotein free radical oxidation induced by Fe^{2+}. To simulate the processes of protein ageing samples of lactoglobulin preparation was incubated under conditions similar to physiological ones (0.85% NaCl, pH 7.0, 35°C) for 7, 14, and 28 days. Samples of recently produced lactoglobulin (0 days of incubation) were used for "control" experiment. The injection of fresh lactoglobulin into the medium of lipoprotein peroxidation led to the decrease in the intensity of total chemiluminescence emission by 33%. We observed a strong downward trend in antioxidative capacity of lactoglobulin that correlated with the decrease in amide group content as well as lactoperoxidase activity in the preparation.

Expert Commentary B – Extensive knowledge about the protein processing has resulted in the paradox caused by the inexplicability of extreme process stages - structure formation of de-novo synthesized molecules and their elimination after the performance of the physiological functions. A huge amount of material about molecular mechanisms of formation and maintenance of proteins in native conformation as a complex of the processes caused by the interaction of initial protein sequence with numerous components of the cellular folding system is saved up. Thus there is an open question - how the correct folding of protein molecules is provided at a huge variety of possible alternatives? The obvious failure of the providing of each protein with a personal chaperone system makes the considered part of the question as though unsolvable. On the other hand, how the eliminable proteins, and the ones that have accomplished their biological functions, or denaturated proteins are recognized? In what way do quite few ubiquitin-ligases E3 differentiate molecules which are subjected to the marking against a background of an unimaginably large number of the conformational conditions among great variety of proteins? Similar question can be set in respect of the immune molecular mechanisms. How the differentiation of a great number of self proteins in the native conformation from denaturated or foreign proteins is provided? In case of biological macromolecules it looks like too many of them owe too much to the few. The classification of the data on structural maintenance of mutual complementarity of protein molecules allows simplifying essentially the problem, making it reduced to the formation and recognition of small superficial microclusters of the amino-acid residues which is necessary and sufficient condition of the protein processing.

Chapter 1 - Protein phosphatase 2A (PP2A) is a large family of serine/threonine phosphatases that are differentially regulated, have restricted substrate specificity, and display tissue- and cell-specific distribution. PP2A enzymes are believed to be heterotrimeric enzymes in vivo. The core enzyme comprises a catalytic C subunit, associated with a scaffolding subunit, called A. Distinct classes of B-type regulatory subunits, termed B, B', B'' and B''', can bind to the (AC) dimer, resulting in formation of a wide variety of heterotrimeric complexes. B-type subunits are in part responsible for PP2A targeting and functional specificity. The C subunit is also subjected to reversible phosphorylation, which results in transient PP2A inactivation, and to carboxyl methylation, which critically modulates PP2A holoenzyme assembly and substrate specificity. A body of evidence has shown that PP2A enzymes are crucial components of signaling pathways that regulate cell growth and proliferation. Indeed, deregulation of PP2A contributes to cell transformation. For instance, okadaic acid, a potent PP2A inhibitor, is a tumor promoter in mice. Mutations in the genes encoding selective PP2A subunits have been found in some cancers. Changes in endogenous PP2A subunit composition and expression may contribute to tumorigenesis. Specific PP2A isoforms also modulate the Wnt signaling pathway that controls cell growth and polarity, and tissue specification. Interestingly, viral proteins from transforming DNA tumor viruses, such as polyomavirus middle T and SV40 small t antigens directly modulate PP2A targeting, activity and substrate specificity. Interaction of SV40 small t with PP2A leads to deregulation of gene transcription, changes in the architecture of the actin and microtubule cytoskeleton, defects in tight junction biogenesis, alterations in the cell cycle, and deregulation of cell growth and survival. In this review, the authors will examine in-depth the multiple roles played by PP2A in the regulation of cell transformation and tumorigenesis.

Chapter 2 - Protein is an important component of adequate nutrition which is essential to promote optimal growth and neurodevelopment. Principles which guide protein nutrition during infancy have not been well established and they are based on limited data from neonates and extrapolation from studies undertaken in animals and older children and adults. Advances in our understanding of protein turnover, synthesis, catabolism, and deposition in the neonatal period and both short- and long-term consequences of early protein deficiency, contribute to the ongoing development of these principles. This chapter explores past, present and future trends in dietary protein research in the neonatal period to help us understand the impact of dietary protein intake in relation to health and chronic disease. We will follow the debate with regards to protein requirements for preterm infants, share our current understanding of the unique relationship between levels of dietary protein, carbohydrates and minerals, and reveal the potential of tailoring dietary protein intake to specific neonatal conditions. We aim to summarize results of noteworthy studies, identify principles, and stimulate ideas about areas for future research.

Chapter 3 - In supervised learning it is commonly assumed that labeled data available for learning a prediction model is a random sample from the underlying distribution. In real-life scenarios, however, this assumption is often violated due to sample selection bias. Consequently, a prediction model learned from such labeled data could be suboptimal and may not generalize well on unseen examples from the same population. In this chapter, we provide a brief review of the sample selection bias problem and existing methods for detecting and correcting such bias. We then describe our contrast classifier framework for addressing this problem and illustrate its effectiveness in several bioinformatics applications related to learning from biased protein structure databases.

Chapter 4 - The present chapter emphasizes the multifunctional properties of milk proteins and the new technologies to obtain biological active proteins and peptides. New approaches have recently been applied in dietary protein science and technology due to improved knowledge regarding the performance of peptides in the modulation of biological systems. Milk proteins have been the object of special attention not only because milk is the first and the most complete food for newborns, but also because it continues to be part of the diet of many adults. Besides the nutrients that are present in milk, this product contains bioactive or biofunctional peptides with effects and activities that resemble those produced by hormones or pharmacological preparations. These peptides result from the hydrolysis of casein and whey proteins during digestion, cheese-making and milk fermentation processes. Reports on immunomodulatory, hypocholesterolemic, hypotensive, antimicrobial, sleep-inducing, epidermal growth factor, transforming growth factor, anticancer, opioid, mineral-binding and bone-formation activities and properties associated with milk proteins have been published in recent literature. Phosphopeptides were found to exert positive effects on the solubility and absorbability of minerals, in addition to being capable of improving bone mineralization. The immunomodulating potential of whey proteins is unquestionable and is in part attributed to their capacity to stimulate glutathione synthesis that promotes the synthesis of immunoglobulins in lymphocytes. Glutathione is necessary for lymphocyte proliferation. Some milk protein hydrolysates carry information concerning the organism from which they were originated. These milk hydrolysates contain opioid peptides having pharmacological similarities to opium (morphine), which, when introduced into the bloodstream, induce an analgesic and sedative effect due to their action on the nervous system. Secondarily, they act on the circulatory system, the passage rate through the digestive tract and the rate of intestinal absorption of foods. Milk whey contains lactoferrin and lactoperoxidase, both of which are known to have antimicrobial and antiviral activity, inhibiting the proliferation of Gram-positive and Gram-negative bacteria, moulds and viruses. Several scientific communications report the inhibition of the hepatitis C virus and H. pylori (bacterium that causes gastritis and peptic ulcers). As a nutritive and tasty food, milk remains a constituent of the human diet after childhood, whether in its original form or as dairy products. Besides, this paper discusses the functional properties of whey proteins, the advantages of their use as food supplements in sports nutrition, as well as the possible benefits of the several growth factors present in whey.

Chapter 5 - Milk is one of the most common causes of food allergies among children under one year of age. No specific therapy exists for this allergy, and thus the only feasible response is to avoid assumption of milk and derived products. Studies conducted on the serum of children with hypersensitivity to milk have shown that caseins are the proteins with the greater allergenic potential. However, in some cases, children have also shown hypersensitivity to the β-lactoglobulines and to the α-lactoalbumines.

When food intolerance is diagnosed in an infant, it is often necessary to impose a period of total parenteral feeding, followed by breast feeding, considered the most correct method of re-feeding; when human milk cannot be given, alternative food sources must be sought.

Clinical studies have demonstrated that donkey milk could substitute breast feeding in infants affected by severe Ig-E mediated milk allergies. In these subjects, donkey milk is not only useful, but also safer than other types of milk. In fact donkey milk composition in lipids (high levels of linoleic and linolenic acid) and proteins (low caseins content) is very close to human milk.

In their studies on donkey milk the authors tried to deepen the knowledge about this product especially regarding the protein fractions. Among caseins, the auhtors identified mainly α_{S1}- and β-caseins; other types of caseins in such as κ- and γ-caseins were not found probably because of their low amount. Lysozyme content in donkey milk resulted to be very high (mean value 1.0 mg/ml) if compared to bovine (traces), caprine (traces) and human milk and varied at different stages of lactation. The high lysozyme content of donkey milk may be responsible of the low bacterial count reported in literature and also makes this milk suitable to prevent intestine infection to infants.

Among seroproteins, β-lactoglobulin and α-lactoalbumin content in donkey milk was 3.7 and 1.8 mg/ml respectively and remained invariable during the different stages of lactation.

The low caseins fraction amount as well as the high lysozyme content of donkey milk associated with its good palatability make this milk a valid alternative to powdered milk and soybean milk for infant affected by dairy cows milk allergy. Nutritional properties of donkey milk have also been compared with sheep milk, one of the best milk used for cheese-making from thousands of years, in order to evaluate the differences between characteristics of protein fractions in three different milks: human, donkey and sheep milk.

Chapter 6 – The increase of intensive production systems in dairy sheep flocks necessitates specifications on feeding strategies and housing structures to optimize milk production and animal welfare, and to control the impact of animal husbandry on the environment. The nutritional and processing properties, and thereby the market value of milk, largely depends on its protein content. The efficiency of dietary nitrogen (N) utilization for milk protein synthesis is rather low in dairy animals so farmers are driven to use high protein level diets for sustaining milk production. High protein level diet can result in a number of deleterious events for animals, farmers, and environment, such as: i) an increase in the milk production cost, ii) an overloading of renal draining activity, iii) an increase of N output to the environment, iv) high levels of ammonia pollution in animal houses, which originates from animals' urine and feces and can be injurious to both livestock and stock persons' health. In particular, an increased N excretion in urine and feces, as a consequence of poor dietary N absorption, can result in high microorganism and ammonia concentrations in animal houses. This requires high ventilation rate for sustaining the performance and health of farmed livestock. In addition, it has been postulated that poor welfare conditions can modify the digestive tract physiology, producing a modification in mobility and nutrient utilization and, as a result, higher amount of incompletely digested material. Based on these considerations, increasing the protein supply to the ewes' mammary gland by reducing total protein degradability in the rumen seems to be a more suitable strategy than increasing the protein level of the diet, because it improves the efficiency of dietary protein utilization by the lactating ewe, especially when specific limiting amino acids are encapsulated to prevent bacterial deamination in the rumen to make them available for absorption in the small intestine. A number of recent studies have addressed the control of micro-climate and of air pollution in animal houses as well as the specification of good management practices aimed at optimizing milk production. Little is known about the influence of protein diet on cheese quality and proteolysis during ripening as well as on the effects of sheep housing conditions on ovine cheese manufactured from raw milk, which is a focus aspect to link flock management and the quality of milk to be processed into dairy products. In this review new achievements on the effects of both feeding treatment and flock management on ovine milk and cheese features are reported.

Chapter 7 - An understanding of the structure and matrix of the whole protein is often vital to understanding its digestive behavior, utilization and availability in both humans and animals. Traditional approach to study dietary protein is to focus on total protein composition and amino acid profile using "wet" chemical analysis as suggested by NRC, called "a chemical approach". However, using the NRC chemical approach to determine energy and nutritive values is not accurate in comparison with biological approach. The fatal shortcoming of this approach is that the NRC chemical approach does not consider food/seed/feed intrinsic structure, matrix and molecular chemical make-up. It only considers total chemical composition. Traditional "wet" chemical analysis can determine total protein composition and amino acid profile, but fails to detect protein inherent structure, molecular chemical make-up and protein matrix due to destruction of feed protein tissues during the processing for chemical analysis. In this chapter, a novel approach is introduced to reveal protein structure, molecular chemical make-up, and protein matrix within intact tissue at cellular and subcellular level with advanced synchrotron-based analytical technology (SFTIRM). This technology provides new concept and methodology for dietary protein research.

In: Protein Research Progress
Editor: A. B. Boscoe and C. R. Listow, pp. 1-7
ISBN 978-1-60456-663-3

Expert Commentary A

CHANGES IN ANTIOXIDATIVE PROPERTIES OF LACTOGLOBULIN PREPARATION INDUCED BY AGE-RELATED DEAMIDATION

Michael Y. Bibov*, Irina A. Sorokina and Alexander I. Lukash
ROSTOV STATE UNIVERSITY, DEPT. OF BIOCHEMISTRY AND MICROBIOLOGY
ROSTOV-ON-DON, BOLSHAYA SADOVAYA, 105, 344006 RUSSIA

ABSTRACT

The influence of deamidation processes in lactoglobulin preparation on its antioxidative properties was studied in a model system of yolk lipoprotein free radical oxidation induced by Fe^{2+}. To simulate the processes of protein ageing samples of lactoglobulin preparation was incubated under conditions similar to physiological ones (0.85% NaCl, pH 7.0, 35°C) for 7, 14, and 28 days. Samples of recently produced lactoglobulin (0 days of incubation) were used for "control" experiment. The injection of fresh lactoglobulin into the medium of lipoprotein peroxidation led to the decrease in the intensity of total chemiluminescence emission by 33%. We observed a strong downward trend in antioxidative capacity of lactoglobulin that correlated with the decrease in amide group content as well as lactoperoxidase activity in the preparation.

Keywords: lactoglobulin, deamidation, chemiluminescence, lactoperoxidase.

INTRODUCTION

Posttranslational modifications are capable to bring about a wide range of structural and functional changes in protein molecules via chemical modifications of amino acid residues included into a polypeptide chain. Therefore, the spontaneous nonenzymatic modifications

* Correspondence: Tel.+7-8632-63-92-96, Fax:+7-8632-64-52-55. E-mail: mbibov@yahoo.com

are considered to be especially important. It is suggested that such modifications can work as an adaptive mechanism evolutionally developed to regulate the lifetime of protein molecules (Robinson A. *et al.,* 1991; Gillery R. *et al.,* 1991; Wright H.T. 1991).

The central place among multitude of nonenzymatic posttranslational modifications belongs to the reaction of deamidation of asparaginyl (Asn) and glutaminyl (Gln) residues (Wright H.T. 1991). It is suggested that Asn and Gln residues incorporated into the polypeptide chain work as a "molecular clock" timing the functional activity of a protein (Robinson A. *et al.,* 1991). To differ from other posttranslational modifications, which result from either enzymatic reaction or the affect of different outer factors, deamidation is a hydrolytic reaction requiring only water for its fulfillment. Both products of this reaction are natural amino acids – aspartate (Asp) and glutamate (Glu). The rate of nonenzymatic deamidation depends not only on primary structure but also on protein conformation and properties of solution surrounding the amide groups (Robinson A. *et al.,* 1991). For example, the rate of amide hydrolysis can be hastened by the increase in temperature and ionic strength of a solution (Robinson A. *et al.,* 1991). The spontaneous cleavage of amide groups was formerly observed in different proteins such as human's serum albumin, immunoglobulins, RNAase, lysozyme and lactoferrin under conditions similar to physiological ones (0.85% NaCl, pH 7.0, 35°C) (Pushkina N.V. 1988; Belizy S. *et al.,* 2001).

The processes of nonenzymatic deamidation and consequent autofragmentation participate in ageing and degradation of protein molecules (Olekhnovich L.P. *et al.,* 1985). Changes in the extent of amidation lead to certain local damages in protein molecules that cause the occurrence of modified proteins with defected biological functions (Pushkina N.V. 1979). This consequently leads to the decrease in specific and nonspecific biological activity of protein molecules. The object of present research was lactoglobulin preparation known as a complex immunobiological system that contains both factors of specific immunity (IgG, IgM, IgA) and nonspecific antimicrobial resistance (lactoperoxidase, xanthinoxidase, superoxide dismutase, lactoferrin) (Bos C. *et al.,* 2000).

The present research was aimed to investigation of the role of posttranslational deamidation process in changing of antioxidative properties of lactoglobulin preparation.

Materials and Methods

Lactoglobulin preparation was obtained from fresh cow's whey using tangential flow ultrafiltration technology.

The dynamics of lactoglobulin deamidation was studied during incubation at 37°C in neutral saline containing 5 mg/ml protein and 0.85% NaCl for 7, 14 and 28 days under sterile conditions provided with 0.22 μ bacterial filters (Schleicher Schull, Germany). On the 7th, 14th, and 28th day of incubation the proteins of lactoglobulin preparation were precipitated by trichloracetic acid (TCA), centrifuged at 3000 g for 15 min and the obtained pellet was twice washed centrifugically in icy cold acetone and diethyl ether. Finally, the obtained pellet was dried at 37°C and used for quantitative determination of total amide group (TAG) content as well as easily and poorly hydrolysable amide fractions (EHA and PHA, respectively) that released ammonia within 15 min and 3 h of hydrolysis in 0.5 H_2SO_4 at 100°C, respectively

(Pushkina N.V. *et al.,* 1976). The concentration of ammonia released after hydrolysis of amide groups was determined fluorometrically according to the method described by Sugawara K. *et al* (1981) and expressed in μmole ammoniacal nitrogen per g protein, the latter was quantitatively determined by the method of Lowry O.H. *et al* (1951). Spectrofluorimetric assay was carried out using a spectrofluorimeter Fluorat-02 (Russia) (λ_{ex}=413 nm, λ_{em}=476 nm). The content of PHA was calculated as the difference between TAG and EHA. These amide fractions, especially EHA and PHA approximately correspond to the contents of asparagine and glutamine in the protein (Pushkina N.V. 1988).

The extent of destruction of whey proteins was estimated by the increase in the medium weight molecule (MWM) amount that was determined spectrophotometrically (Bobrov V.M. *et al.,* 1999). The ratio of light absorption with the wavelengths 280 nm (A_{280}) and 254 nm (A_{254}) was used s a measure of MWM content. The ratio A_{280}/A_{254} was called the "distributional index" of MWM.

Antioxidative activity of lactoglobulin was determined by its effect on Fe^{2+}-indused chemiluminescence of yolk lipoproteins from hen's egg (Sherstnev M.P. *et al.,* 1989). Chemiluminescence was studied using a regular type chemiluminometer – 0.03 (Russia). The yolk lipoprotein suspension was prepared by dilution of hen's egg yolk with distilled water 1 : 10. The yolk suspension (400μl), 6 ml of 25 mM potassium phosphate buffer (pH 7.7, 37°C), and 200 μl of lactoglobulin solution (5 mg/ml protein) were placed into a thermostatted measuring vessel. The mixture was kept at 37°C with stirring for 100 sec. and the background chemiluminescence was recorded. Then the system was supplemented with 1 ml of 25 mM $FeSO_4$ and the development of chemiluminescence was recorded. The antioxidant effect of lactoglobulin was evaluated via total intensity of chemiluminescence emission during the analysis (3 min). The suspension of lipoproteins without lactoglobulin addition was used for the "Control Experiment".

The activity of lactoperoxidase was measured spectrophotometrically as described by Alperovich D.V. *et al.* (1990).

Statistical analysis was carried out by one-way and repeated measures ANOVA with the Student-Newman-Keuls post hoc test (Winer B.J. *et al.,* 1991). All data presented in this work are mean values of 8 – 10 determinations. To evaluate the extent of correlation between different parameters the correlation coefficients were calculated. The data were considered to be reliable providing that $p<0.05$.

RESULTS AND DISCUSSION

Spontaneous deamidation of lactoglobulin during its incubation was observed as a gradual decrease in the content of amide groups. The data presented in the table quantitatively demonstrate the dynamics of changes in the content of amide groups during the 7, 14 and 28-day incubation under simulated physiological conditions. The content of amide groups in the lactoglobulin of the "Control Group" was about 733.5±2.8 μmole/g protein, where 276±3.1 μmole/g protein was referred to EHA and 457.5±4.7 μmole/g protein – to PHA.

Table 1. Changes in content of amide groups in proteins of lactoglobulin preparation during incubation

Day of incubation	Amide group content (M±m μM of ammonia nitrogen per g protein)		
	TAG	EHA	PHA
0	733.5±2.8	276±3.1	457.5±4.7
7	510.5±4.3 $p_1<0.001$	123.1±2.0 $p_1<0.001$	387.4±3.4 $p_1<0.001$
14	465.9±3.5 $p_1<0.001$ $p_2<0.001$	106±1.9 $p_1<0.001$ $p_2<0.001$	360±3.6 $p_1<0.001$ $p_2>0.05$
28	346.6±5.4 $p_1<0.001$ $p_2<0.001$	79.4±3.2 $p_1<0.001$ $p_2<0.001$	267.2±3.5 $p_1<0.001$ $p_2<0.001$

p_1 – changes reliable with respect to 0 days of incubation.
p_2 – changes reliable with respect to the previous analysis.

During ageing of lactoglobulin the spontaneous deamidation was taking place that was followed by gradual decrease in the content of amide groups. After the first week of incubation the total content of amide groups in proteins of lactoglobulin preparation was decreased by 30%. Within fortnight it fell by 37%, and by the 28th day of incubation this decrease had reached the peak of 53% compared to the initial values. The rate of EHA cleavage definitely exceeded that of PHA. After 7, 14, and 28 days of incubation the content of EHA was 55%, 62% and 71% lower than in "fresh" preparation, whereas PHA level was decreased only by 15%, 21% and 42%, respectively.

The data of amide group content negatively correlated with the values of "distributional index" of MWM, which by the 7th, 14th and 28th days of incubation had been increased by 64%, 75% and 81% (Figure 1). The correlation coefficient between these parameters was about -0.85 ($p<0.001$).

Therefore, it appears quite obvious that nonenzymatic deamidation was followed by certain damages of protein structure that consequently could not leave its biological properties unaffected. It was formerly reported that such proteins as catalase, lactoperoxidase, and human's serum albumin were steadily loosing their antioxidative properties during deamidation (Umanskaya N.Y. 1995).

The influence of spontaneous deamidation on antioxidative properties of lactoglobulin preparation was studied in model system of free radical lipid peroxidation in hen's yolk lipoproteins, and measurement of total chemiluminescence emission was used as a criterion of the intensity of its proceeding (Figure 2). The injection of "fresh" lactoglobulin into the reaction mixture led to decrease in the intensity of total emission by 33% compared to the "Control Experiment". After the incubation for 7, 14 and 28 days the sharp decrease in antioxidative properties of lactoglobulin preparation was observed. The supplementation with lactoglobulin incubated even for 7 days demonstrated no reliable effect on total chemiluminescence emission that meant total loss of its antioxidative properties.

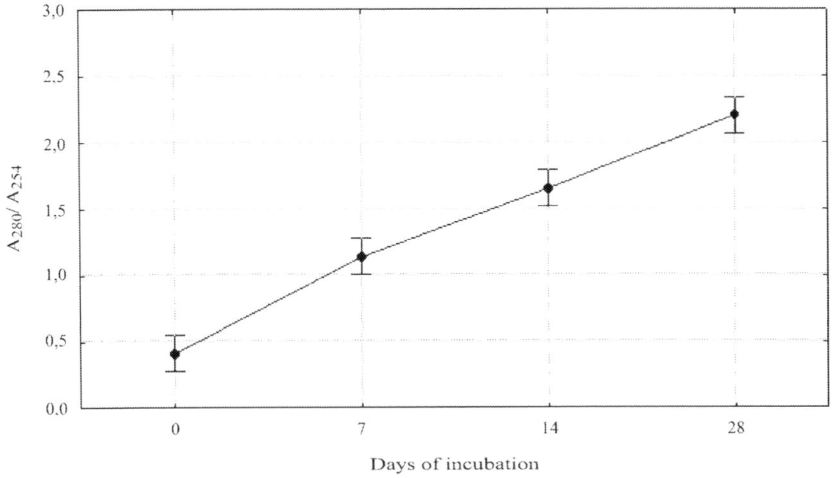

Current effect: F(3,36)=128.40, p<0.001.
Effective hypothesis decomposition.
Vertical bars denote 0.95 confidence intervals.

Figure 1. Changes in the content of medium weight molecules in lactoglobulin preparation during its incubation.

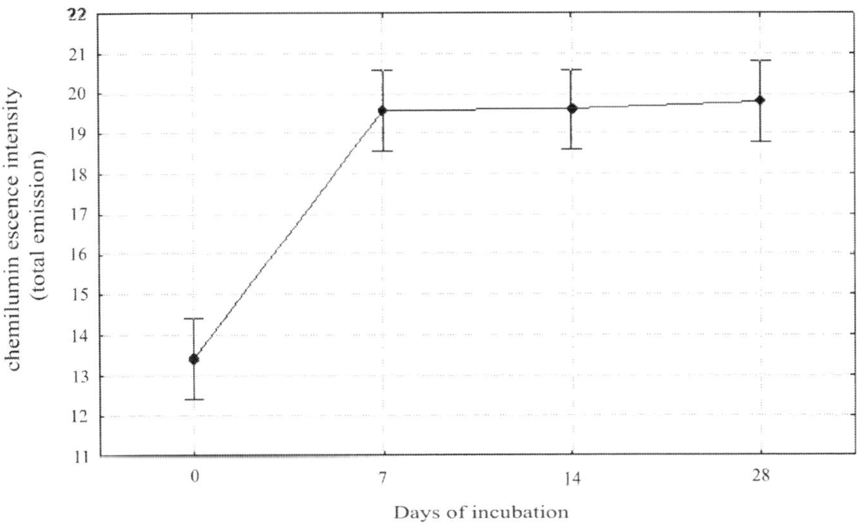

Current effect: F(3,36)=40.219, p<0.001.
Effective hypothesis decomposition.
Vertical bars denote 0.95 confidence intervals.

Figure 2. Changes in antioxidative effect of lactoglobulin preparation during its incubation.

On the other hand, a sharp downward trend in lactoperoxidase activity of lactoglobulin during the incubation was also observed (Figure 3). After one-week incubation the activity of lactoperoxidase in lactoglobulin preparation was totally lost. The correlation coefficient between the content of amide groups and lactoperoxidase activity in lactoglobulin preparation was about 0.78 ($p<0.001$). The highest correlation was observed for amount of EHA and lactoperoxidase activity, where the correlation coefficient reached 0.96 ($p<0.001$), whereas that for PHA was only 0.76 ($p<0.001$).

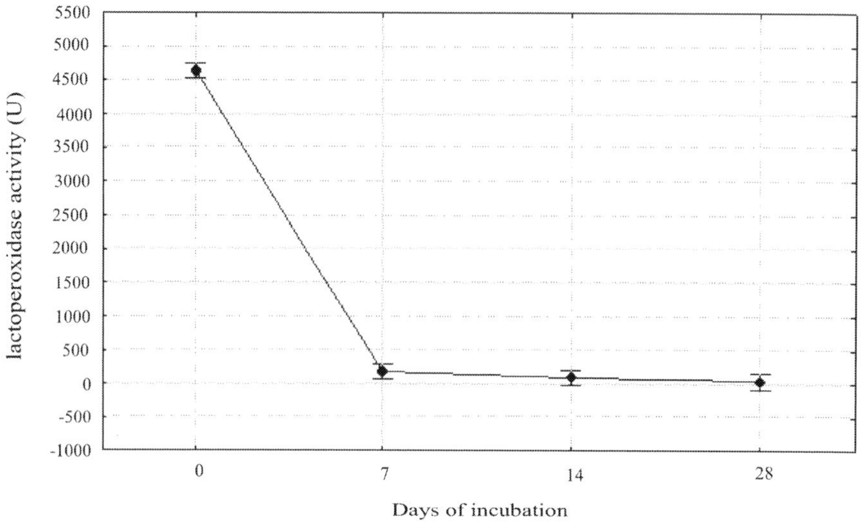

Current effect: $F(3,36)=1628.3$, $p<0.001$.
Effective hypothesis decomposition.
Vertical bars denote 0.95 confidence intervals.

Figure 3. Changes in lactoperoxidase activity of lactoglobulin preparation during its incubation.

To sum it up, we suggest that the loss of antioxidant effect of "aged" lactoglobulin preparation could be explained by the decrease in activity of antioxidative enzyme systems, such as lactoperoxidase, due to deamidation of its molecules and subsequent structural changes that lead to the final destruction of proteins.

Conclusions

As a result of our investigations we have found out that processes of spontaneous posttranslational deamidation participate in the ageing of proteins of lactoglobulin preparation and at least partly responsible for the loss of activity of its non-specific antimicrobial resistance system. Also, the high values of correlation coefficients between changes in amide group content, the amount of medium weight molecules and lactoperoxidase activity can be considered as an additional proof of the capacity of deamidation to induce protein fragmentation in a number of ways that was previously reported by Olekhnovich L.P. *et al.,* (1985).

REFERENCES

Alperovich D.V., Keselman E.V. (1990) *Methods of preparation, analysis and usage of enzyme*. p. 114, Riga.

Belizy S., Nazarova I.N., Procof'ev V.N., Sorokina I.A., Pushkina N.V. and Lukash A.I. (2001) *Biokhimiya.;* 66(5): 576 – 580.

Bobrov V.M., Shishkin S. A. (1999) *Proc. in oto(rhino)laryngology*.; 1: 33 – 34.

Bos C., Gaudichon C., and Tomé D. (2000) *Journal of American College of Nutrition*.; 19(90002): 191 – 205.

Gillery R., Monloisse I., Magart F. (1991) *Diabet. Metab.;* 17: 1 – 16.

Lowry O.H., Rosenbrough N.J., Farr A.L. (1951) *J. Biol. Chem*.; 193(1): 265.

Olekhnovich L.P., Pushkina N.V., Zhdanov A.Y., Lukash A.I., Krichevskaya A.A., Alakhov Y.B. (1985) *Reports of Acad. of Sci. USSR*.; 281(1): 217 – 220.

Pushkina N.V. (1979) *The Ukrainian Biochemical Journal*.; 51(6): 680 – 683.

Pushkina N.V. (1988) *The Ukrainian Biochemical Journal.;* 60(4): 9 – 14.

Pushkina N.V., Lukash A.I. (1976) Proc. of Northern Caucasus Scientific Center of Higher School. *Natur. Sci*.; 2: 95.

Robinson A., Robinson R. (1991) *Proc. Natl. Acad. Sci.; USA*, 15: 8880 – 8884.

Sherstnev M.P., Atanaev T.V., and Vladimirov Y. A. (1989) *Biofizika*.; 34: 684 – 687.

Sugawara K., Oyama F. (1981) *J. Biochem.;* 5: 771 – 774.

Umanskaya N.Y. (1995) *The effect of deamidation and glycation on activity of some proteins with antioxidative properties*. p.74, Rostov State University Press, Rostov-on-Don.

Winer B.J., Brown D.R., Michels K.M. (1991). *Statistical principles in experimental design*, (3rd ed.). McGraw-Hill: New York.

Wright H.T. (1991) *Crit. Rev. Biochem. Mol. Biol.;* 26: 1 – 52.

In: Protein Research Progress
Editor: A. B. Boscoe and C. R. Listow, pp. 9-15
ISBN 978-1-60456-663-3

Expert Commentary B

FORMATION AND RECOGNITION OF SUPERFICIAL MICROCLUSTERS AS THE INTEGRAL PART OF PROCESSING OF PROTEINS

Sergey V. Verevka*
Palladin Institute of Biochemistry,
Ukrainian National Academy of Sciences,
Leontovich, 9 str., Kiev 30, 01601, Ukraine

ABSTRACT

Extensive knowledge about the protein processing has resulted in the paradox caused by the inexplicability of extreme process stages - structure formation of de-novo synthesized molecules and their elimination after the performance of the physiological functions. A huge amount of material about molecular mechanisms of formation and maintenance of proteins in native conformation as a complex of the processes caused by the interaction of initial protein sequence with numerous components of the cellular folding system is saved up. Thus there is an open question - how the correct folding of protein molecules is provided at a huge variety of possible alternatives? The obvious failure of the providing of each protein with a personal chaperone system makes the considered part of the question as though unsolvable. On the other hand, how the eliminable proteins, and the ones that have accomplished their biological functions, or denaturated proteins are recognized? In what way do quite few ubiquitin-ligases E3 differentiate molecules which are subjected to the marking against a background of an unimaginably large number of the conformational conditions among great variety of proteins? Similar question can be set in respect of the immune molecular mechanisms. How the differentiation of a great number of self proteins in the native conformation from denaturated or foreign proteins is provided? In case of biological macromolecules it looks like too many of them owe too much to the few. The classification of the data on structural maintenance of mutual complementarity of protein molecules allows

* Phone: 38(044)2245974 ; FAX: 38(044)2296365 ; E-mail: verevka@biochem.kiev.ua

simplifying essentially the problem, making it reduced to the formation and recognition of small superficial microclusters of the amino-acid residues which is necessary and sufficient condition of the protein processing.

Keywords: protein processing, inter-protein recognition, complementarity, protein folding, protein elimination.

Mutual recognition of the functionally complementary biomolecules makes a basis for the regulation of biochemical processes, which in any measure allows characterizing biochemistry as a science about intermolecular recognition [1]. Special interest here is caused by zones of direct inter-molecular contact, since the research of their structure not only allows explaining the course of the processes caused by these interactions, but also creates preconditions for the purposeful intervention in them. Comparison of the data about direct intermolecular contact dimension has allowed note the important mechanism of distribution of areas of dimension maskable at formation of the diversified complexes. Distribution of dimension of the masked areas shows the group discreteness and the expressed similarity within the limits of group. Two groups of dimension - "standard" (about $1600 \pm 400Å^2$) and "large" (about $3500 \pm 200Å^2$) are observed [2]. Such distribution is hardly casual and, apparently, testifies to existence of thermodynamically necessary and sufficient minimum providing effective mutual binding of the functionally complementary proteins [3]. Proteinase-inhibitor and antigene-antibody complexes form the basis of the "standard" group. According to the formation speed these complexes concern to the fastest processes of inter-protein association, which they comply with necessary and sufficient conditions for fast recognition and binding the complementary proteins. The direct inter-protein contact surfaces can be divided into the binding sites and the bound determinants (ligand groups, epitopes, etc.). In case of «proteinaceous inhibitor - proteinase» and «antigene - antibody» interactions, the dimensions of the bound determinants are practically identical and approximately correspond to 600-800 $Å^2$, that comply with a small group of the compactly located amino-acid residues. The case of the proteinaceous inhibitors complex-forming with corresponding enzymes is more understood in detail. It is shown, that six residues (P3-P3' according to Schechter-Berger nomenclature [4]) of the inhibitor reactive center take up to 77 % of an inhibitor surface, maskable at complexation. Moreover, the residues in positions P1 and P2' have the increased degree of maskability and the abnormally high number of contacts with the amino-acid residues of an enzyme [5,6]. These residues are adequate according to ligand specificity binding (S1-) and effectory (S2'-) sites of an enzyme and are located in optimal, so-called canonical, conformation. Synchronous interaction of the binding and effectory sub-sites of the active center of enzymes with corresponding rigidly fixed amino-acid residues of an inhibitor appears to be a necessary and sufficient condition for highly elective and effective complexation [7, 8]. This kind of complex-forming includes the fastest processes of inter-protein association [5]. The conformational stability of the bound structure makes a necessary condition for stable inter-protein complex formation, as its derangement causes the reduction of protein inhibiting properties [8]. In the considered system "serine proteinase - proteinaceous inhibitor" the basic form of a complex is placed in a noncovalently bound state. Inactivated in the hydrolytic center, anhydro-derivatives of enzymes form the complexes with inhibitors, commensurable according to the stability of complexes with active enzymes [9, 10]. The exception is comprise by the proteinaceous blood inhibitors (serpins), whose reactive centers differ in the conformational mobility with transition through the canonical conformation [11]. For this group of inhibitors, the dynamic of their association with proteinases is determined according to the mobility of the reactive center. In other words, the speed is limited by the formation of canonical conformation in the reactive center in crucial condition [12]. The basic form of complexes with proteinases for the serpins is in a state of the covalent additive product, but when formed with anhydro-derivatives of enzymes these complexes are unstable [13, 14]. The recognition and splitting of the proteinaceous pro-forms activation sites by activating proteinases are provided in a similar way, as well as a number of other highly effective interactions of the complementary proteins. They can be characterized as 1-X-3 motives of the inter-protein recognition at which two functionally important residues are divided by the third, insignificant [8].

In all considered cases the connecting component "sees" not all the fixed molecule, but only the exposed pair of the amino-acid residues of 1-X-3 motif on a surface, in a sense of the superficial microcluster, which provides the recognition and binding of a protein-target by a complementary protein. The derangement of the conformation of a similar microcluster derives a protein from the members that are mediated by the corresponding process complexation. At the same time, the formation of the bound group provides the "visualization" of a protein and transforms the one into the member of the corresponding interactions. In particular, for the majority of the serine proteinase inhibitors of phytogenous nature there is no a little distinct explanation of their functional role as the inhibitors [15]. On the one hand, the inhibited enzymes are not at all in a functional connection with these proteins, on the other – there are a great number of high homologous proteins from the related sources deprived of the inhibitor properties. Apparently, in case of the proteinaceous phytogenous inhibitors, the assumption about casual character of formation in rigidly structured reserve proteins of superficial microclusters – which are the structural analogues of the reactive centers of proteinaceous inhibitors of the functionally foreign enzymes, is allowable. The same assertion may be supposed about the inhibitory properties of numerous ovomucoids.

The maskable areas of «antigene-antibody» interactions are similar to the case of «proteinaceous inhibitor-proteinase» group [2, 3]. Extraordinary high affinity of immunoglobulins to two-ligand compounds intensifies this similarity [16]. Meanwhile, the conformational demands for this group are carried out just the opposite way. Being inert in relation to the proteins in the native conformation, an immunoglobulin effectively binds the denaturated or foreign ones. In contrast to the proteinases, the binding properties of an immunoglobulin are subjected to the essential changes. Modification of the native immunoglobulin by the chaotropic agents results in the "washing out" of the immunoglobulin ligand specificity and display of the affinity in relation to the diversified antigenes. It is supposed, that such changes are caused by the mobility of sub-domain structures of the blood immunoglobulin active centers that is functionally necessary for fine adjustment to the structure of various antigens [17, 18]. Simultaneously, introduction in the system of low-molecular ligands with two - three hydrophobic groups which are located distantly providing the overlapping of the area that is typical for antigene-antibody interactions, results in the essential increase of binding properties of the native and chaotropically modified immunoglobulins [19]. Similar effect can be considered as a resulting display of two processes: induced by the effector structurization of immunoglobulins in highly bounding conformation and the competition for the formed highly bounding macro-domain group of immunoglobulins between low-molecular effectors and statistically formed ligand groups on the surface of an immobilized protein. In other words, remaining inert in relation to its «own» proteins, immunoglobulins show the expressed ability to be arranged to the different «foreign» superficial groups. As it is a question of the microcluster recognition from a pair of the compactly located amino acids, similar reactivity can be brought to the classical division of the incalculable set of amino acid clusters into their «own» and «foreign» clusters, where the «foreign» clusters are represented by all ones which are not «own». With reference to the elementary case of 1-X-3 motif, any superficial fragment of the polypeptide strands (... ABCDEFGHI ...) can be considered as a set of microclusters. The visualization degree of each set is determined by the nature of the exposed on the surface amino acid pairs of their conformational arrangements and the degrees of their stabilization. Thus, each of accessible for the inter-protein contact of the amino acid residues appears to be a component of two microclusters (AXC, CXE, EXG, GXI) - in the third and first positions, accordingly. Recognition and splitting of the bait site α_2-macroglobulin by proteinases is provided in a similar way. Depending on the substrate specificity of an enzyme, the fragment ...-Arg*-Leu-Val*-His-Val-... is splitted on marked arginine, or on valine [20, 21]. Thus, the marked fragment can be considered as a pair of superficial microclusters -Arg-Leu-Val- and -Val-His-Val-. In similar image the surface of any protein molecule appears to be represented by a set of either one or another way stabilized microclusters, to a greater or lesser extent adequate or not adequate to the set of structural rules determined for the particular biological system. The considered problem essentially becomes simpler by the similarity of amino acids within the limits of one group (hydrophobic, polar not-charged, positively or negatively charged).

The stated considerations allow simplifying significantly the concepts about structural maintenance of the protein processing. It is known, that formation of the native structures of a protein comprises an extremely complex process, which takes place in interaction of the initial structure of the formed protein molecule and the cell-producer chaperone components. In this connection the formation of native conformation is surely provided only with the "own" cells. Meantime in the case of the recombinant proteins production, the formation of inactive iso-form and units makes one of the basic unsolved problems of genetic engineering to this day [22, 23]. The task of the structural maintenance of the appropriate conformation formation and its recognition can be significantly simplified. It can be provided by the assumption of formation of the superficial microclusters on the structured molecule surface, submitting to the determined set of the exclusive rules. In other words, the incomprehensible "super knowledge" of the uniquely correct conformation by the cellular chaperone system for each set of synthesized proteins is replaced mostly with more modest function of exclusion any microclusters, which are conformationally distinct from rather small set of the allowed conditions, on the surface of the formed molecule. Such superficial "flattening" is connected to the formation of local power minimums and should be energy-consumptive, as it is actually observed. On the other hand, chaperone proteins have the strong ability to recognize the native and denaturated forms of the "own" protein molecules, remaining inert with respect to the first and vigorously cooperating with the second [24]. Apparently, the formation of mixed aggregates of HSP70 and pathogenic iso-form of prion protein can serve as an extreme case of a similar sort of reactivity [25]. The presence of the stabilized microcluster on the surface of a protein, which breaks the rules determined for a particular cell, appears a necessary and sufficient condition for its recognition by proteins of the chaperone systems. Similarly, the ability of not numerous families of ubiquitin-ligases to differentiate subject to marking and to the subsequent elimination molecules against the huge pool of other proteins can be explained. The presence of wrong superficial microcluster for a particular cellular system appears a sufficient condition for recognition of a corresponding protein molecule against the background of the set of other molecules, which do not have a similar "black label".

Conclusion

The resulted reasons essentially simplify the problem of structural maintenance of the protein processing by means of the formation and recognition of the "own" and the "foreign" microcluster groups which are located on the surface of a protein molecule. The nature of amino acids, forming the microcluster and their conformational condition, appears to be determinative for the molecular differentiation. Of course, the number of "own" clusters is also great, but it is incommensurably less than an unimaginable variety of possible amino acid residues conformational combinations on the surface of a protein. The "foreign" clusters are those which are not corresponding to the exclusive rules, established in a particular cellular system for the "own" ones. The similar assumption significantly simplifies the understanding of mechanisms of formation, functioning and elimination of the proteinaceous molecules.

References

[1] Edelman, G. (2004). Biochemistry and the science of recognition. *J. Biol. Chem., 279*, 7361-7369.

[2] Lo Conte, L., Chothia, C. and Janin, J. (1999). Atomic structure of protein-protein recognition sites. *J. Mol. Biol., 285*, 2177-2198.

[3] Janin, J. (1995). Principles of protein-protein recognition from structure to thermodynamics. *Biochimie, 77*, 497-503.

[4] Schechter, I. and Berger, A. (1968). On the active site of proteases. III. Mapping the active site of papain : specific inhibitor of papain. *Biochem. Biophys. Res. Communs., 32*, 888-902.

[5] Blow, D., Wright, C., Kulka, D., et al. (1972). A model for the assotiation of bovine pancreatic trypsin inhibitor with chymotrypsin and trypsin. *J. Mol. Biol., 69*, 137-144.

[6] Janin, J. and Chothia, C. (1976) Stability and specificity of protein-protein interactions: the case of trypsin-trypsin inhibitor complex. *J. Mol. Biol., 100*, 197-211.

[7] Verevka, S. (1997). On the structure and function of effectory sites in serine proteinases. *Comments on Mol. and Cell. Biophys., 9*, 153-165.

[8] Verevka, S. and Miroshnichenko, O. (2001). 1-X-3 motif in inter-protein recognition: structures, widespreading and possible practical application. *J. Mol. Recognit., 14*, 315-318.

[9] Ako, H., Ryan, C. and Foster, R. (1972). The purification by affinity chromatography of a proteinase inhibitor binding species of anhydro-chymotrypsin. *Biochem. Biophys. Res. Communs., 46*, 1639-1645.

[10] Ako, H., Foster, R. and Ryan, C. (1972). The preparation of anhydro-trypsin and its reactivity with naturally occurring proteinase inhibitors. *Biochem. Biophys. Res. Communs., 47*, 1402-1407.

[11] Elliott, P., Lomas, D., Carrell, R. and Abrahams, J. (1996). Inhibitory conformation of the reactive loop of alpha 1-antitrypsin. *Nat. Struct.Biol., 3,* 676-681.

[12] Kaslik, G., Kardos, J., Szabo, E., et al. (1997). Effects of serpin binding on the target proteinase: global stabilization, localized increased structural flexibility, and conserved hydrogen bonding at the active site. *Biochemistry, 36,* 5455-5464.

[13] Moroi, M., Yamasaki, M. and Aoki, N. (1975). Association of human α1-antitrypsin with anhydro-chymotrypsin. *J. Biochem., 78,* P.925-928.

[14] Enghild, J., Valnickova, Z., Thogersen, I. and Pizzo, S. (1994). Complex between serpins and inactive proteinases are not thermodinamically stable but are recognized by serpin receptors. *J. Biol. Chem., 269,* 29159-20166.

[15] Azzouz, H., Cherqui, A., Campan, E., et al. (2005). Effects of plant protease inhibitors, oryzacystatin I and soybean Bowman-Birk inhibitor, on aphid Macrosiphum europoribae (Homoptera, Aphididae) and its parasitoid Aphelinus abdominalis (Hymenoptera, Aphelinidae). *Journ. Insect. Physiol., 51*, 75-86.

[16] Teng, S., Sproule, K., Hussain, A. and Lowe, C. (1999). A strategy for the generation of biomimetic ligands for affinity chromatography. Combinatorial synthesis and biological evaluation of an IgG binding ligand. *J. Mol. Recognit., 12,* 67-75.

[17] Bobrovnik, S. (1999). Polyreactive immunoglobulins: molecular properties and functions. *Comments on Mol. and Cell. Biophys., 9,* 323-356.

[18] Bobrovnik, S., Petrova, Iu. and Efetov, K. (1997) Transformation of serum immunoglobulins to polyreactive antibodies. *Ukr. Biochim. Zh., 69(3),* 36-42.

[19] Ilyina, L. and Verevka, S. (2003). Ligand-induced structuring of polyreactive immunoglobulins. *Ukr. Biochim. Zh., 75(6),* 56-61.

[20] Sottrup-Jensen, L., Petersen, T. and Magnusson, S. (1980). A thiol ester in α2-macroglobulin cleaved during proteinase complex formation. *FEBS Lett., 121*, 275-279.

[21] Roberts, R. (1985). Alpha-2-macroglobulin. *J. Med., 16,* 129-219.

[22] Horowitz, P. (1999). Ironing of the protein folding problem? *Nat. Biotechnol., 17,* 136-137.

[23] Baneyx, F. and Mujacic, M. (2004). Recombinant protein folding and misfolding in Escherichia coli. *Nat. Biotechnol., 22,* 1399-1408.

[24] Demchenko, A. (2000). Protein folding and molecular chaperones: stochastic process under control. *Biophysics, 45,* 404-410.

[25] Ma, J. and Lindquist, S. (2001). Wild-type PrP and a mutant associated with prion disease and subjected to retrograde transport and proteasome degradation. *Proc Natl Acad Sci USA, 98,* 14955-14960.

In: Protein Research Progress
Editor: A.B. Boscoe and C. R. Listow, pp. 17-60
ISBN 978-1-60456-663-3

Chapter 1

PP2A: MULTITASKING AND MAKING DECISIONS IN CELL GROWTH AND TRANSFORMATION

Jean-Marie Sontag and Estelle Sontag*
Department of Pathology,
University of Texas Southwestern Medical Center,
Dallas, TX 75390-9073.

ABSTRACT

Protein phosphatase 2A (PP2A) is a large family of serine/threonine phosphatases that are differentially regulated, have restricted substrate specificity, and display tissue- and cell-specific distribution. PP2A enzymes are believed to be heterotrimeric enzymes *in vivo*. The core enzyme comprises a catalytic C subunit, associated with a scaffolding subunit, called A. Distinct classes of B-type regulatory subunits, termed B, B', B'' and B''', can bind to the (AC) dimer, resulting in formation of a wide variety of heterotrimeric complexes. B-type subunits are in part responsible for PP2A targeting and functional specificity. The C subunit is also subjected to reversible phosphorylation, which results in transient PP2A inactivation, and to carboxyl methylation, which critically modulates PP2A holoenzyme assembly and substrate specificity. A body of evidence has shown that PP2A enzymes are crucial components of signaling pathways that regulate cell growth and proliferation. Indeed, deregulation of PP2A contributes to cell transformation. For instance, okadaic acid, a potent PP2A inhibitor, is a tumor promoter in mice. Mutations in the genes encoding selective PP2A subunits have been found in some cancers. Changes in endogenous PP2A subunit composition and expression may contribute to tumorigenesis. Specific PP2A isoforms also modulate the Wnt signaling pathway that controls cell growth and polarity, and tissue specification. Interestingly, viral proteins from transforming DNA tumor viruses, such as polyomavirus middle T and SV40 small t antigens directly modulate PP2A targeting, activity and

* Address all correspondence to: Estelle Sontag, Ph.D., Department of Pathology, University of Texas Southwestern Medical Center, 5323 Harry Hines Blvd, Dallas, TX 75390-9073. Phone: (214) 648 2327; fax: (214) 648 2077; e-mail: Estelle.Sontag@UTSouthwestern.edu

substrate specificity. Interaction of SV40 small t with PP2A leads to deregulation of gene transcription, changes in the architecture of the actin and microtubule cytoskeleton, defects in tight junction biogenesis, alterations in the cell cycle, and deregulation of cell growth and survival. In this review, we will examine in-depth the multiple roles played by PP2A in the regulation of cell transformation and tumorigenesis.

INTRODUCTION

Reversible phosphorylation is widely recognized as the most important mechanism in the regulation of protein functions in eukaryotic cells. It is involved in switching cellular activities from one state to another, thereby modulating gene expression, and cell proliferation, differentiation and survival. It is utilized by cells to respond to various environmental stresses and extracellular stimuli by hormones and growth factors. Reversible protein phosphorylation relies entirely on the concerted action of protein kinases and protein phosphatases. These enzymes usually display quite broad substrate specificity *in vitro*, but are required to perform defined functions *in vivo*. In the past, much attention has been paid to studying protein kinases, but it is now becoming increasingly apparent that protein phosphatases play an equally important role in the control of protein phosphorylation.

Phosphatases are divided into two major families of proteins, the phosphotyrosine phosphatases (the PTP family) and the phosphoserine/threonine phosphatases (the PPP and PPM family). Both families originate from two distinct ancestral genes [1-3]. The PPP family comprises the PP1, PP2A, PP2B, PP4, PP5, PP6 and PP7 enzymes, whereas PP2C and bacterial enzymes, like SpoII or PrpC, belong to the PPM family [3]. Among these enzymes, protein phosphatase 2A (PP2A) accounts for as much as 1% of total cellular proteins, and for the major portion of Ser/Thr phosphatase activity in most tissue and cells [4]. The term "PP2A" encompasses a large family of enzymes that are differentially regulated, have restricted substrate specificity, and display tissue- and cell-specific distribution. As detailed below, PP2A regulation is highly complex and dependent on the sophisticated modulation of its subunit composition. Moreover, PP2A subunit interaction, as well as multifaceted protein-protein interactions between PP2A and a wide array of cellular/viral proteins, play a crucial role in specifying PP2A functions. Disruption of these normal interactions is a major contributor to cell transformation.

PP2A STRUCTURE AND REGULATION

PP2A Is a Multimeric Protein Complex

Several holoenzyme complexes of PP2A have been isolated from a variety of tissues and extensively characterized. The core enzyme comprises a ~36kD catalytic subunit (C) that is associated with a 65kD scaffolding subunit called A (or PR65), which modulates PP2A enzymatic properties [5]. The A subunit serves as a scaffolding molecule to coordinate the assembly of PP2A holoenzymes, generating functionally diverse heterotrimers [6]. Two isoforms, α and β, exist for both the A and C subunit, and are expressed at variable levels in different tissues and organisms [7]. Distinct classes of B-type regulatory subunits, termed B (PR55/B55 family), B' (PR61/B56 family), B'' (PR72 family) and B''' (PR93/PR110 family), are encoded by different genes and can bind to the (AC) core enzyme to form a wide variety of heterotrimeric complexes [8, 9] (Figure 1). Different B subunits interact via the same or overlapping sites within the A subunit, which explains why binding of B subunits is mutually exclusive [10, 11]. Prevalent PP2A enzymes are believed to be primarily ABC heterotrimers *in vivo* [12]. Yet, abundant (AC) dimers can be detected during PP2A purification from tissue, maybe as a result of the dissociation of B subunit from the core enzyme. Whereas it is still unclear whether (AC) species occur in cells, individual subunits do not exist as stable monomers *in vivo*.

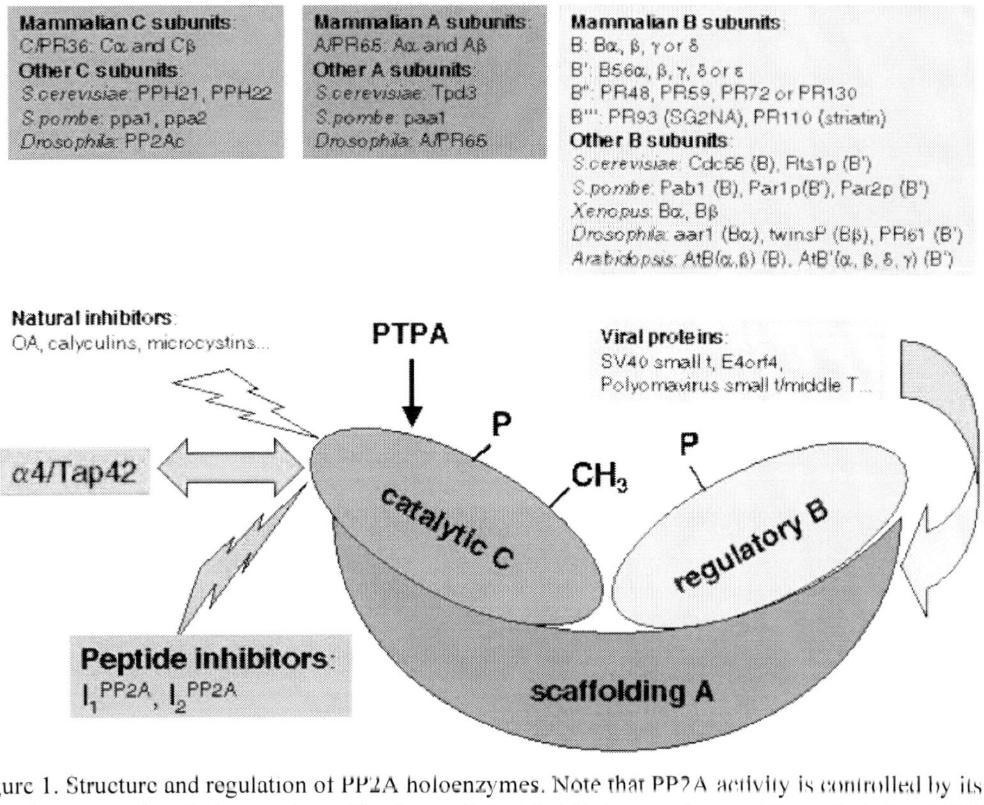

Figure 1. Structure and regulation of PP2A holoenzymes. Note that PP2A activity is controlled by its subunit composition, its interaction with other cellular/viral factors, and reversible methylation ($-CH_3$) and phosphorylation (-P).

PP2A Regulation Is Highly Complex

PP2A Is Regulated by Its Subunit Composition

The expression of multiple A, B and C subunit isoforms provides for a great variety of possible PP2A holoenzymes. It is now well recognized that B-type subunits are in part responsible for PP2A functional specificity. They display different affinities for the core enzyme, and their binding to AC modulates PP2A catalytic activity. The extent of this effect varies greatly according to the substrate and type of subunit present in the enzymatic complex [13]. In addition to modulating PP2A activity, the nature of the B subunit greatly influences substrate specificity. Holoenzymes containing the Bα and PR72 subunits dephosphorylate distinct sites on SV40 Large T antigen [14]. Likewise, ABαC are more potent than AB'C holoenzymes in dephosphorylating tau proteins, whereas AC and C are nearly inactive towards this substrate [15]. In PC6-3 cells, growth arrest and neuronal differentiation is induced by overexpression of Bγ, but not of Bα π [16]. Studies in yeast have elegantly demonstrated that B-type subunits are not functionally redundant, but direct the phosphatase to distinct evolutionary conserved cellular functions [17,18].

A unique feature of the B subunits is their ability to compete for binding to (AC), and to differentially exchange and replace each other in the PP2A complex. For instance, Bα can replace Bβ, but not B', whereas B' can exchange with both Bα/Bβ *in vitro* [13]. Overexpression of B' subunits modulates endogenous PP2A activity [19]. Likewise, overexpressed A subunit mutants that bind to C, but not to B, induce deregulation of intracellular PP2A [20]. Therefore, changing the ratio between endogenous levels of the core enzyme and holoenzymes exerts a tremendous influence on PP2A regulation [19, 20]. Particular cellular stimuli also modulate B subunit expression, which in turn affects the composition of endogenous pools of PP2A [16]. Specific B subunits also influence PP2A localization. Specificity of action is mainly derived from subcellular targeting of distinct pools of phosphatases through dynamic binding with relevant partners. Pools of PP2A have been identified on the plasma membrane [21], Golgi [22], microtubules [15] intermediate filaments [23], mitochondrial membrane [24], centrosomes [25], nucleus [26], and caveolae [27].

PP2A Is Regulated by Phosphorylation

PP2A catalytic subunit ($PP2A_C$) undergoes reversible phosphorylation on Tyr^{307} in the conserved C-terminal sequence [28]. PP2A Tyr phosphorylation has been observed in growth factor-stimulated fibroblasts, and in cells overexpressing $pp60^{v\text{-}src}$ [28,29]. $PP2A_C$ also undergoes Thr phosphorylation in response to an autophosphorylation-activated protein kinase [30]. Phosphorylation on either Tyr or Thr residues results in inhibition of PP2A catalytic activity, while its reactivation occurs through autodephosphorylation at these residues. Phosphorylation of certain B subunits, i.e. B"δ/PR74 and B'/B56α/PR61, that could affect PP2A activity and substrate specificity, has also been documented [31,32].

PP2A Is Regulated by Methylation

$PP2A_C$ undergoes reversible carboxymethylation on the Leu^{309} residue in the conserved C-terminal $TPDYFL^{309}$ motif [33-35]. Methylation is catalyzed by a specific carboxyl methyltransferase [33,34,36-38], whereas the methyl group is removed by a specific methylesterase [39-41]. Several studies have demonstrated that methylation of $PP2A_C$

critically affects PP2A holoenzyme assembly and substrate specificity [42-48]. Interestingly, methylation-defective mutants of $PP2A_C$ become unable to bind to the Bα subunit in yeast and mammalian cells [43-49].

PP2A Is a Target of Viral Proteins

Remarkably, PP2A is the target of many viruses [50]: Papovae viruses (SV40, Polyomavirus), Papilloma viruses (HPV16), Lentiviruses (HIV), Adenoviruses, Cytomegaloviruses (Herpes virus). Among them, the DNA tumor viral antigens, Polyoma middle T and SV40 small t, have been the most studied. They form stable complexes with PP2A [51-53]. SV40 small t directly interacts with the A subunit [54], and may compete with B subunits for binding to this subunit[10]. Indeed, small t is able to replace the B/PR55 subunit, but not the B'/PR61 subunit from trimeric PP2A complexes [54-56]. Interaction of PP2A with small t alters PP2A substrate specificity and activity towards numerous substrates [9,54].

PP2A Is a Target of Naturally-Occurring Inhibitors and Toxins

Two cellular PP2A inhibitory proteins, termed I_1^{PP2A}/PHAP-I and I_2^{PP2A}/SET/TAF-Iβ have been isolated [57-60]. They inhibit all PP2A isoforms, probably by directly binding to $PP2A_C$. Interestingly, in acute non-lymphocytic myeloid leukemia, SET is fused to CAN (nucleoporin Nup24). The formation of this fusion protein may impair the normal regulation of PP2A and contribute to leukaemogenesis [60]. PP2A is also the target of many toxins produced by bacteria and marine dinoflagellates, e.g. fostriecin, tautomycin, nodularins, microcystins, calyculins, okadaic acid [9]. All these compounds inhibit PP2A activity with variable specificity. The best studied and most widely used is okadaic acid (OA), which is responsible for diarrhetic shellfish poisoning [61]. Although OA directly binds to and has the highest affinity for $PP2A_C$ [62], it also inhibits other phosphatases, i.e. PP1, PP4/PPX, PP5 and PP2B, in a concentration-dependent manner [63-66].

PP2A Is Regulated by PTPA

It has initially been reported that, despite being primarily a Ser/Thr phosphatase, PP2A can display a low, but detectable level of phosphotyrosine phosphatase (PTPase) activity *in vitro* [67,68]. This PTPase activity appears to depend on a protein factor called the PTPase activator (PTPA). PTPA is most effective in stimulating the PTPase activity of $PP2A_C$ and AC dimers, and to a lesser extent, those of AB"C and ABC trimers [69,70]. However, recent data in *S.cerevisiae* rather suggest that RRD1 and RRD2 (the yeast equivalents of human PTPA) are required for biogenesis of active, and phosphoserine/threonine-specific SIT4 (yeast-related PP2A phosphatase) [71]. Indeed, deletion of the yeast PTPA homologs generate a $PP2A_C$ with a conformation different from the wild-type enzyme, as indicated by its altered substrate specificity, reduced protein stability, and metal dependence. Moreover, PTPA fulfills an essential function conserved from yeast to man [71].

Role of PP2A in Cell Transformation

PP2A and Tumorigenesis

Deregulation of cell growth and adhesion, and altered gene expression are hallmarks of the transformed cell phenotype. There is strong evidence supporting a role for PP2A in regulation of tumor formation. Inhibition of PP2A activity by OA, calyculin A, microcystins and nodularins, induces many types of cancer [72]. However, the PP2A inhibitor, fostriecin, blocks cell proliferation and displays significant anti-tumor activity [73,74]. Furthermore, PP2A activity and expression are notably decreased in drug-resistant breast cancer cells [75,76]. Mutations in genes expressing the Aα and Aβ subunits, that may affect B subunit binding, have been found in lung and colon tumors, melanomas, breast carcinomas, and primary lung and colon tumor-derived cell lines [77-80]. In the B16 mouse melanoma model, the B'/B56γ subunit was highly expressed in metastatic cells, but not in corresponding nonmetastatic or normal melanocytes [81]. Recently, a retrotransposon insertion in the gene encoding B'/B56γ1, causing the expression of an N-terminally truncated form of this subunit, was associated with a higher metastatic state of melanoma cells [82,83]. On the other hand, inactivating mutations or deletions of $PP2A_C$ have never been detected in human tumors, which likely reflects its essential importance for cellular viability [84].

PP2A Is Targeted by Transforming DNA Tumor Viruses

PP2A is absolutely required for transformation of quiescent cells by SV40 and Polyomavirus [9,85-88]. Interaction of viral proteins with specific PP2A enzymes is utilized by DNA tumor viruses to deregulate normal signaling cascades that control gene expression, the cell cycle, cell growth and survival, in order to promote cell transformation. For instance, in fibroblasts, SV40 small t activates several signal transduction pathways, including the stress-activated protein kinase (SAPK) and mitogen-activated protein kinase (MAP kinase) cascades, and induces transcriptional activation of Sp1, AP-1, CREB and NF-κB. Indeed, SV40 small t stimulates serum-independent cell growth, cell cycle progression, cell survival, and cellular and viral promoter activation [9,86-89]. Like SV40 small t, polyomavirus small t stimulates cell transformation and enhances viral DNA replication through its ability to transactivate *c-Fos* and promote cell cycle progression in a PP2A-dependent manner [90]. Polyomavirus middle T is present in a multi-protein complex comprising PP2A, phosphoinositide 3-kinase (PI 3-kinase), protein tyrosine kinases of the src family, phospholipase Cγ, Shc and 14-3-3 proteins, which are known to regulate apoptosis and mediate mitogenic signaling [90].

PP2A and the Cell Cycle

In cancer, there are fundamental alterations in the genetic control of the cell cycle and cell division, resulting in unrestrained cell proliferation. Progression of cells through the various phases of the cell cycle is governed in part by the sequential activity of different cyclin-dependent kinases (CDKs). Genetic evidence in yeast and *Drosophila* points towards a critical role for PP2A in the regulation of cell cycle progression, since its dysfunction induces cell cycle defects and abnormal cytokinesis [73,91-98].

G_1-S Transition

PP2A has stimulatory effects during the early phases of the cell cycle, by affecting the activity of G_1-specific CDK complexes controlling the G_1-S transition. Exposure of mammalian cells to OA causes inhibition of DNA synthesis and G_1 arrest [99-102]. OA can induce a decrease in cyclin D_2, E, and A expression, which results in severely impaired CDK activity, and G_1 arrest [102,103]. In contrast, deregulation of PP2A by SV40 small t stimulates G1-S cell cycle progression by upregulating cyclin A, D_1 and E expression [9,86-89].

S Phase and DNA Replication

PP2A also regulates the initiation of DNA replication at the S phase of the cell cycle in yeast, viral and vertebrate systems [94,104,105]. Immunodepletion of PP2A from a *Xenopus* egg extract results in strong inhibition of DNA replication [105]. PP2A is required for initiation of replication, but not for elongation of previously engaged replication forks [105]. It regulates the loading of Cdc45 onto chromatin, by dephosphorylating and subsequently activating a soluble factor required to recruit Cdc45 to the pre-replication complex [106]. The role of PP2A in Cdc45 loading can be linked either to growth-stimulatory signaling pathways [107] or to DNA damage control checkpoints [108]. Another target of PP2A is the protein Cdc6, which is required for establishment of the pre-replication complex [109]. Cdc6 interacts with the nuclear PP2A B"/ PR48 subunit [110]. Overexpression of PR48, but not of Bα or B'α1 subunits, induces G_1 arrest in Hela cells [110]. Furthermore, the B"/PR72 subunit is implicated in the control of viral DNA replication [14].

G_2-M Transition

Genetic and biochemical studies have revealed that PP2A inhibits the activation of the M-phase specific CDK or M-phase promoting factor (MPF). The MPF complex consists of the $p34^{cdc2}$ kinase (Cdc2 or Cdk1) and cyclin B [111-115], which is essential for entry and progression from G_2 to M [116]. It phosphorylates specific substrates, such as histone H1, lamins, vimentin, cyclins and microtubule-associated proteins, explaining the initiation of mitotic processes such as nuclear breakdown, chromosome condensation and spindle formation. In G_2/M, Cdc2 associates with freshly synthesized cyclin B and is phosphorylated on Thr^{161}, Thr^{14} and Tyr^{15}. This triple-phosphorylated complex is inactive and called pre-MPF. Activation occurs when Thr^{14} and Tyr^{15} are dephosphorylated by the dual-specificity phosphatase, Cdc25.

It has been proposed that PP2A dephosphorylates Cdk-activating kinase (CAK or cdk7-cyclin H complex) and thereby inhibits the pathway that leads to phosphorylation of Cdc2 on the Thr^{161} residue and its activation [117]. PP2A can indirectly inactivate Cdc2, by dephosphorylating and activating the Wee1 kinase, which phosphorylates Cdc2 on the Tyr^{15} residue [112,118,119]. Moreover, *in vitro* data indicate that PP2A can dephosphorylate and inactivate Cdc25 [114,118-120]. In fission yeast, loss of one of the two genes that encode $PP2A_C$ causes a semi-Wee1 phenotype. It is lethal when combined with mutations in the *wee1*+ gene, and partially suppresses the growth defect of Cdc25 mutants [112].

The role of PP2A in G_2/M is further supported by data from viral studies. Deregulation of PP2A by SV40 small t results in the accumulation of cells in the G_2/M [9,86-89]. PP2A interacts with VPR, an accessory protein from HIV-1, and is required for the ability of VPR to induce G_2/M cell cycle arrest [121,122].

Mitosis, Cytokinesis and Transition to G_1

Whereas PP2A enzymes have to be inactivated for cells to progress into mitosis, they have to be reactivated at the exit of mitosis. In *Xenopus* embryos and kidney cells, OA treatment blocks mitosis at the metaphase-anaphase transition [113,123,124]. At the end of mitosis, MPF is inactivated by cyclin B destruction and dephosphorylation of Cdc2 at Thr^{161}. Several experiments suggest that PP2A regulates the anaphase-promoting complex, which is involved in degradation of mitotic cyclins and Cdc2 inactivation [125]. The exit from mitosis is also promoted by dephosphorylation of MPF substrates [116]. PP2A is the major enzyme that dephosphorylates several physiological substrates of Cdc2, e.g. histone H1, the high mobility protein I(Y), caldesmon, vimentin, and the cis-golgi matrix protein GM130 [119].

Genetical studies in yeast and *Drosophila* confirm that PP2A plays a positive role in regulation of the metaphase-anaphase transition [119]. Indeed, PP2A mutants induce delayed and anomalous septation and cytokinesis [91,95,126-129]. In *S. pombe*, disruption of the *par1* and *par2* genes (homologues of B'/PR61) show abnormal septum positioning and growth and cytokinesis defects [129,130]. In *S. cerevisiae*, mutations in Cdc55 (B/PR55 homologue), result in highly elongated and multiple budded cells indicative of delayed cytokinesis [126]. These cells also lack a functional kinetochore/spindle assembly checkpoint [131,132]. Mutations in PPH21/PPH22 ($PP2A_C$ homologues) also induce defects in *S. cerevisiae* bud morphogenesis and actin [128]. In *S.cerevisiae*, activation of the mitotic exit network triggers both septin ring splitting and actomyosin ring contraction [133]. The PP2A B'/Rts1p subunit [17,18] is involved in the regulation of septin dynamics, through dephosphorylation of the septin protein Shs1 [134].

Lastly, Aurora is a family of Ser/Thr protein kinases that regulate many processes during cell division, i.e., chromosome segregation, centrosome maturation, mitotic spindle function, and cytokinesis [135]. Interestingly, overexpression of human Aurora kinases (A, B and C) has been detected in tumor cell lines [135]. Recent data suggest that, in *Drosophila*, PP2A is a negative regulator of Aurora B kinase, and controls its function during the cell cycle [136].

PP2A and DNA Damage-Sensing Pathways

There is a general consensus that DNA is the main target for most, if not all genotoxic agents. These include environmental alkyliting carcinogens, ultraviolet (UV)-light and ionizing radiation (IR), as well as many anticancer drugs. If not repaired prior to replication, DNA damage is generally harmful, resulting in effects detectable at the level of chromosomal aberrations, recombinations, as well as gene mutations and malignant transformation. In response to DNA damage, there are checkpoints to arrest the cell cycle, in order to provide time for DNA repair and to set up an adaptative response to counteract further damage. According to whether or not damaged DNA is repaired, cells either resume growth, or enter a status of permanent growth arrest and become apoptotic. Major DNA damage checkpoints are positioned before the cell enters S phase (G_1-S checkpoint), or after DNA replication (G_2-M checkpoint). Two major components of DNA damage pathways include the retinoblastoma family of proteins (pRb, p107 and Rb2/p130), and p53, which are tumor suppressors.

Rb

Members of the Rb family regulate cell growth in normal conditions and after exposure to genotoxic stimuli. They maintain order at the G_1 checkpoint through their ability to bind and sequester members of the E2F family of transcription factors [137], which control the expression of genes involved in cell cycle progression. Under stress conditions (UV-light, oxidative stress), PP2A associates with p107 and p130 and allows their interaction with E2F proteins, thereby blocking cells in G_1/S [138-141]. p107 dephosphorylation occurs through specific interaction with PP2A B''/PR59 subunit [139]. Although it has initially been reported that PP2A regulates pRb through control of G_1 CDK activity [103], recent data rather support a direct interaction between pRb and PP2A [138,140]. Indeed, in cells under oxidative stress, pRb is not dephosphorylated following treatment with OA, calyculin A or SV40 small t[140]. In NRK-52E cells, elevated cAMP causes a marked reduction of pRb phosphorylation. This dephosphorylation is blocked by OA at concentrations selective for PP2A [138]. Furthermore, it has been shown that pRb, p107 and p130 are specifically associated with early S-phase chromatin in primary cells [142]. SV40 small t and OA inhibit the stable association of hypophosphorylated pRb with chromatin [143]. Thus, in S phase, PP2A may normally license the recruitment of pRb to chromatin sites from which, after DNA damage, it relocalizes to selected replication control sites and suppresses abnormal, postdamage replicative activity.

The ability of PP2A to regulate S-phase progression via pRb suggests that cells that are deficient in PP2A-dependent pRb dephosphorylation may become genetically unstable following oxidative injury. This is illustrated by OA-induced tumor formation [144] and the transforming activity of SV40 small t [86]. Moreover, as previously mentioned, mutations in the genes encoding the two isoforms of the A subunit, have been found in multiple human tumors [77,79,80,145,146]. Under those circumstances, PP2A can be regarded as a tumor suppressor, and it is possible that this activity is driven by modulation of pRb, p107, p130 dephosphorylation following genotoxic damage. Such a tumor-suppressive role for PP2A via p130 regulation has been suggested [147]. Ovarian adrenocarcinoma cells treated with retinoic acid stop growing at the G_1 phase and accumulate p130 and PP2A [148]. These studies also show direct interaction between PP2A and p130. Inhibition of PP2A with OA results in hyperphosphorylation of p130 and its degradation through an ubiquitination pathway.

p53

The cell cycle regulator, p53, plays a critical role in cellular response to environmental stresses. Upon a cell's exposure to hypoxia or genomic damage, the stress-activated protein kinases, ataxia-telangiectasia-mutated (ATM) and ataxia and rad3 related (ATR), target p53 for phosphorylation and activation [149]. In this activated state, p53 exists as a DNA-binding tetramer [150], and controls the transcription of a diverse array of target genes. Those include: p21, 14-3-3τ, and MGC10, which induce cell cycle arrest; Bax, Puma and Killer/Dr5, which induce apoptosis; and p53R2 and Gadd45, which participate in DNA repair. The induction of the CDK inhibitor p21 results in G_1 arrest, thereby preventing the replication of damaged DNA [151]. After cellular damage is repaired, activated p53 becomes harmful to normal cell growth and needs to be eliminated. The phosphorylation of p53 on Ser $^{15, 20, 33, 37}$ critically modulates its activation and degradation [152]. MDM2, a p53 target gene, negatively regulates p53 and contributes to the proteolytic degradation of p53 by facilitating its ubiquitination [153]. However, the p19 (ARF) protein binds to MDM2 and prevents MDM2-mediated p53 degradation [154].

Data show multiple levels of PP2A involvement in p53-mediated cell cycle arrest and DNA damage repair. For instance, PP2A dephosphorylates MDM2 and modulates MDM2-p53 interaction via recruitment through cyclin G_1 [155-157]. Data suggest that, like cyclin G_1, cyclin G_2 also controls p53 activity via PP2A [156,157]. Cyclin G_2 can change PP2A targeting and substrate specificity by binding to the B' and $PP2A_C$ subunits, forming heterotrimers that lack the A subunit. The interaction between PP2A and cyclin G_2 leads to nuclear aberrations and G_1/S cell cycle arrest [158]. PP2A also directly interacts with and dephosphorylates p53 at Ser^{37}, thereby inactivating p53 [159]. Analysis of γ-irradiated Molt-4 cells shows that transient phosphorylation on Ser^{37} is also important for p53 transcriptional activity [159]. Dephosphorylation events in these cells is regulated by increased expression and interaction of $PP2A_C$ with p53 in a cyclin G_1-independent manner [159]. Moreover, PP2A could indirectly be the target of ATM kinase activity [160]. In Jurkat cells exposed to ionizing radiation, ATM kinase activity leads to dissociation of B/PR55 from corresponding PP2A heterotrimers. The AC dimer or a newly formed ABC heterotrimer may induce cell cycle checkpoint activation and cell cycle arrest. Alternatively, ATM may counterbalance the cyclin G_1-dependent dephoshorylation of MDM2 by PP2A and subsequent inhibition of p53. This way, ATM would stop PP2A-dependent cell cycle progression and promote DNA repair through p53 activation. A direct role for PP2A in p53 regulation is consistent with reports that PP2A regulates cellular genomic stability [161].

Lastly, another key component in the regulation of DNA damage repair is the DNA-dependent protein kinase (DNA-PK), which has structural homologies with ATM, ATR and phosphoinositide 3-kinase (PI 3-kinase). There is strong evidence for a pro-apoptotic role of DNA-PK via its interaction with p53 [162]. PP2A-dependent DNA-PK dephosphorylation induces its activation [163], adding another tier to the involvement of PP2A in DNA damage-sensing pathways.

PP2A and Apoptosis

Apoptosis, or programmed cell death, is an intrinsic process that is essential for animal development and tissue homeostasis. Dysregulation of this tightly controlled mechanism of cell suicide may result in cancer, neurodegenerative diseases, or other pathological conditions. A variety of extracellular and intracellular signals can trigger an apoptotic response, including cross-linking of so-called 'death receptors' (such as the Fas and tumor necrosis factor (TNF) receptors), UV and ionizing radiation, anticancer drugs, growth factor deprivation, and overexpression of certain oncogenes or tumor suppressor genes. Like many physiological processes, commitment to apoptosis is controlled by the reversible phosphorylation of apoptotic signaling proteins. As described below, PP2A plays a critical and multifaceted role in this regulation.

Bcl-2 and Caspases

The Bcl-2 family consists of pro- and anti-apoptotic proteins that integrate signals from survival factors and death stimuli at the outer mitochondrial membrane. They regulate the release of apoptotic activators and subsequent activation of effector caspases, which are the executioners of the apoptotic response. Caspases are an evolutionarily conserved family of cysteine proteases that catalyze the cleavage of target proteins. They mediate the morphological changes characteristic of apoptotic cell death, i.e. membrane blebbing and formation of apoptotic vesicles, cytoplasmic shrinkage, nuclear condensation, and DNA fragmentation.

In a leukemia-derived cell line, toxic concentrations of the lipid second messenger, ceramide, activates and promotes the translocation of PP2A to the mitochondrial membrane. There, PP2A interacts with and dephosphorylates Bcl-2 phosphorylated on Ser^{70}, resulting in inhibition of its anti-apoptotic function [164-166]. The PP2A enzymes involved in that process are B/PR55 or B'/PR61-containing heterotrimers [167]. Notably, Bβ2, a recently identified B subunit isoform, also binds to the mitochondrial membrane. It accelerates neuronal cell death after survival factor deprivation, possibly through Bcl-2 dephosphorylation [168]. Thus, induction of apoptosis may involve multiple PP2A holoenzymes. Bad, another pro-apoptotic member of the Bcl-2 family, is also dephosphorylated by PP2A *in vitro* [169].

PP2A is also a target of caspases. Caspase-3 can cleave PP2A A/PR65 subunit in its C-terminal domain. *In vitro*, this event results in increased PP2A catalytic activity. *In vivo*, it correlates with caspase-3 activation and subsequent commitment of cells to apoptosis. The partial or total loss of the A/PR65 subunit may enhance PP2A activity towards yet unidentified pro-apoptotic substrates involved in the cellular apoptotic response [170,171].

Apoptotic Signaling Pathways

Although the role of PP2A in the modulation of apoptotic cascades cannot be denied, it is far from being completely understood. Indeed, PP2A displays both pro- and anti-apoptotic functions that are cell type- and signal-specific. Such apparent discrepancies probably reflect the involvement of multiple and defined PP2A isoforms that act at different levels in specific cellular apoptotic signaling cascades. For instance, PP2A participates in Fas, TNF-α and

TRAIL apoptotic signaling. In stimulated U937 cells, apoptosis is blocked by the PP2A inhibitor, calyculin A [172]. Likewise, expression of the PP2A inhibitor I_2^{PP2A} protects Hela cells from Fas-mediated apoptosis by increasing MAPK activity [172]. However, PP2A has anti-apoptotic functions in T lymphocytes; in these cells, integrin α2β1 inhibits Fas-mediated apoptosis by PP2A-dependent activation of the MAPK pathway [173]. PP2A is also an essential factor for survival and growth of myeloma cells, via regulation of intracellular reactive oxygen intermediates and anti-apoptotic factors [174]. RNAi studies in *Drosophila* have shown that the loss of the B'/B56 subunit causes apoptotic cell death, pointing to an anti-apoptotic role for PP2A in this model [175,176].

Viral-Induced Cellular Apoptosis

Remarkably, deregulation of PP2A-dependent apoptotic signaling is a preferred target of many viruses. Trigerring of cell death pathways is often utilized by the infected host cell as a defense mechanism to eliminate infected cells and limit viral replication. However, viruses have evolved strategies to evade this initial pro-apoptotic response in the host. For instance, interaction of SV40 small t and Polyomavirus small t and middle T antigens promotes viral replication and cell survival in infected kidney cells [50]. On the other hand, at later stages of infection, some viruses induce apoptosis as a method of cell killing to enhance virus spread. For instance, the adenoviral protein E4orf4 targets PP2A as a strategy to activate apoptotic signal transduction pathways [177-179]. E4orf4 directly associates with the B/PR55 subunit and induces apoptosis in transformed cells. E4orf4 can also interact with the B'/PR61 subunit, but this interaction does not induce cell death. In the human osteosarcoma cell line U2OS, interaction of SV40 small t with PP2A also induces apoptosis [180].

PP2A and Cell Adhesion

Cell-Matrix Adhesion and Migration

The process of metastatis requires migration of tumor cells from the primary tumor into the vasculature and the target tissue [181]. Studies in a variety of tumor types have shown that metastatic tumor cells have increased motility compared to non-metastatic cells [182]. Cellular migration is influenced by the degree of adhesiveness to a substratum, signaling through integrins, and active turnover of focal adhesions [183]. There is, however, a fine balance between adherence/detachment and motility. A strong shift toward focal adhesion formation results in strong adherence with little migration; in contrast, excessive loss of adherence results in the inability to navigate across the substratum. Interactions with a substratum leads to the formation and stabilization of focal adhesions, where the actin skeleton converges with integrins and an interconnection of protein complexes [184,185]. Paxillin is a key adaptor protein involved in the formation of stable adhesions. The ability of paxillin to facilitate interaction between molecules of the focal adhesion is regulated by phosphorylation [186]. Paxillin phosphorylation at discrete Ser/Thr sites modulates both cell adhesion and spreading [187]. It is dephosphorylated by both PP2A and PTPases [188]. The PP2A B'/B56γ1 subunits interact with, co-localizes with, and dephosphorylates paxillin[82]. Other studies performed in Lewis lung carcinoma cells, human HMEC-1 microvascular cells, and human head-and neck squamous carcinoma cells [189,190], also support a key role for PP2A in regulation of cell migration through paxillin. In these cells, inhibition of PP2A leads to dissolution of focal adhesion kinase (FAK)/Src/paxillin complexes, activation of Src, and increased Src-dependent cell motility [189,190].

Cell-Cell Adhesion

Cell-cell adhesion participates in the maintenance of tissue architecture and cell polarity. Reduced cell-cell adhesiveness is a hallmark of human cancers. It allows cancer cells to disobey the social order in tissue, resulting in destruction of the histological structure.

Adherens Junctions

E-cadherin, and its undercoat proteins, catenins, which connect cadherins to actin filaments, are concentrated at adherens junctions (AJs) at the lateral borders of epithelial cells [191]. In mouse embryos, PP2A Cα subunit colocalizes with and dephosphorylates the AJ proteins E-cadherin and β-catenin at the plasma membrane[84]. PP2A may regulate the membrane targeting of E-cadherin and β-catenin, since those proteins are mainly localized in the cytosol in PP2A $C\alpha^{-/-}$ transgenic mice[84]. In this study, PP2A-AJ protein codistribution was analyzed at a very early developmental stage (E6.5), and the PP2A holoenzyme involved was not identified. However, in mature and polarized epithelial cells, and in human colon tissue, the PP2A ABαC heterotrimer interacts with the tight junction (TJ) complex at the apical membrane, but not with E-cadherin at the AJ [192]. Moreover, deregulation of PP2A does not greatly influence E-cadherin localization and AJ formation [89,192]. Therefore, it is possible that the PP2A B'/B56-regulated Wnt signaling responsible for embryonic patterning is involved in the early colocalization of PP2A with AJ proteins [193]. Later on during development and tissue/cell polarization, PP2A may become segregated from AJs, as observed with other TJ proteins.

Tight Junctions

TJs are intercellular junctions adjacent to the apical end of the lateral membrane surface. They perform two important and specialized functions. First, the barrier function of TJs regulates the passage of ions, water and various macromolecules, through paracellular spaces. Secondly, the fence function maintains cell polarity, by preventing intermixing of molecules in the apical membrane with those in the lateral domain. This function is deeply involved in cancer biology, since loss of polarity is a critical step in tumorigenesis. For instance, loss of the TJ proteins, ZO-1 and claudin-7, is associated with certain breast cancers [194,195]. Claudins are responsible for the formation of TJs strands, whereas ZO-1 serves as a scaffold to organize other TJ proteins and is connected to the actin cytoskeleton. Early evidence for a role of PP2A in the control of TJ assembly and structure comes from studies where OA and calyculin A were used on epithelial cells. High concentrations of OA or prolonged incubation of MDCK cells with this inhibitor induce TJ leakiness [196-198]. It is interesting to note that OA and calyculin A are naturally occurring toxins that modulate intestinal paracellular permeability and are responsible for diarrheic shellfish poisoning in humans [196,198]. However, at high concentrations, these inhibitors inhibit PP1, PP2A and PP2A-like phosphatases, so they cannot be utilized to define the biological role of a specific PP2A isoform in the regulation of TJs. Recently, for the first time, it was established that the PP2A ABαC isoform is part of the TJ complex and negatively regulates TJ assembly[192]. Indeed, overexpression of PP2A$_C$ in MDCK cells severely prevents calcium-dependent TJ formation by inhibiting the recruitment of the TJ proteins, ZO-1, occludin and claudin-1, to the membrane during junctional biogenesis. Conversely, inhibition of PP2A activity promotes TJ protein phosphorylation and TJ formation. PP2A activity negatively regulates TJ assembly without appreciably affecting the organization of F-actin and E-cadherin. On the other hand, PP2A may regulate TJ biogenesis by regulating the phosphorylation and activation of PKCζ, an atypical protein kinase C isoform critically required for TJ formation and the establishement of cell polarity [192]. Interestingly, deregulation of PP2A following expression of SV40 small t in MDCK cells is associated with severe defects in the biogenesis and barrier properties of TJs[89]. These effects are associated with a loss of PP2A from areas of cell-cell contact, and altered distribution and reduced cellular amounts of ZO-1, occludin, and claudin-1. They also correlate with deregulation of cell growth, acute morphological changes and multilayering[89]. These data suggest that deregulation of PP2A-dependent cell adhesion is an important contributor to the transforming activity of SV40 small t in epithelial cells.

Desmosomes

Desmosomes are the most basal epithelial adhesions, and are connected with the cytokeratin filamentous network. Inhibition of PP2A following treatment of hepatocytes with microcystin-LR induces disruption of desmoplakin organization at the cell surface and dissociation of desmosomes[199]. However, the precise desmosomal proteins targeted by PP2A remain unknown.

Gap Junctions

Gap junctions (GJs) are aqueous channels connecting adjacent cells with selective permeability for small molecules [200]. These channels are formed when oligomers of connexin proteins in each of the apposing plasma membranes become aligned, providing a continuous passageway between cells. GJs are thought to maintain tissue homeostasis and are implicated in cell growth, development and differentiation [201,202]. They are frequently absent or there expression is reduced in tumors. Expression of recombinant GJ proteins in cancer cell lines slows down growth and even suppresses tumorogenicity [201,202]. Alterations in the phosphorylation state of connexins are involved in a broad variety of GJ functions, i.e. trafficking, assembly/disassembly, and gating. Connexin43 (Cx43) is the major and most studied connexin protein [203]. Its phosphorylation can induce decreased gap junctional intercellular communication [204]. The utilization of inhibitors on different cell types suggest that Cx43 phosphorylation is regulated, either directly or indirectly, by PP2A [205,206].

PP2A and Transcription

Phosphorylation of transcription factors regulates their trans-activating potential, DNA-binding properties and intracellular location. OA treatment of cells suggests a role for PP2A in regulation of their transcriptional activity and gene expression [207]. Inhibition of PP2A by SV40 small t can regulate the transcriptional activity of transcription factors as diverse as CREB, Sp1, AP-1 and NF-κB, and promote cell transformation [9,86]. A growing number of transcription factors appears to be directly regulated by PP2A-dependent dephosphorylation events. These include CREB, AP1, Sp1, STAT3, HOX11, NF-κB/RelA and retinoid X receptor.

CREB

CREB is activated by cAMP-dependent kinase phosphorylation at Ser^{133}. Co-purification experiments and study of cells transfected with SV40 small t suggest that PP2A interacts with CREB and induces CREB inactivation by dephosphorylation [208,209]. CREB mediated transcription is stimulated by coactivators of the four-and-a-half-LIM-only protein family (FHL). While a prerequisite for transcriptional activation of CREB is phosphorylation at Ser^{133}, FHL proteins induce CREB-mediated transcription in a $phosphoSer^{133}$-independent fashion [210]. In NIH 3T3 cells, specific inhibition of PP2A by SV40 small t enhances CREB-dependent FHL2-induced transcription and stimulates intrinsic FHL2 activity [211]. Thus, PP2A is a negative regulator of CREB and FHL proteins.

AP1 (c-Jun, c-Fos)

Microinjection studies with purified PP2A enzymes, and I_2^{PP2A} expression analysis have demonstrated that PP2A enhances *c-Jun* activity through dephosphorylation of the Ser^{63} residue [58,212]. This induces *c-Jun* and activator protein-1 (a complex of *c-Jun* and *c-Fos*) transcriptional activity. PP2A indirectly controls *c-Fos* function by regulating the *c-Fos* promoters, and therefore *c-Fos* expression. This is achieved through PP2A regulation of the transcription factors Sp1, CREB, STAT3 and even AP-1, which all can bind the *c-Fos* promoter and drive *c-Fos* expression [213].

Sp1

PP2A has been implicated in the regulation of Sp1 activity in CV-1 cells[19] and T lymphocytes [214]. In CV-1 cells, both AC dimers and PR55/B-containing heterotrimers dephosphorylate Sp1[19]. Overexpression of PP2A B'/B56 subunits inhibits Sp1-dependent promoter transactivation. Expressed SV40 small t induces Sp1-dependent promoter transactivation. Moreover, in SV40-infected cells, early SV40 DNA replication is promoted by small t-stimulated Sp1 activity; in turn, this may promote the transformation of resting cells by SV40[19]. In contrast to CV-1 cells wherein PP2A interacts with, dephosphorylates and inactivates Sp1, Sp1 dephosphorylation by PP2A in dividing T lymphocytes induces Sp1 activation and is cell cycle dependent [214]. These discrepancies may result from differential Sp1 regulation through different pathways and maybe through different PP2A heterotrimers.

STAT3/STAT6

PP2A dephosphorylates STAT3 and down-regulates STAT3 trans-activating potential [215,216]. There is also evidence that PP2A regulates IL4-induced transcriptional activity of STAT6, either through direct binding to STAT6, or indirectly, by regulating a STAT6 Ser kinase downstream of JAK kinases [217].

HOX11

$PP2A_C$ interacts with and has been implicated in the regulation of HOX11, an homeobox transcription factor controlling genesis of the spleen formation, and displaying oncogenic properties [111]. *In vitro*, HOX11 suppresses PP2A activity. *In vivo,* it disrupts a G_2/M cell cycle checkpoint, since microinjection of HOX11 into *Xenopus* oocytes induces premature G_2-M progression. It remains to be established whether $PP2A_C$ controls HOX11 transcriptional activity or DNA binding.

NF-κB (RelA)

The Nuclear factor-κB (NF-κB)/Rel transcription factors are key regulators of a variety of genes involved in inflammation, growth, differentiation, apoptosis and development [218]. This family of proteins includes RelA, c-Rel, NF-κB1 (p50/p105), NF-κB2 (p52/p100) and RelB. NF-κB is phosphorylated by I-κB kinase. *In vitro* studies indicate that phosphorylation of NF-κB enhances its DNA binding ability [219]. Studies with OA have shown that PP2A is a critical regulator of NF-κB [220,221]. This property is utilized by viruses to deregulate NF-κB-controlled survival pathways. Inhibition of PP2A by SV40 small t induces NF-κB-dependent transactivation [222]. Human T-lymphotropic virus type I tax activates I-κB kinase by inhibiting I-κB kinase-associated PP2A [223]. In melanocytes, PP2A directly interacts with RelA through the A/PR65 subunit, resulting in inhibition of its transcriptional activity [224]. Interestingly, when compared to normal melanocytes, PP2A activity in melanoma cells is reduced and correlates with enhanced RelA transcriptional activity. Thus, PP2A may have tumor-suppressive properties in the cells.

Retinoid X Receptor

T lymphocyte activation signals regulate the expression and transactivation function of the retinoid X receptor (RXR) α through the interplay of complex signaling cascades. RXRs play a central role as nuclear transcription factors by homodimerizing or heterodimerizing with many other members of the retinoid and steroid receptor family of transcription factors [225]. PP2A negatively regulates the levels and transcriptional activation function of RXRα in T cells [226].

PP2A and the Cytoskeleton

Cell morphology is maintained and controlled through three major cytoskeletal structures: F-actin, microtubules and intermediate filaments. They play important functions in maintenance of cell architecture, regulation of cell-cell adhesive junctions, organelle transport, cell migration and differentiation. It is increasingly evident that protein phosphatases play a key role in regulating the assembly/disassembly of these cytoskeletal structures, either via direct interaction with cytoskeletal proteins, or indirectly, through the modulation of signal transduction pathways. PP2A mutant analyses in yeast [119], *Drosophila* [119] and *Xenopus* [227] show that PP2A regulates actin and microtubule dynamics during mitosis, meiosis and cytokinesis. Biochemical data also suggest a role for PP2A in regulation of intermediate filaments [228-231]. There are clear correlations between certain cancer phenotypes and changes in the properties of cytoskeletal proteins. Some of these phenotypes could result in part from PP2A deregulation. For instance, deregulation of PP2A leads to increased cell motility and invasiveness in endothelial cells through disorganization of the actin and microtubule cytoskeleton [82,232,233].

Actin

OA has been widely reported to induce changes in the F-actin cytoskeleton [234-238]. Usually, inhibition of phosphatase activity leads to disorganization and depolymerzation of F-actin. This well documented effect of OA is even used as a diagnostic tool to determine whether patients have been poisoned by diarrheic shellfish toxins [239]. Some data suggest that OA-induced F-actin depolymerization could be the result of actin phosphorylation by kinases no longer controlled by PP2A. Indeed, treatment of renal epithelial LLC-PK1 cells or sea urchin eggs with calyculin A leads to increased F-actin phosphorylation and subsequent depolymerization [240,241]. These observations are confirmed by studies done with Lewis lung carcinoma (LLC) cells. Comparative analysis shows that LLC-N7 cells, which are highly motile and invasive, have a deficiency in PP2A compared to nonmetastatic LLC-C8 cells. Conversely, OA-treated LLC-C8 cells become migratory and invasive. In these cells, OA causes the redistribution of F-actin towards the periphery of the cells, and eventually a loss of the filamentous actin network[242]. These data suggest an active role for PP2A in cell migration.

Cell-matrix adhesion and migration are important processes that control biological mechanisms such as embryogenesis, angiogenesis, tissue repair, inflammation, invasion, and metastasis. Cell migration is dependent on the extensive remodeling of the actin cytoskeleton that leads to formation of membrane structures called lamellipodia and filopodia [188]. Polymerization of new actin filaments provides the framework for generating these structures. The spatio-temporal modulation of actin polymerization, depolymerization and branching is required for maintenance of actin-containing structures at the leading edge of the cells. These processes are tightly controlled by actin-binding proteins, including the actin depolymerizing factor (ADF)/cofilin family of proteins. ADF and cofilin are actin-binding proteins that have filament disassembly/severing activity and are required for breaking down older parts of the actin network so that the resulting globular (G)-actin subunits can be recycled and reused [243]. In T lymphocytes, PP2A associates with and dephosphorylates cofilin, inducing its activation. Dephosphorylated cofilin associates with actin cytoskeleton, confirming an active role for PP2A in F-actin remodeling [244]. Interestingly, in autonomously proliferating T lymphoma cells, cofilin is spontaneously dephosphorylated and translocates to the nucleus [245]. This may contribute to autonomous cell division and prolonged survival of malignant cells. Indeed, T lymphoma cell incubation with OA is accompanied by apoptosis [245].

F-actin is also involved in the structures that establish and maintain cell-matrix and cell-cell contacts, such as focal adhesions, AJs, and TJs. PP2A is part of these complexes through its interaction with paxillin[82] and TJ proteins[192]. PP2A also plays a role in F-actin organization at the TJs[89]. Indeed, PP2A functional disruption by SV40 small t induces a marked disorganization of the actin cytoskeleton in MDCK cells. Small-t-mediated F-actin rearrangements encompass increased Rac1-induced membrane ruffling and lamellipodia,

Cdc42-initiated filopodia and loss of RhoA-dependent stress fibers. These F-actin changes coincide with elevated levels of Rac1 and Cdc42 and decreased amounts of RhoA in small t-expressing cells[89]. Rac1, Cdc42 and RhoA are small GTPases essential in the control of cell migration, owing to their effects on actin remodeling. These results suggest that PP2A could have a direct role in small GTPase-regulated F-actin polymerization and rearrangement at the cell periphery. It is worth mentioning that Rac1 and Cdc42 are at the core of signaling pathways that lead to junctional assembly. Since alterations in the actin cytoskeleton have been linked to loss of cell polarity and tumor invasiveness, the PP2A-dependent disorganization of the F-actin cytoskeleton induced by SV40 small t in selective cell types [89,246] likely contributes to the role of SV40 in cell transformation.

Tubulin

The importance of PP2A in the regulation of microtubule structure and dynamics during the cell cycle has been unveiled in genetic studies. In *S.cerevisiae,* $PP2A_C$ mutants induce a blockage of cells into G_2 because mitotic spindles are not able to form or extend [94,95], and B/PR55 homologue mutants lack a functional kinetochore/spindle assembly checkpoint [131,132]. Disruption of the *paa1* gene (A/PR65 homologue) in *S.pombe* causes anomalies in microtubule distribution [91]. During embryogenesis, *Drosophila* $PP2A_C$ mutants exhibit blocks in mitosis between prophase and anaphase[97]. These $PP2A_C$-negative embryos possess multiple centrosomes with disorganized, elongated, star-shaped microtubule arrays. These data suggest that PP2A is required for the attachment of microtubules to chromosomal DNA at the kinetochore, and the proper initiation of anaphase. In *Xenopus* eggs, PP2A is required to maintain the short-steady length of microtubules during mitosis, in part by regulating Op18/stathmin, a molecule involved in microtubule dynamics [247].

In mammalian cells, it has been reported that a pool of ABαC heterotrimers directly binds to microtubules *in vitro* and *in vivo* [232,248-250], and this microtubule-associated PP2A pool is regulated during the cell cycle [248]. Besides ABαC [248], the PP2A holoenzyme comprising the PR130 subunit has also been found associated with the centrosome [251]. Interestingly, deregulation of PP2A by SV40 small t blocks centrosome functions [252]. PP2A targeting to the microtubule cytoskeleton in part involves the direct interaction of $PP2A_C$ with assembled tubulin, and is associated with a pronounced inhibition of its catalytic activity [249,250]. Cytoskeletal anchoring may serve to sequester selective intracellular pools of enzymes in an inactive state, and promote kinase-mediated phosphorylation of microtubule-associated proteins [249]. PP2A also regulates microtubule stability in hepatocytes [253], and during podocyte process formation in the newborn kidney [254]. PP2A thus plays an active role in microtubule dynamics, especially during development.

Intermediate Filaments

Intermediate filaments (IFs) are major components of the cytoskeleton and nuclear envelope [255]. Cytoplasmic IF proteins include vimentin in fibroblasts, cytokeratins in epithelial cells, neurofilaments and/or glial fibrillary acidic protein in neuronal tissues, and desmin in muscle cells. IF proteins are among the most prominent cellular phosphoproteins, and reversible phosphorylation plays a key role in their dynamic rearrangement [256]. *In vitro* and *in vivo* experiments show that IF proteins can be phosphorylated by various protein kinases, resulting in IF disassembly. PP2A maintains IF structural organization. Inhibition of PP2A by OA or related compounds leads to hyperphosphorylation and disassembly of vimentin IF networks in fibroblasts, 9L rat brain tumor cells, and oligodendrocyte precursor cells [228-231]. Incubation of hepatocytes with microcystin-LR induces a dramatic reorganization of the IF proteins keratin 8 and keratin 18, resulting in desmosomal disassembly [199]. While it is not clear whether PP2A regulates keratin 8/18 phosphorylation by direct protein-protein interaction, recent studies clearly show that vimentin dephosphorylation occurs through direct binding to PP2A [23,257], probably via the B/PR55 subunit[23].

PP2A and Cell Growth Signaling

Extracellular stimuli produce a whole set of cellular responses, such as proliferation, differentiation, and apoptosis. The utilization of specific inhibitors and viral proteins has demonstrated that PP2A controls all major cellular signaling pathways in an isoform-dependent manner, and by acting at multiple levels in these cascades (Figure 2). PP2A deregulation results in subsequent alterations of these pathways, which explains its association with many cancer phenotypes. Some of the better characterized roles of PP2A in signal transduction derive from studies of the Wnt signaling, MAP kinase cascade (Figure 2) and mTOR pathway.

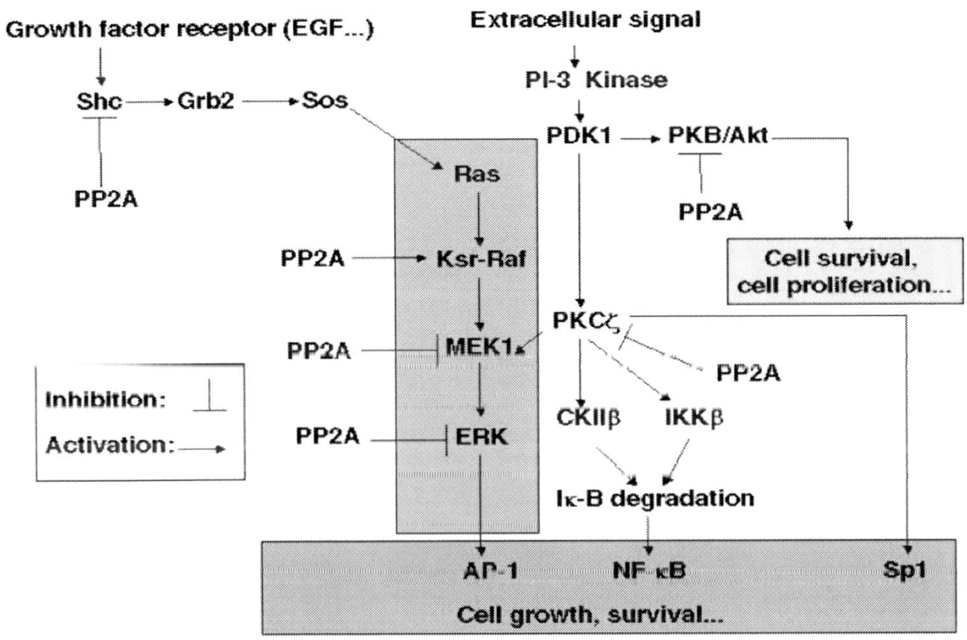

Figure 2. Simplified model of the regulation of MAP kinase signaling by PP2A.

WNT Signaling

The adenomatous polyposis coli (APC) protein, axin and glycogen synthase kinase 3β (GSK3β) form a Wnt-regulated signaling complex that mediates the phosphorylation-dependent degradation of β-catenin [258]. Dysregulation of Wnt/β-catenin signaling disrupts axis formation in vertebrate embryos, and underlies multiple human malignancies, including colon, skin and liver cancers. Axin promotes the phosphorylation of β-catenin by GSK-3β, leading to its elimination. β-catenin removal and subsequent degradation results in inhibition of Wnt signaling. The presence of Wnt normally leads to GSK3β inactivation and accumulation of unphosphorylated β-catenin. β-catenin can then translocate to the nucleus and transactivate Wnt target genes in cooperation with the Lef/TCF transcription factors [258]. PP2A regulates Wnt signaling at several levels in an isoform-dependent manner [193,259-261]. Yeast-two hybrid studies have identified axin as the target of the $PP2A_C$ [262], and the B'/B56β and B'/B56γ subunits [260], and APC as the target of the B'/B56ε subunit[193]. B56ε specifically inhibits Wnt signaling and appears to be essential for β-catenin degradation [260,261]. B56β also inhibits Wnt signaling by affecting β-catenin-induced TCF activation, but without changing β-catenin stability. In contrast, microinjection experiments in *Xenopus* eggs show that $PP2A_C$ has a positive role in Wnt signal transduction, downstream of β-catenin [259].

MAP Kinase Pathway

The mitogen activated protein kinase (MAPK) cascade plays a central role in mediating transcriptional regulation in response to stimulation of cell surface growth factor receptors. The core structure of each MAPK signaling pathway is a module of three kinases: MAP kinase kinase kinase, MAP kinase kinase and MAP kinase. There are four subgroups in the MAP kinase family: ERK, p38, JNK/SAPK and ERK5 [263,264]. ERK is primarily activated by mitogenic stimuli, whereas p38 and JNK/SAPK are activated by stress stimuli or inflammatory cytokines. ERK5 is activated by EGF, NGF, osmotic stresses and oxidative stresses. Activation of these kinases leads to variable cell-type specific responses, e.g. gene expression, proliferation, differentiation, cell cycle arrest, apoptosis and early development.

The Ras/Raf/MEK/ERK MAP kinase cascade plays a central role in mediating cell cycle re-entry and transcriptional responses downstream of many cell-surface receptors. Activation of this pathway is normally transient; its constitutive activation is sufficient to cause tumorigenic transformation of susceptible cells [265]. Numerous observations indicate that PP2A plays a major regulatory role at multiple levels in the cascade (Figure 2). B/PR55-containing PP2A holoenzymes can directly dephosphorylate and inactivate MEK and ERK[55,176]. In wound-healing experiments, β2-adrenergic receptor stimulation inhibits keratinocyte migration; this process involves PP2A activation and interaction with ERK2, and subsequent ERK2 inhibition [266]. A cholesterol-regulated PP2A-containing complex in caveolae is also implicated in the regulation of ERK activity in fibroblasts[27]. Upon Ras stimulation, PP2A also interacts with and dephosphorylates Raf-1 and the protein Kinase suppressor of Ras (Ksr) [267,268]. This causes a relocation of both Raf-1 and Ksr from the cytosol to the plasma membrane, and the activation of MEK by Raf-1 [267,268]. Analysis of insulin, EGF, or insulin-like growth factor 1-stimulated signaling cascades indicates that PP2A interacts with the protein Shc and inhibits the Ras/Raf-1/MEK/ERK pathway [269].

Upon insulin or EGF stimulation, Shc becomes phosphorylated at Tyr^{317}, dissociates from PP2A, then interacts with the protein Grb2, leading to stimulation of the MAPK pathway. In this case, PP2A appears to play the role of a scaffolding protein; its catalytic activity is not required. Atypical PKCζ also regulates mitogenic MAPK signaling through Raf-independent activation of MEK [222]. PP2A can control MEK activation through its ability to directly regulate PKCζ phosphorylation and activation [222]. The association of calmodulin-kinase II with PP2A modulates activation of the MAPK cascade [270]. The kinase PKB/Akt regulates the MAPKpathway through phosphorylation and inhibition of Raf-1[9]. PKB/Akt kinase activity is inhibited by PP2A[118]. All these studies point to a critical role for PP2A in regulation of the MAPK cascade. However, it is important to note that the differential regulatory effects of PP2A in this cascade likely result from the presence of cell-type specific PP2A holoenzymes and regulatory factors.

PP2A and the Nutrient Pathway (TOR)

The TOR (target of rapamycin) pathway controls cellular functions necessary for growth of yeast and larger eukaryotes. mTOR (mammalian TOR) functions in a mitogen-inducible pathway and causes phosphorylation of the translation inhibitor 4E-BP1 and $p70^{S6}$kinase, resulting in phosphorylation of 40S ribosomal protein S6 and initiation of translation [271]. ABC heterotrimers dephosphorylate and inactivate mTOR targets, including $p70^{S6}$kinase and 4E-BP1. Yeast and mammalian TOR proteins regulate PP2A activity through a protein called TAP42 in yeast and alpha4 in mammals. When $PP2A_C$ interacts with TAP42/α4, it becomes unable to dephosphorylate 4E-BP1 [272]. In yeast, TAP42/α4 is phosphorylated by TOR, and phosphorylated TAP42/α4 competes with the Aα and Bα subunits for binding to $PP2A_C$. On the other hand, PP2A heterotrimers dephosphorylate TAP42/α4 and inhibit its binding to $PP2A_C$. Thus, PP2A subunit composition critically regulates TOR signaling. Recent data in yeast also suggest a role for the $PP2A_C$-TAP42 complex in controlling the polarized distribution of actin during the cell cycle [273].

Integrin Signaling

Integrins are a large family of heterodimeric transmembrane receptors composed of α and β subunits. In addition to mediating cell-matrix interactions, integrins activate intracellular signaling pathways which, in collaboration with growth factor-induced signals, regulate cellular functions like migration, growth, survival and differentiation [274]. Integrin signaling is complex and depends on which cellular receptor is stimulated [275]. Activated targets include protein tyrosine kinases of the Src and FAK families, ERK (via Fyn and Shc stimulation), PI-3 kinase, and p38. There is increasing evidence for PP2A involvement in integrin signaling at different levels. PP2A binds to and inhibits Src activity *in vitro* [276]. PP2A also interacts with the Src kinase substrate, Cas [277], a major tyrosine phosphorylated protein in cells transformed by *c-Src* or *v-Crk* [278]. Src/Cas proteins also interact with FAK in a phosphorylation-dependent manner, and regulate the formation/dissociation of focal adhesion complexes [279]. In NIH 3T3 cells, PP2A may be involved in the cell cycle-specific dephosphorylation of Cas and mitotic arrest during integrin-mediated signaling [277]. Integrins signaling for cell survival activates a PI 3-kinase/PKB pathway that does not involve the MAPK cascade. PP2A interacts with integrins and directly regulates PKB and

GSK3β activities [280,281]. In HEK293 cells, PP2A co-localizes with integrins and regulates PKCε, which serves as a signal integrator between cytokine and integrin signaling pathways [282].

PP2A in Development and Differentiation

Proper signaling is a prerequisite for adequate embryonic patterning and cell differentiation. Wnt signaling is involved in *Drosophila*, *Xenopus* and mammalian development. Mutations in components of this signaling cascade leads to defects in morphogenesis and cancer phenotypes, i.e. colorectal cancer, hepatocellular carcinoma, kidney cancer, medulloblastoma [258]. Many studies performed in different organisms show that PP2A plays a crucial role in morphogenesis and development through its modulation of Wnt signaling and additional pathways. Inhibition of PP2A activity can induce severe malformations in embryos, defects in cell differentiation, and cancer. Inhibition of PP2A in *Cassiopea andromeda* can produce an additional head or foot [283]. In *C.elegans*, PP2A B subunit regulates vulval induction [284]. In *Xenopus*, dorsal injection of PP2A B56α, but not the Bα subunit, inhibits Wnt signaling and ventralises the embryo by disruption of the primary axis [193,261]. PP2A is also implicated in many aspects of development in *Drosophila* (wings, eyes) and chicken embryo limb patterning [285]. In mice, knock-out of PP2A Cα subunit is embryonic lethal [286]. In rat, PP2A has a significant role in early lung development, by regulating the mesenchymal cell cycle and differentiation of the respiratory epithelium [287]. In human, defects in the PP2A B55β subunit gene are associated with spino cerebellar ataxia [288]. Mutations in the PP2A Aα subunit gene are found in lung and breast carcinoma, and melanoma[145]. Interaction of the alpha4 subunit with PP2A also participates in the pathogenesis of Opitz BBB/G syndrome, a genetical disorder characterized by malformations of the ventral midline and developmental defects including craniofacial and cardiac anomalies [289].

PP2A regulates embryo patterning by modulating signal transduction pathways and gene expression that are exquisitely controlling the balance between cell proliferation, differentiation and death. Among the signaling cascades affected by PP2A are: the TGFβI pathway [285]; distinct MAP kinase pathways involving TAK1, MKK4-JNK, MKK6-p38 [285], Raf[16,] [290] and other proteins of the Ras/Raf/MEK/ERK pathway [291]; the PI-3 kinase/PKB/Bad pathway [285]; and, not surprisingly, the planar cell polarity signaling pathway [292], which includes the Wnt signaling pathway [293-295]. Examples of transcription factors directly regulated by PP2A during development are numerous. For example, in the Hedgehog signal transduction pathway, PP2A is believed to dephosphorylate a putative transcription factor involved in the COUP-TFII, Islet-1 and Gli regulation [296]. In *Drosophila*, PP2A B'/PR61 subunit positively modulates *Sex combs reduced* (*SCR*), a HOX protein that determines the identity of the labial and prothoracic segments [297]. Recent data show that Zinc finger homeodomain enhancer-binding protein or *Zfhep* is the target of PP2A *in vitro* [298]. *Zfhep* is member of an unusual family of transcription factors. It is essential for early embryonic T cell precursor differentiation, craniofacial development, and skeletal patterning [299]. The PP2A B56δ subunit interacts and dephosphorylates the bHLH factors

HAND1 and *HAND2*, which are required for heart, vascular, neuronal, limb, and extraembryonic development [300]. During RCHOI trophoblast stem cell differentiation, B56δ expression is downregulated and *HAND1* phosphorylation increases [300]. Chondrogenesis is a fundamental process of limb formation that is also regulated by PP2A through its ability to dephosphorylate CREB [301]. Lastly, as mentioned previously, STAT6, a transcription factor involved in T and B cell maturation, is also regulated by PP2A[217].

CONCLUSIONS

Studies from the past decade decisively point to a critical role of PP2A in cell transformation. Clearly, by interacting with specific targets (Table 1), PP2A enzymes regulate multiple pathways and cellular functions that are implicated in tumorigenesis. Mutations in certain PP2A subunit genes are associated with cancer. In addition, changes in the endogenous expression and/or subunit composition of PP2A by cellular/viral factors all induce profound changes in PP2A activity, targeting and substrate specificity. In turn, PP2A dysfunction is sufficient for inducing deregulation of major signaling cascades involved in the control of gene expression, cell growth and survival.

Table 1. List of major cellular targets that have been identified to interact with Specific PP2A subunits or holoenzymes, and may contribute to its role in cell transformation

PP2A	Interacting protein	Major cellular functions
A	NF-κB/RelA	Growth, differentiation...
	Caspase-3	Apoptosis
C	α4/Tap42	Growth, metabolism, development...
	Axin	Development, Wnt signaling...
B subunit-containing PP2A enzymes	KSR, Raf-1	Growth, proliferation...
	PKCζ	Growth, adhesion, survival...
	Microtubules	Cell cycle, differentiation...
	Sp1	Growth, differentiation...
	Vimentin	IF dynamics...
B' subunit-containing PP2A enzymes	Bcl-2	Apoptosis
	Axin	Development, Wnt signaling...
	APC	Development, Wnt signaling...
	Paxillin	Adhesion, migration...
	Cyclin G2	Cell cycle...
	Cyclin G1	Cell cycle...
	HAND1/HAND2	Development
B" subunit-containing PP2A enzymes	Cdc6	DNA replication
	p107, p130	DNA repair

Thus, it is becoming clear that PP2A is far from being a simple housekeeping enzyme as previously thought. However, many questions remain to be solved to fully understand the role PP2A in cell regulation. Firstly, much remains to be learned about the unique mechanisms that regulate PP2A activity, expression and holoenzyme assembly. Secondly, additional studies are needed to assign defined cellular functions to specific PP2A holoenzymes. Thirdly, growing evidence indicates that protein-protein interactions among PP2A subunits and their cellular/viral partners play a crucial role in establishing PP2A functional specificity (Table 1). Yet, the nature of many PP2A-interacting parners is still unknown, and novel PP2A-containing multi-protein complexes are still waiting to be uncovered. Yet, recent discoveries unveil the exciting possibility that specific PP2A holoenzymes may be crucial therapeutic targets for future strategies aimed at preventing tumorigenesis or counteracting cancer progression.

REFERENCES

[1] Shenolikar S. Protein serine/threonine phosphatases--new avenues for cell regulation. *Annu Rev. Cell Biol.* 1994;10:55-86.

[2] Wera S, Hemmings BA. Serine/threonine protein phosphatases. *Biochem. J.* 1995 Oct;311 (Pt 1):17-29.

[3] Barford D. Molecular mechanisms of the protein serine/threonine phosphatases. *Trends Biochem. Sci.* 1996 Nov;21(11):407-12.

[4] Cohen PT. Novel protein serine/threonine phosphatases: variety is the spice of life. *Trends Biochem. Sci.* 1997 Jul;22(7):245-51.

[5] Turowski P, Favre B, Campbell KS, Lamb NJ, Hemmings BA. Modulation of the enzymatic properties of protein phosphatase 2A catalytic subunit by the recombinant 65-kDa regulatory subunit PR65alpha. *Eur. J. Biochem.* 1997 Aug;248(1):200-8.

[6] Groves MR, Hanlon N, Turowski P, Hemmings BA, Barford D. The structure of the protein phosphatase 2A PR65/A subunit reveals the conformation of its 15 tandemly repeated HEAT motifs. *Cell* 1999 Jan;96(1):99-110.

[7] Janssens V, Goris J. Protein phosphatase 2A: a highly regulated family of serine/threonine phosphatases implicated in cell growth and signalling. *Biochem. J.* 2001 Feb;353(Pt 3):417-39.

[8] Janssens V, Goris J. Protein phosphatase 2A: a highly regulated family of serine/threonine phosphatases implicated in cell growth and signalling. *Biochem. J.* 2001 Feb;353(Pt 3):417-39.

[9] Sontag E. Protein phosphatase 2A: the Trojan Horse of cellular signaling. *Cell Signal* 2001 Jan;13(1):7-16.

[10] Ruediger R, Hentz M, Fait J, Mumby M, Walter G. Molecular model of the A subunit of protein phosphatase 2A: interaction with other subunits and tumor antigens. *J. Virol.* 1994 Jan;68(1):123-9.

[11] Hemmings BA, Adams-Pearson C, Maurer F, Muller P, Goris J, Merlevede W, Hofsteenge J, Stone SR. alpha- and beta-forms of the 65-kDa subunit of protein

phosphatase 2A have a similar 39 amino acid repeating structure. *Biochemistry* 1990 Apr;29(13):3166-73.

[12] Kremmer E, Ohst K, Kiefer J, Brewis N, Walter G. Separation of PP2A core enzyme and holoenzyme with monoclonal antibodies against the regulatory A subunit: abundant expression of both forms in cells. *Mol. Cell Biol.* 1997 Mar;17(3):1692-701.

[13] Kamibayashi C, Estes R, Lickteig RL, Yang SI, Craft C, Mumby MC. Comparison of heterotrimeric protein phosphatase 2A containing different B subunits. *J. Biol. Chem.* 1994 Aug;269(31):20139-48.

[14] Cegielska A, Shaffer S, Derua R, Goris J, Virshup DM. Different oligomeric forms of protein phosphatase 2A activate and inhibit simian virus 40 DNA replication. *Mol. Cell Biol.* 1994 Jul;14(7):4616-23.

[15] Sontag E, Nunbhakdi-Craig V, Lee G, Bloom GS, Mumby MC. Regulation of the phosphorylation state and microtubule-binding activity of Tau by protein phosphatase 2A. *Neuron* 1996 Dec;17(6):1201-7.

[16] Strack S. Overexpression of the protein phosphatase 2A regulatory subunit Bgamma promotes neuronal differentiation by activating the MAP kinase (MAPK) cascade. *J. Biol. Chem.* 2002 Nov;277(44):41525-32.

[17] Shu Y, Yang H, Hallberg E, Hallberg R. Molecular genetic analysis of Rts1p, a B' regulatory subunit of Saccharomyces cerevisiae protein phosphatase 2A. *Mol. Cell Biol.* 1997 Jun;17(6):3242-53.

[18] Zhao Y, Boguslawski G, Zitomer RS, DePaoli-Roach AA. Saccharomyces cerevisiae homologs of mammalian B and B' subunits of protein phosphatase 2A direct the enzyme to distinct cellular functions. *J. Biol. Chem.* 1997 Mar;272(13):8256-62.

[19] Garcia A, Cereghini S, Sontag E. Protein phosphatase 2A and phosphatidylinositol 3-kinase regulate the activity of Sp1-responsive promoters. *J. Biol. Chem.* 2000 Mar;275(13):9385-9.

[20] Ruediger R, Brewis N, Ohst K, Walter G. Increasing the ratio of PP2A core enzyme to holoenzyme inhibits Tat- stimulated HIV-1 transcription and virus production. *Virology* 1997 Nov;238(2):432-43.

[21] Ludowyke RI, Holst J, Mudge LM, Sim AT. Transient Translocation and Activation of Protein Phosphatase 2A during Mast Cell Secretion. *J. Biol. Chem.* 2000 Mar;275(9):6144-52.

[22] Ito A, Koma Y, Sohda M, Watabe K, Nagano T, Misumi Y, Nojima H, Kitamura Y. Localization of the PP2A B56gamma regulatory subunit at the Golgi complex: possible role in vesicle transport and migration. *Am. J. Pathol.* 2003 Feb;162(2):479-89.

[23] Turowski P, Myles T, Hemmings BA, Fernandez A, Lamb NJ. Vimentin dephosphorylation by protein phosphatase 2A is modulated by the targeting subunit B55. *Mol. Biol Cell* 1999 Jun;10(6):1997 2015.

[24] Dagda RK, Zaucha JA, Wadzinski BE, Strack S. A developmentally regulated, neuron-specific splice variant of the variable subunit B-beta targets protein phosphatase 2A to mitochondria and modulates apoptosis. *J. Biol. Chem.* 2003 Apr.

[25] Takahashi M, Shibata H, Shimakawa M, Miyamoto M, Mukai H, Ono Y. Characterization of a novel giant scaffolding protein, CG-NAP, that anchors multiple

signaling enzymes to centrosome and the golgi apparatus. *J. Biol. Chem.* 1999 Jun;274(24):17267-74.

[26] Tehrani MA, Mumby MC, Kamibayashi C. Identification of a novel protein phosphatase 2A regulatory subunit highly expressed in muscle. *J. Biol. Chem.* 1996 Mar;271(9):5164-70.

[27] Wang PY, Liu P, Weng J, Sontag E, Anderson RG. A cholesterol-regulated PP2A/HePTP complex with dual specificity ERK1/2 phosphatase activity. *EMBO J.* 2003 Jun;22(11):2658-67.

[28] Chen J, Martin BL, Brautigan DL. Regulation of protein serine-threonine phosphatase type-2A by tyrosine phosphorylation. *Science* 1992 Aug;257(5074):1261-4.

[29] Chen J, Parsons S, Brautigan DL. Tyrosine phosphorylation of protein phosphatase 2A in response to growth stimulation and v-src transformation of fibroblasts. *J. Biol. Chem.* 1994 Mar;269(11):7957-62.

[30] Guo H, Reddy SA, Damuni Z. Purification and characterization of an autophosphorylation-activated protein serine threonine kinase that phosphorylates and inactivates protein phosphatase 2A. *J. Biol. Chem.* 1993 May;268(15):11193-8.

[31] Usui H, Inoue R, Tanabe O, Nishito Y, Shimizu M, Hayashi H, Kagamiyama H, Takeda M. Activation of protein phosphatase 2A by cAMP-dependent protein kinase-catalyzed phosphorylation of the 74-kDa B" (delta) regulatory subunit in vitro and identification of the phosphorylation sites. *FEBS Lett* 1998 Jul;430(3):312-6.

[32] Xu Z, Williams BR. The B56alpha regulatory subunit of protein phosphatase 2A is a target for regulation by double-stranded RNA-dependent protein kinase PKR. *Mol. Cell Biol.* 2000 Jul;20(14):5285-99.

[33] Lee J, Stock J. Protein phosphatase 2A catalytic subunit is methyl-esterified at its carboxyl terminus by a novel methyltransferase. *J. Biol. Chem.* 1993 Sep;268(26):19192-5.

[34] Xie H, Clarke S. Protein phosphatase 2A is reversibly modified by methyl esterification at its C-terminal leucine residue in bovine brain. *J. Biol. Chem.* 1994 Jan;269(3):1981-4.

[35] Favre B, Zolnierowicz S, Turowski P, Hemmings BA. The catalytic subunit of protein phosphatase 2A is carboxyl-methylated in vivo. *J. Biol. Chem.* 1994 Jun;269(23):16311-7.

[36] Xie H, Clarke S. Methyl esterification of C-terminal leucine residues in cytosolic 36-kDa polypeptides of bovine brain. A novel eucaryotic protein carboxyl methylation reaction. *J. Biol. Chem.* 1993 Jun;268(18):13364-71.

[37] George RR, Harris R, Nunn CM, Cramer R, Djordjevic S. Chaperonin assisted overexpression, purification, and characterisation of human PP2A methyltransferase. *Protein Expr. Purif.* 2002 Nov;26(2):266-74.

[38] Leulliot N, Quevillon-Cheruel S, Sorel I, Li De La Sierra-Gallay, Collinet B, Graille M, Blondeau K, Bettache N, Poupon A, Janin J, Van Tilbeurgh H. Structure of PPM1, a leucine carboxy methyltransferase involved in the regulation of protein phosphatase 2A activity. *J. Biol. Chem.* 2003 Dec.

[39] Xie H, Clarke S. An enzymatic activity in bovine brain that catalyzes the reversal of the C-terminal methyl esterification of protein phosphatase 2A. *Biochem. Biophys. Res. Commun* 1994 Sep;203(3):1710-5.

[40] Lee J, Chen Y, Tolstykh T, Stock J. A specific protein carboxyl methylesterase that demethylates phosphoprotein phosphatase 2A in bovine brain. *Proc. Natl. Acad. Sci. USA* 1996 Jun;93(12):6043-7.

[41] Ogris E, Du X, Nelson KC, Mak EK, Yu XX, Lane WS, Pallas DC. A Protein Phosphatase Methylesterase (PME-1) Is One of Several Novel Proteins Stably Associating with Two Inactive Mutants of Protein Phosphatase 2A. *J. Biol. Chem.* 1999 May;274(20):14382-91.

[42] Wei H, Ashby DG, Moreno CS, Ogris E, Yeong FM, Corbett AH, Pallas DC. Carboxymethylation of the PP2A Catalytic Subunit in Saccharomyces cerevisiae Is Required for Efficient Interaction with the B-type Subunits Cdc55p and Rts1p. *J. Biol. Chem.* 2001 Jan;276(2):1570-7.

[43] Yu XX, Du X, Moreno CS, Green RE, Ogris E, Feng Q, Chou L, McQuoid MJ, Pallas DC. Methylation of the Protein Phosphatase 2A Catalytic Subunit Is Essential for Association of Balpha Regulatory Subunit But Not SG2NA, Striatin, or Polyomavirus Middle Tumor Antigen. *Mol. Biol. Cell* 2001 Jan;12(1):185-99.

[44] Evans DR, Hemmings BA. Mutation of the C-terminal leucine residue of PP2Ac inhibits PR55/B subunit binding and confers supersensitivity to microtubule destabilization in Saccharomyces cerevisiae. *Mol. Gen. Genet* 2000 Nov;264(4):425-32.

[45] Chung H, Nairn AC, Murata K, Brautigan DL. Mutation of Tyr307 and Leu309 in the protein phosphatase 2A catalytic subunit favors association with the alpha 4 subunit which promotes dephosphorylation of elongation factor-2. *Biochemistry* 1999 Aug;38(32):10371-6.

[46] Ogris E, Gibson DM, Pallas DC. Protein phosphatase 2A subunit assembly: the catalytic subunit carboxy terminus is important for binding cellular B subunit but not polyomavirus middle tumor antigen. *Oncogene* 1997 Aug;15(8):911-7.

[47] Tolstykh T, Lee J, Vafai S, Stock JB. Carboxyl methylation regulates phosphoprotein phosphatase 2A by controlling the association of regulatory B subunits. *EMBO J.* 2000 Nov;19(21):5682-91.

[48] Wu J, Tolstykh T, Lee J, Boyd K, Stock JB, Broach JR. Carboxyl methylation of the phosphoprotein phosphatase 2A catalytic subunit promotes its functional association with regulatory subunits in vivo. *EMBO J.* 2000 Nov;19(21):5672-81.

[49] Wei H, Ashby DG, Moreno CS, Ogris E, Yeong FM, Corbett AH, Pallas DC. Carboxymethylation of the PP2A Catalytic Subunit in Saccharomyces cerevisiae Is Required for Efficient Interaction with the B type Subunits Cdc55p and Rts1p. *J. Biol. Chem.* 2001 Jan;276(2):1570-7.

[50] Garcia A, Cayla X, Sontag E. Protein phosphatase 2A: a definite player in viral and parasitic regulation. *Microbes Infect.* 2000 Apr;2(4):401-7.

[51] Pallas DC, Shahrik LK, Martin BL, Jaspers S, Miller TB, Brautigan DL, Roberts TM. Polyoma small and middle T antigens and SV40 small t antigen form stable complexes with protein phosphatase 2A. *Cell* 1990 Jan;60(1):167-76.

[52] Ulug ET, Cartwright AJ, Courtneidge SA. Characterization of the interaction of polyomavirus middle T antigen with type 2A protein phosphatase. *J. Virol.* 1992 Mar;66(3):1458-67.

[53] Walter G, Ruediger R, Slaughter C, Mumby M. Association of protein phosphatase 2A with polyoma virus medium tumor antigen. *Proc. Natl. Acad. Sci. USA* 1990 Apr;87(7):2521-5.

[54] Yang SI, Lickteig RL, Estes R, Rundell K, Walter G, Mumby MC. Control of protein phosphatase 2A by simian virus 40 small-t antigen. *Mol. Cell Biol.* 1991 Apr;11(4):1988-95.

[55] Sontag E, Fedorov S, Kamibayashi C, Robbins D, Cobb M, Mumby M. The interaction of SV40 small tumor antigen with protein phosphatase 2A stimulates the map kinase pathway and induces cell proliferation. *Cell* 1993 Dec;75(5):887-97.

[56] Pallas DC, Weller W, Jaspers S, Miller TB, Lane WS, Roberts TM. The third subunit of protein phosphatase 2A (PP2A), a 55-kilodalton protein which is apparently substituted for by T antigens in complexes with the 36- and 63-kilodalton PP2A subunits, bears little resemblance to T antigens. *J. Virol.* 1992 Feb;66(2):886-93.

[57] Li M, Guo H, Damuni Z. Purification and characterization of two potent heat-stable protein inhibitors of protein phosphatase 2A from bovine kidney. *Biochemistry* 1995 Feb;34(6):1988-96.

[58] Al Murrani SW, Woodgett JR, Damuni Z. Expression of I2PP2A, an inhibitor of protein phosphatase 2A, induces c-Jun and AP-1 activity. *Biochem. J.* 1999 Jul;341 (Pt 2):293-8.

[59] Li M, Makkinje A, Damuni Z. Molecular identification of I1PP2A, a novel potent heat-stable inhibitor protein of protein phosphatase 2A. *Biochemistry* 1996 Jun;35(22):6998-7002.

[60] Li M, Makkinje A, Damuni Z. The myeloid leukemia-associated protein SET is a potent inhibitor of protein phosphatase 2A. *J. Biol. Chem.* 1996 May;271(19):11059-62.

[61] Bialojan C, Takai A. Inhibitory effect of a marine-sponge toxin, okadaic acid, on protein phosphatases. Specificity and kinetics. *Biochem J.* 1988 Nov;256(1):283-90.

[62] Zhang Z, Zhao S, Long F, Zhang L, Bai G, Shima H, Nagao M, Lee EY. A mutant of protein phosphatase-1 that exhibits altered toxin sensitivity. *J. Biol. Chem.* 1994 Jun;269(25):16997-7000.

[63] Cohen P, Holmes CF, Tsukitani Y. Okadaic acid: a new probe for the study of cellular regulation. *Trends Biochem. Sci.* 1990 Mar;15(3):98-102.

[64] Chen MX, McPartlin AE, Brown L, Chen YH, Barker HM, Cohen PT. A novel human protein serine/threonine phosphatase, which possesses four tetratricopeptide repeat motifs and localizes to the nucleus. *EMBO J.* 1994 Sep;13(18):4278-90.

[65] Brewis ND, Street AJ, Prescott AR, Cohen PT. PPX, a novel protein serine/threonine phosphatase localized to centrosomes. *EMBO J.* 1993 Mar;12(3):987-96.

[66] Hastie CJ, Cohen PT. Purification of protein phosphatase 4 catalytic subunit: inhibition by the antitumour drug fostriecin and other tumour suppressors and promoters. *FEBS Lett* 1998 Jul;431(3):357-61.

[67] Chernoff J, Li HC, Cheng YS, Chen LB. Characterization of a phosphotyrosyl protein phosphatase activity associated with a phosphoseryl protein phosphatase of Mr = 95,000 from bovine heart. *J. Biol. Chem.* 1983 Jun;258(12):7852-7.
[68] Silberman SR, Speth M, Nemani R, Ganapathi MK, Dombradi V, Paris H, Lee EY. Isolation and characterization of rabbit skeletal muscle protein phosphatases C-I and C-II. *J. Biol. Chem.* 1984 Mar;259(5):2913-22.
[69] Van HC, Cayla X, Bosch M, Merlevede W, Goris J. The phosphotyrosyl phosphatase activator of protein phosphatase 2A. A novel purification method, immunological and enzymic characterization. *Eur. J. Biochem.* 1994 Dec;226(3):899-907.
[70] Cayla X, Van Hoof C, Bosch M, Waelkens E, Vandekerckhove J, Peeters B, Merlevede W, Goris J. Molecular cloning, expression, and characterization of PTPA, a protein that activates the tyrosyl phosphatase activity of protein phosphatase 2A. *J. Biol. Chem.* 1994 Jun;269(22):15668-75.
[71] Fellner T, Lackner DH, Hombauer H, Piribauer P, Mudrak I, Zaragoza K, Juno C, Ogris E. A novel and essential mechanism determining specificity and activity of protein phosphatase 2A (PP2A) in vivo. *Genes Dev.* 2003 Sep;17(17):2138-50.
[72] MacKintosh C, MacKintosh RW. Inhibitors of protein kinases and phosphatases. *Trends Biochem. Sci.* 1994 Nov;19(11):444-8.
[73] Cheng A, Balczon R, Zuo Z, Koons JS, Walsh AH, Honkanen RE. Fostriecin-mediated G2-M-phase growth arrest correlates with abnormal centrosome replication, the formation of aberrant mitotic spindles, and the inhibition of serine/threonine protein phosphatase activity. *Cancer Res.* 1998 Aug;58(16):3611-9.
[74] Evans DR, Simon JA. The predicted beta12-beta13 loop is important for inhibition of PP2Acalpha by the antitumor drug fostriecin. *FEBS Lett* 2001 Jun;498(1):110-5.
[75] Ratnasinghe D, Phang JM, Yeh GC. Differential expression and activity of phosphatases and protein kinases in adriamycin sensitive and resistant human breast cancer MCF-7 cells. *Int. J. Oncol.* 1998 Jul;13(1):79-84.
[76] Suzuki K, Takahashi K. Reduced expression of the regulatory A subunit of serine/threonine protein phosphatase 2A in human breast cancer MCF-7 cells. *Int. J. Oncol.* 2003 Nov;23(5):1263-8.
[77] Wang SS, Esplin ED, Li JL, Huang L, Gazdar A, Minna J, Evans GA. Alterations of the PPP2R1B gene in human lung and colon cancer [In Process Citation]. *Science* 1998 Oct;282(5387):284-7.
[78] Tamaki M, Goi T, Hirono Y, Katayama K, Yamaguchi A. PPP2R1B gene alterations inhibit interaction of PP2A-Abeta and PP2A-C proteins in colorectal cancers. *Oncol. Rep.* 2004 Mar;11(3):655-9.
[79] Takagi Y, Futamura M, Yamaguchi K, Aoki S, Takahashi T, Saji S. Alterations of the PPP2R1B gene located at 11q23 in human colorectal cancers. *Gut* 2000 Aug;47(2):268-71.
[80] Calin GA, di Iasio MG, Caprini E, Vorechovsky I, Natali PG, Sozzi G, Croce CM, Barbanti-Brodano G, Russo G, Negrini M. Low frequency of alterations of the alpha (PPP2R1A) and beta (PPP2R1B) isoforms of the subunit A of the serine-threonine phosphatase 2A in human neoplasms. *Oncogene* 2000 Feb;19(9):1191-5.

[81] Francia G, Poulsom R, Hanby AM, Mitchell SD, Williams G, Mckee P, Hart IR. Identification by differential display of a protein phosphatase-2A regulatory subunit preferentially expressed in malignant melanoma cells. *Int. J. Cancer* 1999 Aug;82(5):709-13.

[82] Ito A, Kataoka TR, Watanabe M, Nishiyama K, Mazaki Y, Sabe H, Kitamura Y, Nojima H. A truncated isoform of the PP2A B56 subunit promotes cell motility through paxillin phosphorylation. *EMBO J.* 2000 Feb;19(4):562-71.

[83] Ito A, Koma Y, Watabe K, Nagano T, Endo Y, Nojima H, Kitamura Y. A truncated isoform of the protein phosphatase 2A B56gamma regulatory subunit may promote genetic instability and cause tumor progression. *Am. J. Pathol.* 2003 Jan;162(1):81-91.

[84] Gotz J, Probst A, Mistl C, Nitsch RM, Ehler E. Distinct role of protein phosphatase 2A subunit Calpha in the regulation of E-cadherin and beta-catenin during development. *Mech. Dev.* 2000 May;93(1-2):83-93.

[85] Campbell KS, Auger KR, Hemmings BA, Roberts TM, Pallas DC. Identification of regions in polyomavirus middle T and small t antigens important for association with protein phosphatase 2A. *J. Virol.* 1995 Jun;69(6):3721-8.

[86] Rundell K, Parakati R. The role of the SV40 ST antigen in cell growth promotion and transformation. *Semin. Cancer Biol.* 2001 Feb;11(1):5-13.

[87] Yu J, Boyapati A, Rundell K. Critical role for SV40 small-t antigen in human cell transformation. *Virology* 2001 Nov;290(2):192-8.

[88] Montano X, Rundell K. Role of SV40 small t in cell lysis, transformation, and signaling. *Methods Mol. Biol.* 2001;165:229-42.

[89] Nunbhakdi-Craig V, Craig L, Machleidt T, Sontag E. Simian virus 40 small tumor antigen induces deregulation of the actin cytoskeleton and tight junctions in kidney epithelial cells. *J. Virol.* 2003 Mar;77(5):2807-18.

[90] Mullane KP, Ratnofsky M, Cullere X, Schaffhausen B. Signaling from polyomavirus middle T and small T defines different roles for protein phosphatase 2A. *Mol. Cell Biol.* 1998 Dec;18(12):7556-64.

[91] Kinoshita K, Nemoto T, Nabeshima K, Kondoh H, Niwa H, Yanagida M. The regulatory subunits of fission yeast protein phosphatase 2A (PP2A) affect cell morphogenesis, cell wall synthesis and cytokinesis. *Genes Cells* 1996 Jan;1(1):29-45.

[92] Gomes R, Karess RE, Ohkura H, Glover DM, Sunkel CE. Abnormal anaphase resolution (aar): a locus required for progression through mitosis in Drosophila. *J. Cell Sci.* 1993 Feb;104 (Pt 2):583-93.

[93] Mayer-Jaekel RE, Ohkura H, Ferrigno P, Andjelkovic N, Shiomi K, Uemura T, Glover DM, Hemmings BA. Drosophila mutants in the 55 kDa regulatory subunit of protein phosphatase 2A show strongly reduced ability to dephosphorylate substrates of p34cdc2. *J. Cell Sci.* 1994 Sep;107 (Pt 9):2609-16.

[94] Lin FC, Arndt KT. The role of Saccharomyces cerevisiae type 2A phosphatase in the actin cytoskeleton and in entry into mitosis. *EMBO J.* 1995 Jun;14(12):2745-59.

[95] Evans DR, Stark MJ. Mutations in the Saccharomyces cerevisiae type 2A protein phosphatase catalytic subunit reveal roles in cell wall integrity, actin cytoskeleton organization and mitosis. *Genetics* 1997 Feb;145(2):227-41.

[96] Milczarek GJ, Chen W, Gupta A, Martinez JD, Bowden GT. Okadaic acid mediates p53 hyperphosphorylation and growth arrest in cells with wild-type p53 but increases aberrant mitoses in cells with non-functional p53. *Carcinogenesis* 1999 Jun;20(6):1043-8.

[97] Snaith HA, Armstrong CG, Guo Y, Kaiser K, Cohen PT. Deficiency of protein phosphatase 2A uncouples the nuclear and centrosome cycles and prevents attachment of microtubules to the kinetochore in Drosophila microtubule star (mts) embryos. *J. Cell Sci.* 1996 Dec;109(Pt 13):3001-12.

[98] Mayer-Jaekel RE, Ohkura H, Gomes R, Sunkel CE, Baumgartner S, Hemmings BA, Glover DM. The 55 kd regulatory subunit of Drosophila protein phosphatase 2A is required for anaphase. *Cell* 1993 Feb;72(4):621-33.

[99] Ishida Y, Furukawa Y, DeCaprio JA, Saito M, Griffin JD. Treatment of myeloid leukemic cells with the phosphatase inhibitor okadaic acid induces cell cycle arrest at either G1/S or G2/M depending on dose. *J. Cell Physiol.* 1992 Mar;150(3):484-92.

[100] Mellgren G, Vintermyr OK, Boe R, Doskeland SO. Hepatocyte DNA replication is abolished by inhibitors selecting protein phosphatase 2A rather than phosphatase 1. *Exp. Cell Res.* 1993 Apr;205(2):293-301.

[101] Kim TA, Velasquez BR, Wenner CE. Okadaic acid regulation of the retinoblastoma gene product is correlated with the inhibition of growth factor-induced cell proliferation in mouse fibroblasts. *Proc. Natl. Acad. Sci. USA* 1993 Jun;90(12):5460-3.

[102] Schonthal A, Feramisco JR. Inhibition of histone H1 kinase expression, retinoblastoma protein phosphorylation, and cell proliferation by the phosphatase inhibitor okadaic acid. *Oncogene* 1993 Feb;8(2):433-41.

[103] Yan Y, Mumby MC. Distinct roles for PP1 and PP2A in phosphorylation of the retinoblastoma protein. PP2a regulates the activities of G(1) cyclin-dependent kinases. *J. Biol. Chem.* 1999 Nov;274(45):31917-24.

[104] Scheidtmann KH, Mumby MC, Rundell K, Walter G. Dephosphorylation of simian virus 40 large-T antigen and p53 protein by protein phosphatase 2A: inhibition by small-t antigen. Mol Cell Biol 1991 Apr;11(4):1996-2003.

[105] Lin XH, Walter J, Scheidtmann K, Ohst K, Newport J, Walter G. Protein phosphatase 2A is required for the initiation of chromosomal DNA replication [In Process Citation]. *Proc. Natl. Acad. Sci. USA* 1998 Dec;95(25):14693-8.

[106] Chou DM, Petersen P, Walter JC, Walter G. Protein phosphatase 2A regulates binding of Cdc45 to the prereplication complex. *J. Biol. Chem.* 2002 Oct;277(43):40520-7.

[107] Wilson NJ, Moss ST, Csar XF, Ward AC, Hamilton JA. Protein phosphatase 2A is expressed in response to colony-stimulating factor 1 in macrophages and is required for cell cycle progression independently of extracellular signal-regulated protein kinase activity. *Biochem. J.* 1999 May;339(Pt 3):517-24.

[108] Murakami H, Nurse P. DNA replication and damage checkpoints and meiotic cell cycle controls in the fission and budding yeasts. *Biochem. J.* 2000 Jul;349(Pt 1):1-12.

[109] Lei M, Tye BK. Initiating DNA synthesis: from recruiting to activating the MCM complex. *J. Cell Sci.* 2001 Apr;114(Pt 8):1447-54.

[110] Yan Z, Fedorov SA, Mumby MC, Williams RS. PR48, a novel regulatory subunit of protein phosphatase 2A, interacts with Cdc6 and modulates DNA replication in human cells. *Mol. Cell Biol.* 2000 Feb;20(3):1021-9.

[111] Kawabe T, Muslin AJ, Korsmeyer SJ. HOX11 interacts with protein phosphatases PP2A and PP1 and disrupts a G2/M cell-cycle checkpoint. *Nature* 1997 Jan;385(6615):454-8.

[112] Kinoshita N, Yamano H, Niwa H, Yoshida T, Yanagida M. Negative regulation of mitosis by the fission yeast protein phosphatase ppa2. *Genes Dev.* 1993 Jun;7(6):1059-71.

[113] Felix MA, Cohen P, Karsenti E. Cdc2 H1 kinase is negatively regulated by a type 2A phosphatase in the Xenopus early embryonic cell cycle: evidence from the effects of okadaic acid. *EMBO J.* 1990 Mar;9(3):675-83.

[114] Clarke PR, Hoffmann I, Draetta G, Karsenti E. Dephosphorylation of cdc25-C by a type-2A protein phosphatase: specific regulation during the cell cycle in Xenopus egg extracts. *Mol. Biol. Cell* 1993 Apr;4(4):397-411.

[115] Lee TH, Turck C, Kirschner MW. Inhibition of cdc2 activation by INH/PP2A. *Mol. Biol. Cell* 1994 Mar;5(3):323-38.

[116] Coleman TR, Dunphy WG. Cdc2 regulatory factors. *Curr. Opin. Cell Biol.* 1994 Dec;6(6):877-82.

[117] Lee TH, Solomon MJ, Mumby MC, Kirschner MW. INH, a negative regulator of MPF, is a form of protein phosphatase 2A. *Cell* 1991 Jan;64(2):415-23.

[118] Millward TA, Zolnierowicz S, Hemmings BA. Regulation of protein kinase cascades by protein phosphatase 2A. *Trends Biochem. Sci.* 1999 May;24(5):186-91.

[119] Janssens V, Goris J. Protein phosphatase 2A: a highly regulated family of serine/threonine phosphatases implicated in cell growth and signalling. *Biochem. J.* 2001 Feb;353(Pt 3):417-39.

[120] Borgne A, Meijer L. Sequential dephosphorylation of p34(cdc2) on Thr-14 and Tyr-15 at the prophase/metaphase transition. *J. Biol. Chem.* 1996 Nov;271(44):27847-54.

[121] Masuda M, Nagai Y, Oshima N, Tanaka K, Murakami H, Igarashi H, Okayama H. Genetic studies with the fission yeast Schizosaccharomyces pombe suggest involvement of wee1, ppa2, and rad24 in induction of cell cycle arrest by human immunodeficiency virus type 1 Vpr. *J. Virol.* 2000 Mar;74(6):2636-46.

[122] Re F, Braaten D, Franke EK, Luban J. Human immunodeficiency virus type 1 Vpr arrests the cell cycle in G2 by inhibiting the activation of p34cdc2-cyclin B. *J. Virol.* 1995 Nov;69(11):6859-64.

[123] Yamashita K, Yasuda H, Pines J, Yasumoto K, Nishitani H, Ohtsubo M, Hunter T, Sugimura T, Nishimoto T. Okadaic acid, a potent inhibitor of type 1 and type 2A protein phosphatases, activates cdc2/H1 kinase and transiently induces a premature mitosis-like state in BHK21 cells. *EMBO J.* 1990 Dec;9(13):4331-8.

[124] Vandre DD, Wills VL. Inhibition of mitosis by okadaic acid: possible involvement of a protein phosphatase 2A in the transition from metaphase to anaphase. *J. Cell Sci.* 1992 Jan;101 (Pt 1):79-91.

[125] Deak P, Donaldson M, Glover DM. Mutations in makos, a Drosophila gene encoding the Cdc27 subunit of the anaphase promoting complex, enhance centrosomal defects in

polo and are suppressed by mutations in twins/aar, which encodes a regulatory subunit of PP2A. *J. Cell Sci.* 2003 Oct;116(Pt 20):4147-58.

[126] Healy AM, Zolnierowicz S, Stapleton AE, Goebl M, DePaoli-Roach AA, Pringle JR. CDC55, a Saccharomyces cerevisiae gene involved in cellular morphogenesis: identification, characterization, and homology to the B subunit of mammalian type 2A protein phosphatase. *Mol. Cell Biol.* 1991 Nov;11(11):5767-80.

[127] van Zyl W, Huang W, Sneddon AA, Stark M, Camier S, Werner M, Marck C, Sentenac A, Broach JR. Inactivation of the protein phosphatase 2A regulatory subunit A results in morphological and transcriptional defects in Saccharomyces cerevisiae. *Mol. Cell Biol.* 1992 Nov;12(11):4946-59.

[128] Ronne H, Carlberg M, Hu GZ, Nehlin JO. Protein phosphatase 2A in Saccharomyces cerevisiae: effects on cell growth and bud morphogenesis. *Mol. Cell Biol.* 1991 Oct;11(10):4876-84.

[129] Jiang W, Hallberg RL. Isolation and characterization of par1(+) and par2(+). Two schizosaccharomyces pombe genes encoding b' subunits of protein phosphatase 2a [In Process Citation]. *Genetics* 2000 Mar;154(3):1025-38.

[130] Tanabe O, Hirata D, Usui H, Nishito Y, Miyakawa T, Igarashi K, Takeda M. Fission yeast homologues of the B' subunit of protein phosphatase 2A: multiple roles in mitotic cell division and functional interaction with calcineurin. *Genes Cells* 2001 May;6(5):455-73.

[131] Minshull J, Straight A, Rudner AD, Dernburg AF, Belmont A, Murray AW. Protein phosphatase 2A regulates MPF activity and sister chromatid cohesion in budding yeast. *Curr. Biol.* 1996 Dec;6(12):1609-20.

[132] Wang Y, Burke DJ. Cdc55p, the B-type regulatory subunit of protein phosphatase 2A, has multiple functions in mitosis and is required for the kinetochore/spindle checkpoint in Saccharomyces cerevisiae. *Mol. Cell Biol.* 1997 Feb;17(2):620-6.

[133] Lippincott J, Shannon KB, Shou W, Deshaies RJ, Li R. The Tem1 small GTPase controls actomyosin and septin dynamics during cytokinesis. *J. Cell Sci.* 2001 Apr;114(Pt 7):1379-86.

[134] Dobbelaere J, Gentry MS, Hallberg RL, Barral Y. Phosphorylation-dependent regulation of septin dynamics during the cell cycle. *Dev. Cell* 2003 Mar;4(3):345-57.

[135] Carmena M, Earnshaw WC. The cellular geography of aurora kinases. *Nat. Rev. Mol. Cell Biol.* 2003 Nov;4(11):842-54.

[136] Sugiyama K, Sugiura K, Hara T, Sugimoto K, Shima H, Honda K, Furukawa K, Yamashita S, Urano T. Aurora-B associated protein phosphatases as negative regulators of kinase activation. *Oncogene* 2002 May;21(20):3103-11.

[137] Stiegler P, Giordano A. Role of pRB2/p130 in cellular growth regulation. *Anal. Quant. Cytol. Histol.* 1999 Aug;21(4):363-6.

[138] Feschenko MS, Stevenson E, Nairn AC, Sweadner KJ. A novel cAMP-stimulated pathway in protein phosphatase 2A activation. *J. Pharmacol. Exp. Ther.* 2002 Jul;302(1):111-8.

[139] Voorhoeve PM, Hijmans EM, Bernards R. Functional interaction between a novel protein phosphatase 2A regulatory subunit, PR59, and the retinoblastoma-related p107 protein. *Oncogene* 1999 Jan;18(2):515-24.

[140] Cicchillitti L, Fasanaro P, Biglioli P, Capogrossi MC, Martelli F. Oxidative stress induces protein phosphatase 2A-dependent dephosphorylation of the pocket proteins pRb, p107, and p130. *J. Biol. Chem.* 2003 May;278(21):19509-17.
[141] Voorhoeve PM, Watson RJ, Farlie PG, Bernards R, Lam EW. Rapid dephosphorylation of p107 following UV irradiation. *Oncogene* 1999 Jan;18(3):679-88.
[142] Kennedy BK, Barbie DA, Classon M, Dyson N, Harlow E. Nuclear organization of DNA replication in primary mammalian cells. *Genes Dev.* 2000 Nov;14(22):2855-68.
[143] Avni D, Yang H, Martelli F, Hofmann F, ElShamy WM, Ganesan S, Scully R, Livingston DM. Active localization of the retinoblastoma protein in chromatin and its response to S phase DNA damage. *Mol. Cell* 2003 Sep;12(3):735-46.
[144] Suganuma M, Fujiki H, Suguri H, Yoshizawa S, Hirota M, Nakayasu M, Ojika M, Wakamatsu K, Yamada K, Sugimura T. Okadaic acid: an additional non-phorbol-12-tetradecanoate-13-acetate-type tumor promoter. *Proc. Natl. Acad. Sci. USA* 1988 Mar;85(6):1768-71.
[145] Ruediger R, Pham HT, Walter G. Disruption of protein phosphatase 2A subunit interaction in human cancers with mutations in the A alpha subunit gene. *Oncogene* 2001 Jan;20(1):10-5.
[146] Ruediger R, Pham HT, Walter G. Alterations in protein phosphatase 2A subunit interaction in human carcinomas of the lung and colon with mutations in the A beta subunit gene. *Oncogene* 2001 Apr;20(15):1892-9.
[147] Vuocolo S, Purev E, Zhang D, Bartek J, Hansen K, Soprano DR, Soprano KJ. Protein phosphatase 2A associates with Rb2/p130 and mediates retinoic acid-induced growth suppression of ovarian carcinoma cells. *J. Biol. Chem.* 2003 Oct;278(43):41881-9.
[148] Wu S, Donigan A, Platsoucas CD, Jung W, Soprano DR, Soprano KJ. All-trans-retinoic acid blocks cell cycle progression of human ovarian adenocarcinoma cells at late G1. *Exp. Cell Res.* 1997 May;232(2):277-86.
[149] Levine AJ. p53, the cellular gatekeeper for growth and division. *Cell* 1997 Feb;88(3):323-31.
[150] Funk WD, Pak DT, Karas RH, Wright WE, Shay JW. A transcriptionally active DNA-binding site for human p53 protein complexes. *Mol. Cell Biol.* 1992 Jun;12(6):2866-71.
[151] Ko LJ, Prives C. p53: puzzle and paradigm. *Genes Dev.* 1996 May;10(9):1054-72.
[152] Ashcroft M, Kubbutat MH, Vousden KH. Regulation of p53 function and stability by phosphorylation. *Mol. Cell Biol.* 1999 Mar;19(3):1751-8.
[153] Ko LJ, Prives C. p53: puzzle and paradigm. *Genes Dev.* 1996 May;10(9):1054-72.
[154] Zhang Y, Xiong Y, Yarbrough WG. ARF promotes MDM2 degradation and stabilizes p53: ARF-INK4a locus deletion impairs both the Rb and p53 tumor suppression pathways. *Cell* 1998 Mar;92(6):725-34.
[155] Okamoto K, Kamibayashi C, Serrano M, Prives C, Mumby MC, Beach D. p53-dependent association between cyclin G and the B' subunit of protein phosphatase 2A. *Mol. Cell Biol.* 1996 Nov;16(11):6593-602.
[156] Okamoto K, Li H, Jensen MR, Zhang T, Taya Y, Thorgeirsson SS, Prives C. Cyclin G recruits PP2A to dephosphorylate Mdm2. *Mol. Cell* 2002 Apr;9(4):761-71.
[157] Kimura SH, Nojima H. Cyclin G1 associates with MDM2 and regulates accumulation and degradation of p53 protein. *Genes Cells* 2002 Aug;7(8):869-80.

[158] Bennin DA, Don AS, Brake T, McKenzie JL, Rosenbaum H, Ortiz L, DePaoli-Roach AA, Horne MC. Cyclin G2 associates with protein phosphatase 2A catalytic and regulatory B' subunits in active complexes and induces nuclear aberrations and a G1/S phase cell cycle arrest. *J. Biol. Chem.* 2002 Jul;277(30):27449-67.

[159] Dohoney KM, Guillerm C, Whiteford C, Elbi C, Lambert PF, Hager GL, Brady JN. Phosphorylation of p53 at serine 37 is important for transcriptional activity and regulation in response to DNA damage. *Oncogene* 2004 Jan;23(1):49-57.

[160] Guo CY, Brautigan DL, Larner JM. ATM-dependent dissociation of B55 regulatory subunit from nuclear PP2A in response to ionizing radiation. J. Biol. Chem. 2002 Feb;277(7):4839-44.

[161] Schonthal AH. Role of PP2A in intracellular signal transduction pathways. *Front Biosci.* 1998 Dec;3:D1262-D1273.

[162] Yang J, Yu Y, Hamrick HE, Duerksen-Hughes PJ. ATM, ATR and DNA-PK: initiators of the cellular genotoxic stress responses. *Carcinogenesis* 2003 Oct;24(10):1571-80.

[163] Douglas P, Moorhead GB, Ye R, Lees-Miller SP. Protein phosphatases regulate DNA-dependent protein kinase activity. *J. Biol. Chem.* 2001 Jun;276(22):18992-8.

[164] Leoni LM, Shih HC, Deng L, Tuey C, Walter G, Carson DA, Cottam HB. Modulation of ceramide-activated protein phosphatase 2A activity by low molecular weight aromatic compounds. *Biochem. Pharmacol.* 1998 Apr;55(7):1105-11.

[165] Ruvolo PP, Deng X, Ito T, Carr BK, May WS. Ceramide induces Bcl2 dephosphorylation via a mechanism involving mitochondrial PP2A. *J. Biol. Chem.* 1999 Jul;274(29):20296-300.

[166] Ruvolo PP, Clark W, Mumby M, Gao F, May WS. A functional role for the B56 alpha-subunit of protein phosphatase 2A in ceramide-mediated regulation of Bcl2 phosphorylation status and function. *J. Biol. Chem.* 2002 Jun;277(25):22847-52.

[167] Galadari S, Kishikawa K, Kamibayashi C, Mumby MC, Hannun YA. Purification and characterization of ceramide-activated protein phosphatases. *Biochemistry* 1998 Aug;37(32):11232-8.

[168] Dagda RK, Zaucha JA, Wadzinski BE, Strack S. A developmentally regulated, neuron-specific splice variant of the variable subunit Bbeta targets protein phosphatase 2A to mitochondria and modulates apoptosis. *J. Biol. Chem.* 2003 Jul;278(27):24976-85.

[169] Chiang CW, Harris G, Ellig C, Masters SC, Subramanian R, Shenolikar S, Wadzinski BE, Yang E. Protein phosphatase 2A activates the proapoptotic function of BAD in interleukin- 3-dependent lymphoid cells by a mechanism requiring 14-3-3 dissociation. *Blood* 2001 Mar;97(5):1289-97.

[170] Santoro MF, Annand RR, Robertson MM, Peng YW, Brady MJ, Mankovich JA, Hackett MC, Ghayur T, Walter G, Wong WW, Giegel DA. Regulation of protein phosphatase 2A activity by caspase-3 during apoptosis. *J. Biol. Chem.* 1998 May;273(21):13119-28.

[171] Talanian RV, Quinlan C, Trautz S, Hackett MC, Mankovich JA, Banach D, Ghayur T, Brady KD, Wong WW. Substrate specificities of caspase family proteases. *J. Biol. Chem.* 1997 Apr;272(15):9677-82.

[172] Harmala-Brasken AS, Mikhailov A, Soderstrom TS, Meinander A, Holmstrom TH, Damuni Z, Eriksson JE. Type-2A protein phosphatase activity is required to maintain death receptor responsiveness. *Oncogene* 2003 Oct;22(48):7677-86.

[173] Gendron S, Couture J, Aoudjit F. Integrin alpha2beta1 inhibits Fas-mediated apoptosis in T lymphocytes by protein phosphatase 2A-dependent activation of the MAPK/ERK pathway. *J. Biol. Chem.* 2003 Dec;278(49):48633-43.

[174] Kang HS, Choi I. Protein phosphatase 2A modulates the proliferation of human multiple myeloma cells via regulation of the production of reactive oxygen intermediates and anti-apoptotic factors. *Cell Immunol.* 2001 Oct;213(1):34-44.

[175] Li X, Scuderi A, Letsou A, Virshup DM. B56-associated protein phosphatase 2A is required for survival and protects from apoptosis in Drosophila melanogaster. *Mol. Cell Biol.* 2002 Jun;22(11):3674-84.

[176] Silverstein AM, Barrow CA, Davis AJ, Mumby MC. Actions of PP2A on the MAP kinase pathway and apoptosis are mediated by distinct regulatory subunits. *Proc. Natl. Acad. Sci. USA* 2002 Apr;99(7):4221-6.

[177] Kleinberger T, Shenk T. Adenovirus E4orf4 protein binds to protein phosphatase 2A, and the complex down regulates E1A-enhanced junB transcription. *J. Virol.* 1993 Dec;67(12):7556-60.

[178] Shtrichman R, Sharf R, Barr H, Dobner T, Kleinberger T. Induction of apoptosis by adenovirus E4orf4 protein is specific to transformed cells and requires an interaction with protein phosphatase 2A. *Proc. Natl. Acad. Sci. USA* 1999 Aug;96(18):10080-5.

[179] Kleinberger T. Induction of apoptosis by adenovirus E4orf4 protein. *Apoptosis* 2000 Jun;5(3):211-5.

[180] Gjoerup O, Zaveri D, Roberts TM. Induction of p53-independent apoptosis by simian virus 40 small t antigen. *J. Virol.* 2001 Oct;75(19):9142-55.

[181] Bernstein LR, Liotta LA. Molecular mediators of interactions with extracellular matrix components in metastasis and angiogenesis. *Curr. Opin Oncol.* 1994 Jan;6(1):106-13.

[182] Nakayama S, Sasaki A, Mese H, Alcalde RE, Matsumura T. Establishment of high and low metastasis cell lines derived from a human tongue squamous cell carcinoma. *Invasion Metastasis* 1998;18(5-6):219-28.

[183] Jones RJ, Brunton VG, Frame MC. Adhesion-linked kinases in cancer; emphasis on src, focal adhesion kinase and PI 3-kinase. *Eur. J. Cancer* 2000 Aug;36(13 Spec No):1595-606.

[184] Greenwood JA, Theibert AB, Prestwich GD, Murphy-Ullrich JE. Restructuring of focal adhesion plaques by PI 3-kinase. Regulation by PtdIns (3,4,5)-p(3) binding to alpha-actinin. *J. Cell Biol.* 2000 Aug;150(3):627-42.

[185] Cary LA, Han DC, Guan JL. Integrin-mediated signal transduction pathways. *Histol Histopathol.* 1999 Jul;14(3):1001-9.

[186] Turner CE. Paxillin and focal adhesion signalling. *Nat. Cell Biol.* 2000 Dec;2(12):E231-E236.

[187] Brown MC, Perrotta JA, Turner CE. Serine and threonine phosphorylation of the paxillin LIM domains regulates paxillin focal adhesion localization and cell adhesion to fibronectin. *Mol. Biol. Cell* 1998 Jul;9(7):1803-16.

[188] Larsen M, Tremblay ML, Yamada KM. Phosphatases in cell-matrix adhesion and migration. *Nat. Rev. Mol. Cell Biol.* 2003 Sep;4(9):700-11.

[189] Young MR, Kolesiak K, Meisinger J. Protein phosphatase-2A regulates endothelial cell motility and both the phosphorylation and the stability of focal adhesion complexes. *Int. J. Cancer* 2002 Jul;100(3):276-82.

[190] Young MR, Liu SW, Meisinger J. Protein phosphatase-2A restricts migration of Lewis lung carcinoma cells by modulating the phosphorylation of focal adhesion proteins. Int *J. Cancer* 2003 Jan;103(1):38-44.

[191] Hirohashi S, Kanai Y. Cell adhesion system and human cancer morphogenesis. *Cancer Sci.* 2003 Jul;94(7):575-81.

[192] Nunbhakdi-Craig V, Machleidt T, Ogris E, Bellotto D, White CL, III, Sontag E. Protein phosphatase 2A associates with and regulates atypical PKC and the epithelial tight junction complex. *J. Cell Biol.* 2002 Sep;158(5):967-78.

[193] Seeling JM, Miller JR, Gil R, Moon RT, White R, Virshup DM. Regulation of beta-Catenin Signaling by the B56 Subunit of Protein Phosphatase 2A. *Science* 1999 Mar;283(5410):2089-91.

[194] Hoover KB, Liao SY, Bryant PJ. Loss of the tight junction MAGUK ZO-1 in breast cancer: relationship to glandular differentiation and loss of heterozygosity. *Am. J. Pathol.* 1998 Dec;153(6):1767-73.

[195] Kominsky SL, Argani P, Korz D, Evron E, Raman V, Garrett E, Rein A, Sauter G, Kallioniemi OP, Sukumar S. Loss of the tight junction protein claudin-7 correlates with histological grade in both ductal carcinoma in situ and invasive ductal carcinoma of the breast. *Oncogene* 2003 Apr;22(13):2021-33.

[196] Okada T, Narai A, Matsunaga S, Fusetani N, Shimizu M. Assessment of the marine toxins by monitoring the integrity of human intestinal Caco-2 cell monolayers. *Toxicol In Vitro* 2000 Jun;14(3):219-26.

[197] Singer KL, Stevenson BR, Woo PL, Firestone GL. Relationship of serine/threonine phosphorylation/dephosphorylation signaling to glucocorticoid regulation of tight junction permeability and ZO-1 distribution in nontransformed mammary epithelial cells. *J. Biol. Chem.* 1994 Jun;269(23):16108-15.

[198] Tripuraneni J, Koutsouris A, Pestic L, De Lanerolle P, Hecht G. The toxin of diarrheic shellfish poisoning, okadaic acid, increases intestinal epithelial paracellular permeability. *Gastroenterology* 1997 Jan;112(1):100-8.

[199] Toivola DM, Goldman RD, Garrod DR, Eriksson JE. Protein phosphatases maintain the organization and structural interactions of hepatic keratin intermediate filaments. *J. Cell Sci.* 1997 Jan;110 (Pt 1):23-33.

[200] Loewenstein WR. Junctional intercellular communication: the cell-to-cell membrane channel. *Physiol. Rev.* 1981 Oct;61(4):829 913.

[201] Trosko JE, Ruch RJ. Cell-cell communication in carcinogenesis. *Front Biosci.* 1998 Feb;3:D208-D236.

[202] Lo CW. The role of gap junction membrane channels in development. *J. Bioenerg Biomembr.* 1996 Aug;28(4):379-85.

[203] Yamasaki H, Naus CC. Role of connexin genes in growth control. *Carcinogenesis* 1996 Jun;17(6):1199-213.

[204] Cruciani V, Mikalsen SO, Vasseur P, Sanner T. Effects of peroxisome proliferators and 12-O-tetradecanoyl phorbol-13-acetate on intercellular communication and connexin43 in two hamster fibroblast systems. *Int. J. Cancer* 1997 Oct;73(2):240-8.

[205] Herve JC, Sarrouilhe D. Modulation of junctional communication by phosphorylation: protein phosphatases, the missing link in the chain. Biol Cell 2002 Nov;94(7-8):423-32.

[206] Cruciani V, Kaalhus O, Mikalsen SO. Phosphatases involved in modulation of gap junctional intercellular communication and dephosphorylation of connexin43 in hamster fibroblasts: 2B or not 2B? *Exp. Cell Res.* 1999 Nov;252(2):449-63.

[207] Schonthal AH. Regulation of gene expression by serine/threonine protein phosphatases. *Semin. Cancer Biol.* 1995 Aug;6(4):239-48.

[208] Wadzinski BE, Wheat WH, Jaspers S, Peruski LF, Jr., Lickteig RL, Johnson GL, Klemm DJ. Nuclear protein phosphatase 2A dephosphorylates protein kinase A-phosphorylated CREB and regulates CREB transcriptional stimulation. Mol Cell Biol 1993 May;13(5):2822-34.

[209] Wheat WH, Roesler WJ, Klemm DJ. Simian virus 40 small tumor antigen inhibits dephosphorylation of protein kinase A-phosphorylated CREB and regulates CREB transcriptional stimulation. *Mol. Cell Biol.* 1994 Sep;14(9):5881-90.

[210] Fimia GM, De Cesare D, Sassone-Corsi P. A family of LIM-only transcriptional coactivators: tissue-specific expression and selective activation of CREB and CREM. *Mol. Cell Biol.* 2000 Nov;20(22):8613-22.

[211] Johannessen M, Olsen PA, Johansen B, Seternes OM, Moens U. Activation of the coactivator four-and-a-half-LIM-only protein FHL2 and the c-fos promoter through inhibition of protein phosphatase 2A. *Biochem. Pharmacol.* 2003 Apr;65(8):1317-28.

[212] Alberts AS, Deng T, Lin A, Meinkoth JL, Schonthal A, Mumby MC, Karin M, Feramisco JR. Protein phosphatase 2A potentiates activity of promoters containing AP-1-binding elements. Mol Cell Biol 1993 Apr;13(4):2104-12.

[213] Piechaczyk M, Blanchard JM. c-fos proto-oncogene regulation and function. *Crit. Rev. Oncol. Hematol.* 1994 Oct;17(2):93-131.

[214] Lacroix I, Lipcey C, Imbert J, Kahn-Perles B. Sp1 transcriptional activity is up-regulated by phosphatase 2A in dividing T lymphocytes. *J. Biol. Chem.* 2002 Mar;277(11):9598-605.

[215] Liang H, Venema VJ, Wang X, Ju H, Venema RC, Marrero MB. Regulation of angiotensin II-induced phosphorylation of STAT3 in vascular smooth muscle cells. *J. Biol. Chem.* 1999 Jul;274(28):19846-51.

[216] Woetmann A, Nielsen M, Christensen ST, Brockdorff J, Kaltoft K, Engel AM, Skov S, Brender C, Geisler C, Svejgaard A, Rygaard J, Leick V, Odum N. Inhibition of protein phosphatase 2A induces serine/threonine phosphorylation, subcellular redistribution, and functional inhibition of STAT3. *Proc. Natl. Acad. Sci. USA* 1999 Sep;96(19):10620-5.

[217] Woetmann A, Brockdorff J, Lovato P, Nielsen M, Leick V, Rieneck K, Svejgaard A, Geisler C, Odum N. Protein phosphatase 2A (PP2A) regulates interleukin-4-mediated STAT6 signaling. *J. Biol. Chem.* 2003 Jan;278(5):2787-91.

[218] Baeuerle PA, Baltimore D. NF-kappa B: ten years after. *Cell* 1996 Oct;87(1):13-20.

[219] Naumann M, Scheidereit C. Activation of NF-kappa B in vivo is regulated by multiple phosphorylations. *EMBO J.* 1994 Oct;13(19):4597-607.

[220] Thevenin C, Kim SJ, Rieckmann P, Fujiki H, Norcross MA, Sporn MB, Fauci AS, Kehrl JH. Induction of nuclear factor-kappa B and the human immunodeficiency virus long terminal repeat by okadaic acid, a specific inhibitor of phosphatases 1 and 2A. *New Biol.* 1990 Sep;2(9):793-800.

[221] Vlach J, Garcia A, Jacque JM, Rodriguez MS, Michelson S, Virelizier JL. Induction of Sp1 phosphorylation and NF-kappa B-independent HIV promoter domain activity in T lymphocytes stimulated by okadaic acid. *Virology* 1995 Apr;208(2):753-61.

[222] Sontag E, Sontag JM, Garcia A. Protein phosphatase 2A is a critical regulator of protein kinase C zeta signaling targeted by SV40 small t to promote cell growth and NF-kappaB activation. *EMBO J.* 1997 Sep;16(18):5662-71.

[223] Fu DX, Kuo YL, Liu BY, Jeang KT, Giam CZ. Human T-lymphotropic virus type I tax activates I-kappa B kinase by inhibiting I-kappa B kinase-associated serine/threonine protein phosphatase 2A. *J. Biol. Chem.* 2003 Jan;278(3):1487-93.

[224] Yang J, Fan GH, Wadzinski BE, Sakurai H, Richmond A. Protein phosphatase 2A interacts with and directly dephosphorylates RelA. *J. Biol. Chem.* 2001 Dec;276(51):47828-33.

[225] Rastinejad F. Retinoid X receptor and its partners in the nuclear receptor family. *Curr. Opin. Struct. Biol.* 2001 Feb;11(1):33-8.

[226] Ishaq M, Fan M, Wigmore K, Gaddam A, Natarajan V. Regulation of retinoid X receptor responsive element-dependent transcription in T lymphocytes by Ser/Thr phosphatases: functional divergence of protein kinase C (PKC)theta; and PKC alpha in mediating calcineurin-induced transactivation. *J. Immunol.* 2002 Jul;169(2):732-8.

[227] Lu Q, Dunn RL, Angeles R, Smith GD. Regulation of spindle formation by active mitogen-activated protein kinase and protein phosphatase 2A during mouse oocyte meiosis. *Biol. Reprod.* 2002 Jan;66(1):29-37.

[228] Eriksson JE, Brautigan DL, Vallee R, Olmsted J, Fujiki H, Goldman RD. Cytoskeletal integrity in interphase cells requires protein phosphatase activity. *Proc. Natl. Acad. Sci. USA* 1992 Nov;89(22):11093-7.

[229] Almazan G, Afar DE, Bell JC. Phosphorylation and disruption of intermediate filament proteins in oligodendrocyte precursor cultures treated with calyculin A. *J. Neurosci. Res.* 1993 Oct;36(2):163-72.

[230] Lee WC, Yu JS, Yang SD, Lai YK. Reversible hyperphosphorylation and reorganization of vimentin intermediate filaments by okadaic acid in 9L rat brain tumor cells. *J. Cell Biochem.* 1992 Aug;49(4):378-93.

[231] Yatsunami J, Fujiki H, Suganuma M, Yoshizawa S, Eriksson JE, Olson MO, Goldman RD. Vimentin is hyperphosphorylated in primary human fibroblasts treated with okadaic acid. *Biochem. Biophys. Res. Commun.* 1991 Jun;177(3):1165-70.

[232] Jackson J, Meisinger J, Patel S, Lim ZC, Vellody K, Metz R, Young MR. Protein Phosphatase-2A Associates with the Cytoskeleton to Maintain Cell Spreading and Reduced Motility of Nonmetastatic Lewis Lung Carcinoma Cells: The Loss of This Regulatory Control in Metastatic Cells. *Invasion Metastasis* 1998 Oct;17(4):199-209.

[233] Gabel S, Benefield J, Meisinger J, Petruzzelli GJ, Young M. Protein phosphatases 1 and 2A maintain endothelial cells in a resting state, limiting the motility that is needed for the morphogenic process of angiogenesis. *Otolaryngol. Head Neck Surg.* 1999 Oct;121(4):463-8.

[234] Baldacini O, Lutun P, Girardot R, Monteil H. Effect of okadaic acid on the cytotoxic activity of Clostridium difficile toxin B and Clostridium sordellii toxin L. *Nat. Toxins* 1993;1(6):361-8.

[235] Yano Y, Sakon M, Kambayashi J, Kawasaki T, Senda T, Tanaka K, Yamada F, Shibata N. Cytoskeletal reorganization of human platelets induced by the protein phosphatase 1/2 A inhibitors okadaic acid and calyculin A. *Biochem. J.* 1995 Apr;307 (Pt 2):439-49.

[236] Niggli V, Djafarzadeh S, Keller H. Stimulus-Induced Selective Association of Actin-Associated Proteins (alpha-Actinin) and Protein Kinase C Isoforms with the Cytoskeleton of Human Neutrophils. *Exp. Cell Res.* 1999 Aug;250(2):558-68.

[237] Leira F, Alvarez C, Vieites JM, Vieytes MR, Botana LM. Study of cytoskeletal changes induced by okadaic acid in BE(2)-M17 cells by means of a quantitative fluorimetric microplate assay. *Toxicol In Vitro* 2001 Aug;15(4-5):277-82.

[238] Macias-Silva M, Garcia-Sainz JA. Inhibition of hormone-stimulated inositol phosphate production and disruption of cytoskeletal structure. Effects of okadaic acid, microcystin, chlorpromazine, W7 and nystatin. *Toxicon* 1994 Jan;32(1):105-12.

[239] Leira F, Alvarez C, Cabado AG, Vieites JM, Vieytes MR, Botana LM. Development of a F actin-based live-cell fluorimetric microplate assay for diarrhetic shellfish toxins. *Anal. Biochem.* 2003 Jun;317(2):129-35.

[240] Gu L, Zhang H, Chen Q, Chen J. Calyculin A-induced actin phosphorylation and depolymerization in renal epithelial cells. *Cell Motil. Cytoskeleton* 2003 Apr;54(4):286-95.

[241] Asano Y, Mabuchi I. Calyculin-A, an inhibitor for protein phosphatases, induces cortical contraction in unfertilized sea urchin eggs. *Cell Motil. Cytoskeleton* 2001 Apr;48(4):245-61.

[242] Young MR. Protein phosphatases-1 and -2A regulate tumor cell migration, invasion and cytoskeletal organization. *Adv. Exp. Med. Biol.* 1997;407:311-8.

[243] Dawe HR, Minamide LS, Bamburg JR, Cramer LP. ADF/cofilin controls cell polarity during fibroblast migration. *Curr. Biol.* 2003 Feb;13(3):252-7.

[244] Dawe HR, Minamide LS, Bamburg JR, Cramer LP. ADF/cofilin controls cell polarity during fibroblast migration. *Curr. Biol.* 2003 Feb;13(3):252-7.

[245] Samstag Y, Dreizler EM, Ambach A, Sczakiel G, Meuer SC. Inhibition of constitutive serine phosphatase activity in T lymphoma cells results in phosphorylation of pp19/cofilin and induces apoptosis. *J. Immunol.* 1996 Jun;156(11):4167-73.

[246] Graessmann A, Graessmann M, Tjian R, Topp WC. Simian virus 40 small-t protein is required for loss of actin cable networks in rat cells. *J. Virol.* 1980 Mar;33(3):1182-91.

[247] Tournebize R, Andersen SS, Verde F, Doree M, Karsenti E, Hyman AA. Distinct roles of PP1 and PP2A-like phosphatases in control of microtubule dynamics during mitosis. *EMBO J.* 1997 Sep;16(18):5537-49.

[248] Sontag E, Nunbhakdi-Craig V, Bloom GS, Mumby MC. A novel pool of protein phosphatase 2A is associated with microtubules and is regulated during the cell cycle. *J. Cell Biol.* 1995 Mar;128(6):1131-44.
[249] Sontag E, Nunbhakdi-Craig V, Lee G, Brandt R, Kamibayashi C, Kuret J, White CL, III, Mumby MC, Bloom GS. Molecular Interactions among Protein Phosphatase 2A, Tau, and Microtubules. Implications for the regulation of tau phosphorylation and the development of tauopathies. *J. Biol. Chem.* 1999 Sep;274(36):25490-8.
[250] Hiraga A, Tamura S. Protein phosphatase 2A is associated in an inactive state with microtubules through 2A1-specific interaction with tubulin. *Biochem J.* 2000 Mar;346(Pt 2):433-9.
[251] Takahashi M, Shibata H, Shimakawa M, Miyamoto M, Mukai H, Ono Y. Characterization of a novel giant scaffolding protein, CG-NAP, that anchors multiple signaling enzymes to centrosome and the golgi apparatus. *J. Biol. Chem.* 1999 Jun;274(24):17267-74.
[252] Gaillard S, Fahrbach KM, Parkati R, Rundell K. Overexpression of simian virus 40 small-t antigen blocks centrosome function and mitotic progression in human fibroblasts. *J. Virol.* 2001 Oct;75(20):9799-807.
[253] Hamm-Alvarez SF, Wei X, Berndt N, Runnegar M. Protein phosphatases independently regulate vesicle movement and microtubule subpopulations in hepatocytes. *Am. J. Physiol.* 1996 Sep;271(3 Pt 1):C929-C943.
[254] Kobayashi N, Reiser J, Schwarz K, Sakai T, Kriz W, Mundel P. Process formation of podocytes: morphogenetic activity of microtubules and regulation by protein serine/threonine phosphatase PP2A. *Histochem. Cell Biol.* 2001 Mar;115(3):255-66.
[255] Fuchs E, Weber K. Intermediate filaments: structure, dynamics, function, and disease. *Annu. Rev. Biochem.* 1994;63:345-82.
[256] Eriksson JE, Opal P, Goldman RD. Intermediate filament dynamics. *Curr. Opin. Cell Biol.* 1992 Feb;4(1):99-104.
[257] Cheng TJ, Lin YL, Chiang AS, Lai YK. Association of protein phosphatase 2A with its substrate vimentin intermediate filaments in 9L rat brain tumor cells. *J. Cell Biochem.* 2000 Jul;79(1):126-38.
[258] Polakis P. Wnt signaling and cancer. *Genes Dev.* 2000 Aug;14(15):1837-51.
[259] Ratcliffe MJ, Itoh K, Sokol SY. A positive role for the PP2A catalytic subunit in Wnt signal transduction. *J. Biol. Chem.* 2000 Nov;275(46):35680-3.
[260] Yamamoto H, Hinoi T, Michiue T, Fukui A, Usui H, Janssens V, Van Hoof C, Goris J, Asashima M, Kikuchi A. Inhibition of the Wnt signaling pathway by the PR61 subunit of protein phosphatase 2A. *J. Biol. Chem.* 2001 Jul;276(29):26875-82.
[261] Li X, Yost HJ, Virshup DM, Seeling JM. Protein phosphatase 2A and its B56 regulatory subunit inhibit Wnt signaling in Xenopus. *EMBO J.* 2001 Aug;20(15):4122-31.
[262] Hsu W, Zeng L, Costantini F. Identification of a domain of Axin that binds to the serine/threonine protein phosphatase 2A and a self-binding domain. *J. Biol. Chem.* 1999 Feb;274(6):3439-45.

[263] Kyriakis JM, Avruch J. Mammalian mitogen-activated protein kinase signal transduction pathways activated by stress and inflammation. *Physiol Rev.* 2001 Apr;81(2):807-69.
[264] Chang L, Karin M. Mammalian MAP kinase signalling cascades. *Nature* 2001 Mar;410(6824):37-40.
[265] Gomez N, Cohen P. Dissection of the protein kinase cascade by which nerve growth factor activates MAP kinases. *Nature* 1991 Sep;353(6340):170-3.
[266] Pullar CE, Chen J, Isseroff RR. PP2A activation by beta2-adrenergic receptor agonists: novel regulatory mechanism of keratinocyte migration. *J. Biol. Chem.* 2003 Jun;278(25):22555-62.
[267] Ory S, Zhou M, Conrads TP, Veenstra TD, Morrison DK. Protein phosphatase 2A positively regulates Ras signaling by dephosphorylating KSR1 and Raf-1 on critical 14-3-3 binding sites. *Curr. Biol.* 2003 Aug;13(16):1356-64.
[268] Yoder JH, Chong H, Guan KL, Han M. Modulation of KSR activity in Caenorhabditis elegans by Zn ions, PAR-1 kinase and PP2A phosphatase. *EMBO J.* 2004 Jan;23(1):111-9.
[269] Ugi S, Imamura T, Ricketts W, Olefsky JM. Protein phosphatase 2A forms a molecular complex with Shc and regulates Shc tyrosine phosphorylation and downstream mitogenic signaling. *Mol. Cell Biol.* 2002 Apr;22(7):2375-87.
[270] Lebrin F, Bianchini L, Rabilloud T, Chambaz EM, Goldberg Y. CK2alpha-protein phosphatase 2A molecular complex: possible interaction with the MAP kinase pathway. *Mol. Cell Biochem.* 1999 Jan;191(1-2):207-12.
[271] Huang S, Houghton PJ. Targeting mTOR signaling for cancer therapy. *Curr. Opin. Pharmacol.* 2003 Aug;3(4):371-7.
[272] Thomas G, Hall MN. TOR signalling and control of cell growth. *Curr. Opin. Cell Biol.* 1997 Dec;9(6):782-7.
[273] Wang H, Jiang Y. The Tap42-Protein Phosphatase Type 2A Catalytic Subunit Complex Is Required for Cell Cycle-Dependent Distribution of Actin in Yeast. Mol Cell Biol 2003 May;23(9):3116-25.
[274] Schwartz MA, Baron V. Interactions between mitogenic stimuli, or, a thousand and one connections. *Curr. Opin. Cell Biol.* 1999 Apr;11(2):197-202.
[275] Clark EA, Brugge JS. Integrins and signal transduction pathways: the road taken. *Science* 1995 Apr;268(5208):233-9.
[276] Yokoyama N, Miller WT. Inhibition of Src by direct interaction with protein phosphatase 2A. *FEBS Lett.* 2001 Sep;505(3):460-4.
[277] Yokoyama N, Miller WT. Protein phosphatase 2A interacts with the Src kinase substrate p130(CAS). *Oncogene* 2001 Sep;20(42):6057-65.
[278] Sakai R, Iwamatsu A, Hirano N, Ogawa S, Tanaka T, Mano H, Yazaki Y, Hirai H. A novel signaling molecule, p130, forms stable complexes in vivo with v-Crk and v-Src in a tyrosine phosphorylation-dependent manner. *EMBO J.* 1994 Aug;13(16):3748-56.
[279] Yamakita Y, Totsukawa G, Yamashiro S, Fry D, Zhang X, Hanks SK, Matsumura F. Dissociation of FAK/p130(CAS)/c-Src complex during mitosis: role of mitosis-specific serine phosphorylation of FAK. *J. Cell Biol.* 1999 Jan;144(2):315-24.

[280] Ivaska J, Nissinen L, Immonen N, Eriksson JE, Kahari VM, Heino J. Integrin alpha 2 beta 1 promotes activation of protein phosphatase 2A and dephosphorylation of Akt and glycogen synthase kinase 3 beta. *Mol. Cell Biol.* 2002 Mar;22(5):1352-9.

[281] Pankov R, Cukierman E, Clark K, Matsumoto K, Hahn C, Poulin B, Yamada KM. Specific beta1 integrin site selectively regulates Akt/protein kinase B signaling via local activation of protein phosphatase 2A. *J. Biol. Chem.* 2003 May;278(20):18671-81.

[282] Ivaska J, Bosca L, Parker PJ. PKCepsilon is a permissive link in integrin-dependent IFN-gamma signalling that facilitates JAK phosphorylation of STAT1. *Nat. Cell Biol.* 2003 Apr;5(4):363-9.

[283] Kehls NE, Herrmann K, Berking S. The protein phosphatase inhibitor cantharidin induces head and foot formation in buds of Cassiopea andromeda (Rhizostomae, Scyphozoa). *Int. J. Dev. Biol.* 1999 Jan;43(1):51-7.

[284] Sieburth DS, Sundaram M, Howard RM, Han M. A PP2A regulatory subunit positively regulates Ras-mediated signaling during Caenorhabditis elegans vulval induction. *Genes Dev.* 1999 Oct;13(19):2562-9.

[285] Viallet J, Garcia A, Weydert A. Protein phosphatase 2A as a new target for morphogenetic studies in the chick limb. *Biochimie* 2003 Aug;85(8):753-62.

[286] Gotz J, Probst A, Ehler E, Hemmings B, Kues W. Delayed embryonic lethality in mice lacking protein phosphatase 2A catalytic subunit Calpha. *Proc. Natl. Acad. Sci. USA* 1998 Oct;95(21):12370-5.

[287] Taylor BK, Stoops TD, Everett AD. Protein phosphatase inhibitors arrest cell cycle and reduce branching morphogenesis in fetal rat lung cultures. *Am. J. Physiol. Lung Cell Mol. Physiol.* 2000 May;278(5):L1062-L1070.

[288] Holmes SE, O'Hearn EE, McInnis MG, Gorelick-Feldman DA, Kleiderlein JJ, Callahan C, Kwak NG, Ingersoll-Ashworth RG, Sherr M, Sumner AJ, Sharp AH, Ananth U, Seltzer WK, Boss MA, Vieria-Saecker AM, Epplen JT, Riess O, Ross CA, Margolis RL. Expansion of a novel CAG trinucleotide repeat in the 5' region of PPP2R2B is associated with SCA12. *Nat. Genet.* 1999 Dec;23(4):391-2.

[289] Schweiger S, Schneider R. The MID1/PP2A complex: a key to the pathogenesis of Opitz BBB/G syndrome. *Bioessays* 2003 Apr;25(4):356-66.

[290] Kao G, Tuck S, Baillie D, Sundaram MV. C. elegans SUR-6/PR55 cooperates with LET-92/protein phosphatase 2A and promotes Raf activity independently of inhibitory Akt phosphorylation sites. *Development* 2004 Feb;131(4):755-65.

[291] Wassarman DA, Solomon NM, Chang HC, Karim FD, Therrien M, Rubin GM. Protein phosphatase 2A positively and negatively regulates Ras1- mediated photoreceptor development in Drosophila. *Genes Dev.* 1996 Feb;10(3):272-8.

[292] Tree DR, Ma D, Axelrod JD. A three-tiered mechanism for regulation of planar cell polarity. *Semin. Cell Dev. Biol.* 2002 Jun;13(3):217-24.

[293] Yang J, Wu J, Tan C, Klein PS. PP2A:B56epsilon is required for Wnt/beta-catenin signaling during embryonic development. *Development* 2003 Dec;130(23):5569-78.

[294] Everett AD, Kamibayashi C, Brautigan DL. Transgenic expression of protein phosphatase 2A regulatory subunit B56gamma disrupts distal lung differentiation. *Am. J. Physiol. Lung Cell. Mol Physiol.* 2002 Jun;282(6):L1266-L1271.

[295] Hannus M, Feiguin F, Heisenberg CP, Eaton S. Planar cell polarization requires Widerborst, a B' regulatory subunit of protein phosphatase 2A. *Development* 2002 Jul;129(14):3493-503.
[296] Krishnan V, Pereira FA, Qiu Y, Chen CH, Beachy PA, Tsai SY, Tsai MJ. Mediation of Sonic hedgehog-induced expression of COUP-TFII by a protein phosphatase. *Science* 1997 Dec;278(5345):1947-50.
[297] Berry M, Gehring W. Phosphorylation status of the SCR homeodomain determines its functional activity: essential role for protein phosphatase 2A,B'. *EMBO J.* 2000 Jun;19(12):2946-57.
[298] Costantino ME, Stearman RP, Smith GE, Darling DS. Cell-specific phosphorylation of Zfhep transcription factor. *Biochem. Biophys. Res. Commun.* 2002 Aug;296(2):368-73.
[299] Takagi T, Moribe H, Kondoh H, Higashi Y. DeltaEF1, a zinc finger and homeodomain transcription factor, is required for skeleton patterning in multiple lineages. *Development* 1998 Jan;125(1):21-31.
[300] Firulli BA, Howard MJ, McDaid JR, McIlreavey L, Dionne KM, Centonze VE, Cserjesi P, Virshup DM, Firulli AB. PKA, PKC, and the protein phosphatase 2A influence HAND factor function: a mechanism for tissue-specific transcriptional regulation. *Mol. Cell* 2003 Nov;12(5):1225-37.
[301] Zakany R, Szucs K, Bako E, Felszeghy S, Czifra G, Biro T, Modis L, Gergely P. Protein phosphatase 2A is involved in the regulation of protein kinase A signaling pathway during in vitro chondrogenesis. *Exp. Cell Res.* 2002 Apr;275(1):1-8.

In: Protein Research Progress
Editor: A.B. Boscoe and C. R. Listow, pp. 61-98
ISBN 978-1-60456-663-3

Chapter 2

Dietary Protein Requirements for Preterm Infants in the Neonatal Period: Past, Present, and Future

Shahirose Sadrudin Premji*[*1], Amuchou Singh Soraisham[2], Lorraine Chessell[3] and Reg Sauve[4]

[1]University of Calgary, Faculty of Nursing
2500 University Drive NW, Calgary, Alberta
Canada T2N 1N4
[2]Department of Pediatrics, University of Calgary
Alberta, Canada T2N 2T9
[3]Registered Dietitian, NICU, McMaster Children's Hospital
Ontario, Canada L8N 3Z5
[4]Paediatrics and Community Health Sciences
The University of Calgary, 3330 Hospital Dr NW
Calgary, Alberta T2N 4N1

Abstract

Protein is an important component of adequate nutrition which is essential to promote optimal growth and neurodevelopment. Principles which guide protein nutrition during infancy have not been well established and they are based on limited data from neonates and extrapolation from studies undertaken in animals and older children and adults. Advances in our understanding of protein turnover, synthesis, catabolism, and deposition in the neonatal period and both short- and long-term consequences of early protein deficiency, contribute to the ongoing development of these principles. This chapter explores past, present and future trends in dietary protein research in the neonatal period to help us understand the impact of dietary protein intake in relation to health and

---

[*] Corresponding author: Telephone: 403-220-8053 Fax: 403-284-4803; Email: premjis@ucalgary.ca

chronic disease. We will follow the debate with regards to protein requirements for preterm infants, share our current understanding of the unique relationship between levels of dietary protein, carbohydrates and minerals, and reveal the potential of tailoring dietary protein intake to specific neonatal conditions. We aim to summarize results of noteworthy studies, identify principles, and stimulate ideas about areas for future research.

INTRODUCTION

Dietary protein provides essential amino acids necessary for protein synthesis [1] and impacts growth, nutrient retention, and metabolic response in preterm infants [2]. Significant body protein deficit can accumulate without any protein intake given the limited body protein stores in preterm infants [3]. Rate of metabolic protein loss is two-fold higher in extremely low birth weight (ELBW) infants compared to term infants [3]. Deficiency of protein leads to growth failure, edema and lower resistance to infection. When protein deficiency is extreme it may compromise pulmonary status secondary to muscle weakness [4]. Estimating protein requirements to achieve protein accretion is difficult since preterm infants have a high rate of protein turnover and breakdown [5] due to significant effect of gestation or illness on protein metabolism [3, 6, 7]. Protein balance is also dependent on caloric intake [3] as sufficient energy is needed to allow protein to be used for anabolism rather than being broken down to enter (i.e., carbon skeleton) the citric acid cycle to be used as an energy source [8]. This protein-energy relationship is not as clearly defined in ELBW infants [3].

Adequacy of protein intake is difficult to evaluate in the clinical setting and it is within this context there has been heightened interest and new insight into the processes of whole-body protein accretion in the preterm infant [9]. Indicators of adequacy of protein intake include plasma proteins such as transthyretin, transferrin, and retinol-binding protein [10-14]. The basis for estimating protein needs is further compromised by the lack of a generally accepted goal for feeding these infants [9].

Both total parenteral and enteral nutrition are provided to preterm infants to optimize long-term growth and development [15-17]. Nutritional interventions which correct protein deficiency without exceeding the recommended protein intake or increasing risk of adverse clinical or metabolic sequelae have been proposed. Early administration of parenteral amino acids (e.g., in the first hours after birth) to achieve rapid, maximal nutrition [16, 18] is an important clinical strategy to preserve body protein stores [3]. Emerging considerations with regards to protein quality and specific amino acids emphasize biological function, splanchnic metabolism, and nutritional adequacy of specific proteins and amino acids.

Early nutrition during critical periods of development may create endocrine and metabolic conflict, that is malprogramming, which ultimately increase risk of disease in adult life (e.g., diabetes, hypertension, coronary heart disease, osteoporosis and obesity) [19-22]. Acceptable range of dietary protein may vary from infant to infant and need to be tailored in response to a given early postnatal nutritional plane to ensure expression of a normal phenotype and prevent disease manifestation later in life.

CURRENT RECOMMENDATIONS FOR DIETARY PROTEIN INTAKE IN PRETERM INFANTS

Protein requirements have been studied since 1869 and periodically reestimated and debated as our understanding of metabolic responses in preterm infants has improved [23]. Estimates of protein requirement for preterm infants is most commonly based on the factorial approach [24], a theory based calculation that sums the requirements for growth and those for replacement of inevitable losses in urine, feces, and skin [25]. Zeigler et al [26, 27] have described the body composition of the reference fetus. Theoretical calculations of preterm infant protein requirements to promote growth and nitrogen accretion based on the factorial approach using this reference fetus is estimated at 4 g/kg/day for infants with a birth weight of less than 1200 grams, and 3.5 g/kg/day for infants with a birth weight of 1200 to 1800 grams [28]. Although the reference fetus is a widely accepted representation of protein accretion, the small sizes of the fetuses included in estimating body composition may somewhat underestimate the fetal accretion rates [3]. The same deficits of existing fetal growth reference as described by Kramer et al [29] including errors in reported gestational age (e.g., rounding gestational age and basing gestational age on onset of the last menstrual period), insufficient sample sizes at low gestational ages and inadequate statistical modeling, may also be applicable to the development of this reference fetus. Hence, this reference fetus may not be entirely accurate in estimating protein requirements for preterm infants. A new "reference fetus" that overcomes these limitations needs to be developed.

Protein requirements can also be estimated by determining the actual amount of protein that will support intrauterine rates of growth and nitrogen accretion [30]. However, estimates based on this approach do not result in the same estimate of protein requirements [24] as studies have shown approximately 3 g/kg/day is sufficient to support intrauterine rates of growth and nitrogen accretion [2, 12, 30-32]. These well established protein requirements are based on clinically stable preterm infants between 2 to 4 weeks of age who are receiving enteral feedings and growing [3, 12, 32]. Consequently, protein requirements of low and ELBW infants are less certain [3]. A potential disadvantage of this alternate approach is that a number of very different nutritional strategies can be used to achieve fetal rate of weight gain without due consideration for "quality" of weight gain [24].

Nutritional intakes differ between infants and nurseries, and account for 45% of the variation in growth [33]. Clinically, recommended protein intakes are difficult to establish and maintain, particularly in sick preterm infants [18, 34-37]. Additionally, since these recommended protein intakes, regardless of which approach is used to make estimates, do not account for protein deficits accrued in the early weeks of life, postnatal malnutrition and growth retardation is inevitable [33]. Data from the NICHD Neonatal Research Network demonstrate that postnatal growth deficiency is common in preterm infants, particularly ELBW infants, and most do not achieve catch-up growth during hospitalization [38]. Current recommendations for protein intakes have been based on nitrogen balance studies which may further underestimate the protein requirements of preterm infants [39]. Future research should aim to determine protein requirements of preterm infants based on birth weight, gestational age, postnatal age, and clinical course.

Past research has focused on maximum protein intakes for preterm infants; however, little attention has been directed to identifying the lower limits of protein needs in preterm infants [37]. This is an important issue given that lower protein intakes are inevitable [33]. Amino acid oxidation methodology offers a number of advantages over the nitrogen balance method used in the past, in expanding our understanding of minimum protein intakes for preterm infants. The amino acid oxidation methodology gives insight into how preterm infants adapt to short-term changes in protein intake making it feasible to study the impact of protein-deficient diet. Additionally, it is a more sensitive method of estimating protein intake requirement [37]. Consequently, this alternate approach offers opportunity to better define minimum and maximum protein intakes for preterm infants.

Although recommendations have been made with regards to protein intakes, it is uncertain how long these specific recommendations should be followed. For example, should they be carried on after a preterm infant reaches term corrected age? A nutritional approach needs to be implemented for a much longer period than a few weeks after term and constantly optimized and adapted to the needs of the growing and developing preterm infant. Each preterm infant has an optimal range of protein to energy ratio; the challenge lies in assessing and meeting this range [40].

PROTEIN ENERGY RATIO

Protein synthesis and breakdown are processes which are energy-dependent; hence energy is required for both protein metabolism and deposition [2, 16]. A curvilinear relationship exists with most of the effect of energy on protein gain occurring at less than 50 to 60 kcal/kg/day and a maximum energy threshold, 100 to 120 kcal/kg/day, beyond which protein accretion is not affected [2, 16]. Sufficient energy is needed to allow protein to be used for anabolism rather than being broken down to enter (i.e., carbon skeleton) the citric acid cycle to be used as an energy source [8]. If an excess of energy is supplied with limited protein intake, energy, regardless of its source, is deposited as fat with impact on weight gain [41, 42].

Kashyap et al [2, 43] found that nitrogen retention is directly related to protein intake and fat deposition is related primarily to energy intake. Anthropometric data demonstrate increased fat deposition with energy intake greater than 120 kcal/kg/day [2, 43] and diet induced thermogenesis in instances where protein was not utilized completely [2]. The benefit of improved thermoregulation with increased fat deposition needs to be appraised against the risk of predisposing infants to later obesity [44]. By regulating the composition of dietary intake, particularly protein and energy, different composition of weight gain can be achieved [41]. Since the relative proportions of protein and energy intake determine the composition of newly deposited tissue, it was hypothesized that an increase in protein to energy ratio may promote accretion rates of fat to fat-free mass closer to those of the fetus [45]. This hypothesis was not supported with a protein to energy ratio of 3.2 g/100 kcal [45]. There is limited available data on the maximum protein to energy ratio for preterm infants. A protein to energy ratio of 2.5 to 3.6 g/100 kcal at a caloric intake of 110 to 135 kcal/kg/day has been recommended for preterm infant formulas based on the provision of quantity of protein needed to achieve fetal rates of lean tissue growth and nitrogen accretion. Following this recommendation limits protein intake to 2.8 to 4.8 g/kg/day [46].

ASSESSMENT OF PROTEIN NUTRITIONAL STATUS

Anthropometric measurements currently used to assess nutritional status do not give insight into protein nutritional status, that is, protein deficiency [11]. Protein accretion is a better indicator of nutritional status [16]. From a clinical perspective, measurement of body composition is not easily attainable [16]. Plasma proteins such as transthyretin, transferrin, and retinol-binding protein have been shown to be suitable to evaluate protein nutritional status in stable, growing, appropriate for gestational age very low birth weight (VLBW) infants [10-14]. Findings suggest that at protein intakes <3g/kg/day, the urinary excretion of several amino acids was stable, but increased with protein intakes above 3 g/kg/day suggesting a threshold above which amino acids are excreted at higher rates. These researchers postulated that evaluation of renal clearance for single amino acids might be valuable to assess protein quality. Given strong correlation between urinary and plasma concentrations of most amino acids, measure of urinary amino acids may offer a less invasive approach to assess protein nutritional status [12].

Although metabolic markers such as plasma amino acid concentrations have been proposed as a means of assessing protein nutritional status, a number of different reference standards or norms have been suggested including pre-prandial plasma amino acid concentrations in: a) the fetus of the same postconceptional age, b) umbilical cord blood at birth, or c) breast-fed, growing term infants at 1 to 3 months of age [12]. One study compared plasma levels of essential amino acids in well-growing infants receiving exclusively human milk proteins with values from the literature for the reference standards or norms identified above. Healthy, growing, breast-fed term infants at 1 to 3 months of age appeared to be a more appropriate reference standard or norm [12]. Given all the factors that influence plasma amino acid concentrations including method and moment of sampling, protein quantity and quality, protein turnover rate, energy supply, and gestational age, it has been proposed that cord blood amino acid concentration may be more appropriate [47].

When evaluating the nutritional value of dietary protein, particularly in preterm infants, it is important to test the preterm infant's ability to handle the protein by measuring postprandial plasma amino acid concentrations [48]. In preterm infants, plasma amino acid concentrations are influenced by different quality of the protein as evident from a study which measured pre- and postprandial plasma amino acid concentrations in preterm infants fed with different dietary proteins at similar protein intakes during the first week of life [49]. The postprandial kinetics of the amino acids is influenced by the ingested protein quantity, the time interval between feedings, and the duration of the feeding [49]; hence it is not surprising why plasma amino acid concentrations have been reported to fluctuate continuously [48]. The optimal time to measure postprandial peak values of amino acids remains to be determined [49]. Assessment of protein nutritional status continues to be a clinical challenge for those caring for preterm infants.

Goal of Protein Nutritional Management

The generally accepted nutritional goal of feeding preterm infants, particularly VLBW and ELBW infants is still a matter of controversy [9, 12]. An appropriate clinical objective is the provision of protein intake from birth to 7 days that will prevent deficiencies and catabolism of nutrient substrate [50] and thereafter "achieve a protein gain that approximates the in-utero protein gain of a normal fetus of the same postconceptional age [9].

Although many neonatologists aim to help preterm infants attain intrauterine growth rates when making decisions related to nutritional management of preterm infants, this approach may not be ideal in the extrauterine environment [12] for medical, practical, and theoretical reasons [16]. In ELBW infants, weight change in the early weeks of life reflects more fluid management than tissue deposition. Medical management to prevent or minimize the adverse consequences of respiratory illness [18] or patent ductus arteriosus may include fluid restriction and/or withholding of feedings. Consequently, from a medical and practical perspective the goal of attaining intrauterine growth rates is unrealistic [16]. Theoretically, the body compartment that can change significantly with nutrition is protein mass as glucose is stored as glycogen, and glycogen stores are limited in all infants [16]. According to Thureen and Heird [24] replicating intrauterine body composition in the postnatal period is a more physiologic approach to growth in preterm infants. As well, whether nutritional evaluation should be intrauterine or extrauterine-based is a matter of contention [51].

Enteral Nutrition

Proteins supplied in enteral nutrition are derived from two different sources, namely human milk or formula (i.e., bovine milk) [9]. Until the 1960s banked human milk was routinely used to provide enteral nutrition for LBW infants. Since then expressed breast milk and modified cow milk protein-based formula preparation have become the predominant source of enteral nutrition [46]. The nutrient profiles of preterm formula, particularly amino acids, are different from human and term formulas as a result of studies from the 1980's (e.g., Atkinson et al [52], Atkinson et al [53], and Lucas et al [54]) which demonstrated that preterm infants grew better when milk feedings such as formula, or human milk was supplemented with protein and minerals [46]. Many earlier sources of enteral nutrition including formula's containing half-skimmed cow milk, acidified milk, evaporated milk, and soy milk are considered unsuitable for preterm infants. In this chapter we will only discuss soy milk formulas given their more recent and popular use in clinical practice.

Formula FED Preterm Infants: Higher Versus Lower Protein Intakes

Fetal body composition data collected by Widdowson and Spray in 1951 formed the basis for developing preterm infant formula. The mid 1960s marked the introduction of protein enriched formula (18% whey: 82% casein) developed specifically for preterm infants by Mead Johnson and Company. Over the course of time, preterm formulas have changed to improve digestibility and absorption, limit metabolic acidosis, and decrease osmolality of formula. Reviews completed by Tsang et al (1985 and 1993) to evaluate the evidence regarding the nutritional requirements of preterm infants prompted further modification of preterm formulas [46]. Preterm formulas currently available in North America contain added whey protein (3.0 g of protein per 100 kcal), glucose polymers, medium-chain triglycerides, calcium, phosphorus, electrolytes, folate, and fat-soluble vitamins. If a maximum energy threshold is maintained, formula fed infants receive about 3.0 to 3.6 g/kg/day of protein.

Higher protein intakes have been proposed, particularly for ELBW infants, since increases in protein intake up to an approximate maximum intake of 4.0 to 4.5 g/kg/day at energy intakes above 30 to 50 kcal/kg/day leads to greater protein accretion [16]. While higher protein intakes have the potential to promote greater protein accretion and growth, diet-induced thermogenesis may result from protein intakes not utilized completely [2]. In preterm infants the amino acid metabolic pathways may not be able to handle higher protein intakes efficiently. The resulting metabolic acidosis and higher plasma level of amino acids such as tyrosine and phenylalanine concentrations [9, 55] could theoretically lead to mental retardation.

A Cochrane systematic review evaluating the risks and benefits of higher ($\geq$ 3.0 g/kg/day) versus lower (< 3.0 g/kg/day) protein intakes during the initial hospital stay of formula-fed preterm infants < 2.5 kilograms ascertained that weight gain and nitrogen accretion can be promoted by regulating protein intake [56]. A meta-analysis of five studies [43, 57-60] revealed accelerated weight gain with protein intake of $\geq$ 3.0 g/kg/day but < 4.0 g/kg/day or high protein intake compared to protein intake of < 3.0 g/kg/day or low protein intake (weighted mean difference (WMD) 2.36 g/kg/day, 95% Confidence Interval (CI) 1.31, 3.40). Given the accompanying increase in nitrogen accretion rates (WMD 143.7, 95% CI 128.7, 158.8) it was deduced that this weight gain mostly likely represents an increase in lean body mass [56]. The statistical heterogeneity in nitrogen accretion suggests that any overall summary statistic could be misleading. Results of acid-base status were conflicting with one study [43] reporting values of metabolic status to be within normal limits while a second study [59] reporting late metabolic acidosis in infants receiving higher protein intake. Limited information could be extracted from this systematic review about other potential adverse effects of high protein intake including neurodevelopment outcomes, necrotizing enterocolitis (NEC), sepsis and diarrhea. It is therefore uncertain whether higher protein intake ($\geq$ 3.0 g/kg/day but < 4.0 g/kg/day) compared to low protein intake (< 3.0 g/kg/day) safely promotes optimal growth and development during initial hospital stay of formula-fed preterm infants < 2.5 kilograms.

The above Cochrane systematic review [56] included two studies [61, 62] with multiple reports [63-67], in a post-facto analysis. These studies did not meet the a priori inclusion criteria as the formulas differed in other nutrients (e.g., energy, sodium, potassium, phosphate, and zinc) by more than 10% relative concentration, aside from protein intakes which were above 4.0 g/kg/day. Given the unique comparisons undertaken in these studies, they could not be included in the meta-analysis [56]. Weight gain was not reported by Goldman et al [61] instead regression curves were calculated for infants < 1500 grams and > 1500 grams to determine how long infants should take to reach 2200 grams. More infants in the very high protein intake group took longer than the calculated period of time to reach

2200 grams ($p<0.01$). Goldman et al [63, 64] documented a significantly increased incidence of low IQ scores (i.e., Stanford-Binet test scores below 90) at three and five to seven years of age among infants of birth weight below 1300 grams who were fed high protein intakes of 6 to 7.2 g/kg/day during their initial hospital stay. The Cochrane systematic review maintains that very high protein intakes (> 4.0g/kg/day) during initial hospital stay of formula-fed preterm infants < 2.5 kilograms be considered experimental in light of the current state of evidence [56].

Two more recent studies [68, 69] compared high (≥ 3.0 g/kg/day but < 4.0 g/kg/day) versus very high protein intakes (≥ 4.0 g/kg/day). Both studies infer that short term increases in protein intake promotes weight gain without evidence of metabolic stress. In the Embleton and Cooke study, [68] preterm infants were given the study formula from the time they were tolerating full enteral feedings (i.e., 150 mL/kg/day) until term plus 12 weeks corrected age. In the Cooke et al study, [69] the intervention began when preterm infants were tolerating full enteral feedings (i.e., 135 or more mL/kg/day) and each study formula was provided for one week. These studies, however, emphasize the need to more precisely define protein requirements [68] and determine the long-term benefits of increased protein intakes [69]. The Cooke et al study [69] does not meet the inclusion criteria for the Cochrane Systematic Review as it is not a randomized controlled trial but rather a balanced cross-over design. The Embleton et al study [68] compared three protein intake groups (4.3 g/kg/day versus 3.8 g/kg/day versus 3.5 g/kg/day) in seventy-seven healthy preterm infants ≤ 1750 grams birth weight (≤ 34 weeks gestation) tolerating full enteral formula feed at ≥ 150 mL/kg/day for 48 hours. The formula did not vary in other nutrients (e.g., energy, sodium, potassium, phosphate, and zinc) by more than 10% relative concentration. Although weight gain was highest in the very high protein intake group, differences were not significant. Three infants in the very high protein intake group (≥ 4.0g/kg/day) developed uremia, nevertheless other metabolic indices (e.g., metabolic acidosis, ammonia level, plasma amino acid profiles) remained normal. Blood urea nitrogen levels normalized without interventions. The inclusion of this study in the Cochrane Systematic Review would not alter the conclusions drawn in this review.

Heterogeneity in certain outcomes reported in the Cochrane Systematic Review may be explained by clinical diversity, specifically variability in participants. Rates of protein loss are two-fold higher in ELBW infants compared to term infants [3]. Throughout gestation there is a decrease in the rate of protein accretion per unit body weight. Efficiency of protein gain is inversely related to body weight [70]. Preterm infants have a high rate of protein turnover and breakdown [5] due to significant effect of illness on protein metabolism [3, 6, 7]. Kalhan and Iben [7] caution about the interpretation of studies examining growth, protein and nitrogen metabolism because of numerous confounding variables particularly for the ELBW infant or micropremie. Protein nutrition management will need to be individualized taking into account the various factors that impact protein accretion (e.g., gestational age, weight, and illness) and modified with changing clinical status over the course of the infants stay in the hospital.

Human Milk Supplementation

Preterm infants fed own mother's milk receive approximately 1.4 g/100 mL [71] or about 2.5 g/kg/day protein [34] which is below established protein requirements of at least 3 g/kg/day to support intrauterine rates of growth and nitrogen accretion [2, 12, 30-32]. From a clinical perspective current knowledge about protein energy ratio, discussed early in this chapter, implies that to increase weight gain, particularly protein component of this weight gain, protein (not exceeding the maximum recommended amount) rather than energy should be supplemented to promote higher growth rate in infants receiving human milk [72]. Similarly, when an infant is not taking adequate amounts of formula, strategies to supplement this intake should include increase in protein intake alone [2, 72] (not exceeding the maximum recommended amount) or concentrating the formula to increase both protein and energy intake as well as other macronutrients if needed [2]. Methods used clinically to supplement human milk include powder fortifiers (multinutrient powdered mixtures), complete liquid fortifiers (liquid formulas which are mixed half and half with expressed breast milk) or single nutrient supplements [73]. Multinutrient powder fortifiers are most commonly used when there is an adequate volume of human milk available for feeds. Using a powder formula is often preferable because it doesn't reduce the amount of human milk received by the infant, whereas liquid fortifiers are mixed half and half with human milk.

Historically, early studies comparing the growth of preterm infants fed human milk or cow's milk formulas suggested that infants fed high protein formulas gained weight more quickly than infants fed banked, human milk. Gordon et al [74] compared infants fed an energy intake of 120 kcal/kg/day using human milk or a formula mixture of evaporated milk or partly skimmed cow's milk. In this study, infants fed cow's milk with a protein intake of approximately 6 g/kg/day gained weight faster than infants fed human milk. Subsequent studies reviewed both the quantity and the quality of proteins added to infant formulas proving that indeed, these infants grew faster on the high protein formulas, but that the improvements in growth may come at a metabolic cost. Goldman et al [61] compared infants fed protein intakes at 3-3.6 g/kg/day or 6-7.2 g/kg/day. They found that infants who were fed the lower protein formula had higher incidence of late onset edema and lower albumin levels at 28 days of age, but that infants fed with the higher protein formula had more fever and lethargy, and nipple fed more poorly. In a follow-up study of the same infants, there was a significant increase in low IQ scores (<90) in infants with birth weights < 1300 grams who received the high protein intakes [63].

In a study by Raiha et al [62] infants were fed one of 4 formulas: low protein (1.5%), casein predominant; high protein (3%), casein predominant; low protein (1.5%), whey predominant; high protein (3%), whey predominant, or pooled human milk. In this study, infants on high protein formulas had higher serum urea levels, which were above the normal range. Ammonia levels were elevated with both casein predominant formulas, and the high whey protein formula, and late metabolic acidosis was more common in infants fed high casein formula. In the same group of infants, plasma amino acids were evaluated. Plasma and urine phenylalanine and tyrosine levels were higher in formula fed infants than those fed breast milk, and were significantly higher in infants fed the 3% casein formula [67]. Similarly, there was an increase in urine and plasma methionine levels in all infants fed formula as compared to human milk, with the highest levels in infants fed the 3% protein formulas [65]. The plasma cystine levels were also elevated in infants fed 3% protein formulas as compared to infants fed human milk, while taurine levels were reduced in infants fed formulas [65]. Casein predominant feeds are not routinely used for preterm infants due to the elevated amino acid concentrations of phenylalanine, tyrosine, methionine which is thought to be due to hepatic immaturity [75].

With evidence for poorer outcomes in babies fed very high intakes of protein, researchers again began to focus on the use of human milk as a source of nutrition for VLBW infants. In 1978, Atkinson et al [76] measured the nitrogen concentration of milk from mothers who delivered prematurely. They found that the nitrogen concentration in preterm mother's milk was significantly higher during the first month of lactation than in full term mother's milk. They determined that for a given volume of milk, the preterm infant would receive 20% more nitrogen than the term infant if fed own mother's milk, and approximately double the nitrogen intake of a preterm infant fed mature, donor milk. A subsequent study to look at the composition of nitrogen in preterm mother's milk confirmed that it is qualitatively similar to term mother's milk and that own mother's milk would provide the additional protein nitrogen and non-protein nitrogen needed for the synthesis of new tissue [77]. Therefore, there was renewed interest in the potential use of breast milk for preterm infants. Several short term studies have suggested that infants may gain weight at or near intrauterine rates when fed their own mother's milk [52, 53, 78], suggesting that own mother's milk is a more appropriate feeding than banked human milk in the first weeks of life. However, other studies have not supported these findings [79]. Because there is a significant decrease in the protein content of preterm human milk between transitional and mature milk, while the protein requirement of the preterm infant remains high [75, 80] it is expected that exclusive feeding of unfortified breast milk will result in a slower growth rate and nutritional deficits that persist beyond the period of hospitalization [80].

A Cochrane Systematic Review [81] was completed to determine whether the addition of protein (average increase by 1.5 g/kg/day) to human milk resulted in improved growth and neurodevelopment outcomes without adverse consequences. Results of 4 studies [10, 32, 82, 83] were combined (n = 90 infants) and showed an increase in short-term growth (WMD 3.6 g/kg/day, 95% CI 2.4 to 4.8 g/kg/day), length (WMD 0.28 cm/week, 95% CI 0.18 to 0.38 cm/week) and head growth (WMD 0.15 cm/week, 95% CI 0.06 to 0.23 cm/week) with protein supplementation of human milk. Higher urea levels (WMD 1 mmol/L, 95% CI 0.8 to 1.2 mmol/L) were seen in the supplemented group, although they were still within the normal range. It is postulated that the higher urea levels might be indicative of a more appropriate intake, and that the low levels generally seen in preterm infants fed unsupplemented human milk reflect a deficient protein intake [81]. A study not included in the systematic review [31] also supports a higher rate of growth in infants fed protein-supplemented human milk over human milk alone. In studies comparing infants fed fortified human milk or preterm, protein-rich formulas, both diets have been shown to support growth and nitrogen retention at estimated third trimester intrauterine rates [84, 85].

Studies [10-12, 86] reviewing the association between protein intakes and concentrations of plasma proteins to assess the adequacy of protein intakes showed differences in serum concentration of plasma proteins, particularly transthyretin, transferrin, retinal binding protein and transferring [12, 14], and urine amino acids in protein supplemented babies. They found that protein intake correlated with changes in the serum concentration of transthyretin, retinal binding protein and transferrin, and that transthyretin and transferrin could be used to evaluate protein nutritional status in VLBW infants.

Infants who received human milk alone had low plasma and urine concentrations of several amino acids as compared with cord or fetal amino acids or healthy term breast-fed infants, which indicated deficient protein intakes [12]. On the other hand, infants fed fortified milk had plasma amino acid levels below fetus or cord levels, but similar to the breast-fed term infant or preterm infants fed preterm, protein-rich formula. Similar findings have been reported in other studies [10, 86]. Kashyap et al [31] also reviewed biochemical indicators of protein status in infants fed human milk with or without added protein. Findings were consistent in that urea, transthyretin and albumin levels were higher in the protein-supplemented group, and albumin and transthyretin levels were below normal in unsupplemented infants. Furthermore, nitrogen retention exceeded intrauterine rates in supplemented infants but was below intrauterine rates in unsupplemented infants [31].

Uncertainty remains with regards to the precise quantity of nutrients which should be added to supplements [73] and the source of protein supplementation. Commercially-produced supplements for human milk, such as fortifiers, provide additional nutrients in the form of protein, calcium, phosphate, carbohydrates, vitamins and trace minerals. Of particular importance are calcium and phosphate which are required to attain adequate postnatal skeletal growth as preterm infants have low skeletal stores of these minerals [87]. A Systematic Cochrane Review was completed to determine if multicomponent nutritional supplements to human milk leads to improved growth, bone metabolism and neurodevelopment outcomes [87]. A meta-analysis of thirteen studies (n = 600 infants) demonstrated short-term increases in weight gain (WMD 2.3 g/kg/day, 95% CI, 1.7 to 2.9 g/kg/day), length gain (WMD 0.12 cm/week, 95% CI 0.07 to 0.18 cm/week), and head growth (WMD 0.12 cm/week, 95% CI 0.07 to 0.16 cm/week) in infants receiving fortification. There appeared to be no effect on long-term growth parameters; however there was insufficient data to adequately assess long-term growth, as well as neurodevelopment outcomes. There appeared to be no effect on alkaline phosphates levels. It was challenging to assess the effect of fortification on bone mineral content as findings varied across studies. Furthermore, the meta-analysis was heavily influenced by the larger trial which demonstrated a significant difference in favor of the fortification group. Large amounts of missing data precluded any conclusions about adverse effects associated with use of multicomponent fortification. To date there are no studies evaluating the effects of individual components, such as carbohydrates [88] and calcium and phosphorus [89]. Two studies [32, 83] examining the effect of adding supplemental fat to human milk demonstrated no effect on growth. One of these studies [32] was included in a Cochrane Systematic Review [90] which concluded that the advantages of fat supplementation, including provision of more calories per gram and lower respiratory quotient secondary to diminished carbon dioxide production [91], are difficult to assess because of insufficient data [90]. Long-term growth, neurodevelopmental outcomes and adverse effects of fat supplementation of human milk such as fat accretion or deposition were not assessed; hence the reviewers were unable to comment on potential adverse effects [90].

In preterm infants protein absorption is directly influenced by protein sources as evident from differences in nitrogen absorption rates noted with varied feeding regimen [70]. Nitrogen absorption was reported to be higher with powdered whey- or casein-predominant formulas compared to human milk fortifier [70]. Comparison of human milk protein versus bovine protein as a source for supplementing human milk, suggest that nitrogen absorption and retention are similar [92], growth does not differ between infants fed supplemental protein from human or bovine sources [92-94], and that biochemical markers of protein nutrition such as urea, transthyretin and transferrin were no different between groups [93, 94].

Although manipulation of milk has the potential to impact the intrinsic host-defense properties of the milk, human milk confers non-nutritional advantages such as protection from infection and NEC [73] and improved cognitive and neurodevelopment outcomes [95]. Current research supports the need for supplemental protein for preterm infants fed human milk. Future research should focus on the most appropriate level of supplementation, the precise quantity of nutrients which should be added to supplements, nutrient interactions, the appropriate length of time supplementation should continue, as well as the best source of protein. More evidence for clinical effects of protein supplementation and ruling out any adverse effects of providing higher protein intakes, especially with prolonged protein supplementation is needed.

Specialised Formulas

In 1929, soy protein-based formulas were introduced as a strategy to manage severe persistent lactose intolerance, and galactosemia in infancy. It wasn't until mid-1960s, however, when there was increasing acceptance of this approach [96, 97]. Soy protein-based formulas were indicated for the nutritional management of preterm infants given its characteristic low osmolality and lack of lactose which would confer the advantage of decreasing the incidence of NEC [98]. Soy protein-based formulas currently in the market use a soy isolate; a practice which commenced in the mid-1960s to overcome side effects of diarrhea and excessive intestinal gas in infants consuming this diet [97]. Research in the 1970s and 1980s focused on improving the biological quality of the protein with supplementation of methionine, carnitine, and taurine as evidenced by improved weight gain and/or urea nitrogen excretion and/or serum concentrations of nitrogen and albumin [99-101]. By the late 1990s use of soy protein-based formula gained popularity despite limited indications [97] prompting a critical review of the role of soy protein-based formulas [96, 98]. It is important to assess the efficacy and safety of soy protein-based formulas [97] in preterm infants given their immaturity and unique nutritional needs.

In preterm infants, studies evaluating the nutritional adequacy of soy protein-based formulas have shown contradictory findings of anthropometric measurements (e.g., growth or length) and serum chemistries (e.g., plasma protein levels and/or nitrogen retention and/or serum phosphorus levels). Two randomized controlled trials [102, 103] reported lower weight gain and serum albumin levels in infants receiving soy protein-based formulas. A third randomized controlled trial [104] reported no difference in postnatal growth between infants enrolled in the soy protein-based formula group and conventional milk-based formula (Similac with Iron 24, Ross Laboratories). Shenai and colleagues [104] enrolled VLBW infants and showed nitrogen retention rates comparable to fetal accretion rates, adequate protein utilization (i.e., normal serum protein and albumin levels), and no issues with calcium metabolism (i.e., no difference in urinary calcium excretion, net calcium retention, and serum calcium). Phosphorus metabolism was affected (i.e., greater loss of fecal phosphorus, and significantly lower serum phosphorus levels) raising the concern of phosphorus deficiency rickets. Hall and colleagues [103] enrolled stable, healthy VLBW infants and found no difference in serum calcium, phosphorus, alkaline phosphatase, and 25-hydroxy vitamin D levels in infants receiving a soy protein infant formula supplemented with calcium, phosphorus, and vitamin D when compared to infants receiving a whey-predominant preterm infant formula. The high aluminum content in soy protein-based formulas [105, 106] can theoretically potentiate albumin toxicity in preterm infants because of low albumin binding capacity, reduced renal function or renal insufficiency, and rapid growth [105]. Soy protein-based formulas can thereby impact bone matrix formation and mineralization and contribute to the development of osteopenia in preterm infants [105]. A retrospective review [107] and prospective observational study [108] both reported radiographic evidence of increased osteopenia in LBW infants receiving soy protein-based formulas as a sole source of nutrition. Preterm infants followed by Kulkarni et al [108] received a multivitamin preparation containing 400 IU of vitamin D with some infants receiving additional calcium lactate and vitamin D supplementation based on attending neonatologist preference.

The constitution of soy protein infant formula differs significantly from cow's milk protein infant formula. Soy protein infant formula has higher protein content (2.25 g/100 kcal vs 1.8 g/100 kcal) given lower digestibility and bioavailability of soy protein [96]. Soy protein infant formula has a different pattern of amino acid (i.e., lower amounts of methionine, branched chain amino acids lysine, and praline and higher quantities of aspirate, glycine, arginine, and cystine) and may play a role in the observed differences in postprandial protein metabolism. Differences in digestion kinetics such as rapid transit of soy protein (i.e., intestinal rate of amino acid absorption) and influx of dietary amino acids into the liver for protein synthesis, primarily account for the principal differences in postprandial metabolism between soy and milk protein [109]. High concentrations of phytate in soy protein-based formulas may impair the absorption of minerals and trace elements and interfere with iodine metabolism [96]. The high concentrations of phytoestrogens found in soy protein-based formulas may adversely affect sexual and reproductive health later in life [96]; however this has not been proven. The potential untoward effect of these differences in preterm infants remains unknown. Soy protein is highly allergenic [23] and many infants sensitized to cow's milk protein may also present with allergic reactions to soy proteins [110].

In summary, the use of soy protein-based formulas in preterm infants may be accompanied by a high degree of risk. Preterm infants fed soy protein-based formulas demonstrated poor growth rates, reduced bone mineralization and lower serum levels of albumin. As a result, soy protein-based formulas are not recommended for routine use in preterm infants [96, 98, 111]. Additionally, there is no evidence to support the efficacy of soy protein-based formulas for the management of bovine protein allergy in preterm infants [96]. Clinicians must therefore base their management decision on clinical wisdom. New alternatives to soy protein-based formulas include rice protein-based formulas [110] which have also been studied to some extent. For example, Lasekan et al [112] evaluated growth, tolerance and biochemical changes in healthy term infants randomized to an experimental partially-hydrolyzed rice protein-based formula or standard cow's milk protein-based formula. There were minor differences in amino acid profiles but the two formulas were comparable. Further work remains to be done to develop appropriate guidelines for the management of bovine protein allergy in preterm infants.

Total Parenteral Nutrition

In the beginning of 20th century, it was known that dietary proteins were hydrolyzed in the intestinal tract before absorption occurred. The special nature of protein and its metabolism made it challenging to find suitable ways of administering protein intravenously. The first study in the administration of protein was made in goats in the form of protein hydrolysates by Herriques and Anderson in 1913 [113]. These hydrolysates were products of naturally occurring proteins such as casein and fibrin. Positive nitrogen balance was achieved, thus demonstrating the role of intravenous protein hydrolysates as a possible alternative to dietary protein in animals. In 1937, Robert Elman published the first successful study evaluating the intravenous (IV) infusion of amino acids in the form of fibrinogen hydrolysates in man. He used the technique of hydrolysis of whole protein by strong acids. This was an indisputable landmark in the development of IV nutrition. Arvind Wretlind referred to him as the "father of IV nutrition." In Europe, the first protein hydrolysate to be marketed was developed by Wretlind and was introduced in 1944 in Sweden. Wretlind hydrolysed casein enzymatically and then dialyzed the product to remove large polypeptides [114]. This solution was given the name Aminosol. The disadvantage of the protein hydrolysates was that the amino acid pattern could not be changed. On the other hand, these solutions contained all amino acids necessary for protein synthesis. It took many years of adjusting formulations before crystalline amino acid solutions with similar biologic activity could be produced [113].

The first to introduce a crystalline amino acid was Bansi in 1964 in Germany [113]. Bansi used the proposed formula from the classical work by Rose found in his 1949 publication "Amino Acid Requirements of Man." In his product, the non essential nitrogen content was primarily glycine, which was soon shown to be ineffective. In the early development of amino acid solutions, it was difficult to include tyrosine, cysteine-cystine and glutamine because of technical reasons. Glutamine could not be included in balanced amino acids solutions because of its low solubility and lack of a stable shelf life. This problem was solved by introducing dipeptide, which enhanced the solubility and stability and made it possible to develop solutions containing both glutamine and tyrosine [115].

In 1968, Wilmore and Dudrick reported the first successful use of total parenteral nutrition (TPN), consisting of protein hydrolysate and glucose, in newborn infants with massive resection of small bowel [116]. Earlier versions of intravenous amino acid solutions were based on fibrin or casein hydrolysates. Use of casein hydrolysate was associated with metabolic acidosis and hyperammonemia [117, 118], because the preparations were of suboptimal quality and resulted in poor amino acid utilization. Crystalline amino acid came to the market for use in children in the early 1980s. Studies in neonates in the early 1980s demonstrated that the percent of nitrogen retained with intravenous crystalline amino acids preparations was significantly greater than with casein hydrolysates [119].

The crystalline amino acid solutions (e.g., TrophAmine™, Kendal-McGaw Laboratories, Inc., Irvine, CA) were refined in the late 1980s to maintain plasma amino acid concentrations in preterm and term infants that were comparable to those seen in postprandial breast fed infants of the same gestational age [120]. Trophamine contains a higher ratio of essential to non-essential amino acids, and branched chain amino acids (leucine, isoleucine and valine) account for approximately 50% of the total essential amino acids. This solution improved nitrogen balance and resulted in normal blood urea nitrogen concentrations in term and preterm infants. Helms [121] demonstrated significantly greater weight gain and nitrogen retention in neonates receiving Trophamine than neonates receiving a standard crystalline amino acid formulation (FreAmine III, Kendal McGaw Laboratories, Inc).

The plasma amino acid concentrations are significantly different between preterm and term infants [122, 123]. Amino acid requirements of the preterm neonates are poorly defined, particularly when fed parenterally [124]. Parenterally fed neonates are at risk of toxicity from excess amino acid intake due to immaturity of the degradative pathways, renal immaturity for nitrogen excretion, lack of small intestinal metabolism and metabolic immaturity.

Currently available parenteral amino acid solutions used in preterm infants were developed from studies in infants older and more mature than today's ELBW infant population. These solutions were designed to produce plasma amino acid patterns observed in normally growing, one month old breast fed infants. The major change in their composition was an increase in the total amount of essential amino acids, as well as their relative fraction versus non essential amino acids. There were specific reductions in certain amino acids (e.g. glycine, phenylalanine and methionine) that were considered to be potentially toxic and that might produce dangerously high plasma concentrations at higher amino acid infusion rates. Recent studies have shown that even Trophamine when infused at rates of 3 g/kg/day in ELBW infants, still produce lower concentration of the essential amino acid lysine and threonine than found in the plasma of normally growing human fetuses of the same gestational age [125]. At present, emphasis is placed on the growth of a reasonably stable infant while formulating parenteral amino acids. Many infants and children who require parenteral amino acids are quite ill and may have vastly different amino acid requirements [126]. Hence, the currently available parenteral amino acid solutions may not be optimal for ELBW infants.

Future research should aim at identifying the amino acid and protein requirement of preterm infants based on birth weight, gestational age and illness. There is considerable opportunity to improve the quality of intravenous amino acid mixtures that might produce even better nitrogen and protein balance and growth of lean body mass in very preterm infants [127]. Addition of essential amino acids such as cystein and tyrosine which are unstable in solution and conditional amino acids like glutamine and arginine will be addressed later in this chapter.

TPN Associated Liver Disease: Role of Protein/Amino Acids

Liver disease due to parenteral and enteral nutrition is a well recognized iatrogenic phenomenon, but its cause and pathogenesis have not been clearly elucidated [128]. Parenteral nutrition associated liver disease (PNALD) is seen in 40-70% of VLBW infants receiving TPN [129-131]. With the advances in the field of neonatology, the incidence of PNALD in surgical neonates has decreased significantly [132]. Peden [133] reported the first case of cholestatic jaundice and cirrhosis developing in a preterm infant (birth weight 1 kg) who received protein and hypertonic glucose infusion. This report was followed by series that described neonates, mostly preterm, developing hyperbilirubinemia within 2 weeks of commencing TPN [134].

The exact pathogenesis of TPN associated cholestasis is unknown but most likely multifactorial. The prevalence of PNALD increases with degree of prematurity, duration of TPN and associated medical and surgical conditions [135]. There are several mechanisms that are shown to contribute to liver disease in patients receiving TPN. These include excessive calorie load, various components of TPN solutions, sepsis, nutritional deficiencies such taurine, choline, carnitine and manganese toxicity. In this chapter, we discuss the various mechanisms postulated for the role of protein and amino acids in the development of PNALD including increased protein load, direct hepatotoxicity of amino acid and deficiency of certain essential and non essential amino acids.

Severity of TPN associated cholestasis depends on the duration of TPN and quantity of protein infused [136]. Volume of amino acid infused or cumulative amino acid intake was associated with higher incidence of cholestasis. A prospective study by Vileisis et al [137] comparing low protein intake (2.3 gm/kg/day) and high protein intake (3.6 gm/kg/day) in 82 neonates showed that none of the infants who received TPN for less than 2 week (39/82) developed cholestatic jaundice. Among infants exposed to more than 2 weeks of TPN, the incidence of cholestatic jaundice was 27% in low protein group and 33% in high protein group, respectively. However, infants prospectively assigned to high protein intake developed cholestatic jaundice earlier than low protein intake (27 days versus 47 days, $p < 0.001$) and achieved a significantly greater peak direct bilirubin level. The authors concluded that increased parenteral protein intake is associated with an earlier onset and increased magnitude of cholestatic jaundice [137]. Amino acid infusion may be associated with decrease bile flow. Animal studies have shown that infusion of amino acids decreases the bile flow in an isolate perfused rat liver model [138]. A similar effect was also noted in rabbits and the effect was more marked in infant than adult rabbits, a finding suggesting an enhanced susceptibility of the neonatal animal [139].

TPN solution is directly toxic to the liver and its toxicity appears to be mediated by one or more amino acids. In an experimental rabbit model, infusion of methionine was found to be hepatotoxic [140]. Merritt et al [141] observed that the intraperitoneal injection of tryptophan was hepatotoxic in suckling and weaning rats. Similarly, histologic abnormalities in the liver were noted in neonatal rats and gerbils receiving light exposed TPN containing tryptophan [142, 143]. However, there is no convincing evidence that specific amino acid is solely responsible for cholestasis. For example, methionine has been implicated, yet serum levels of methionine in infants receiving TPN are similar to those in enterally fed infants [120]. Similarly photo degradation products of amino acids have also been implicated [144].

Taurine deficiency: TPN associated liver disease may de due to lack of certain essential amino acids. Taurine deficiency has been postulated as a mechanism of TPN associated cholestasis. Taurine is an essential amino acid in neonates and is an important substrate for the conjugation of bile acids. Taurine effectively conjugates lithocholic acid sulphates preventing cholestasis. It is protective against bile acid induced liver injury. Taurine deficiency leads to an increase in the less soluble glycol conjugates of bile acids with subsequent cholestasis. However taurine supplementation during first 10 days of age of TPN therapy in 20 preterm neonates did not show any effect on hepatocellular function [145]. In a prospective trial, taurine supplementation has been shown to decrease TPN associated cholestasis especially in ELBW infants and neonates with NEC [146].

Carnitine deficiency: Carnitine is synthesized in liver and kidney from lysine and methionine, both of which are essential amino acids. Carnitine, which is required for efficient oxidation of long chain fatty acids, may also be conditionally essential in patients receiving long term TPN. Low levels of carnitine are found in infants and adults receiving TPN. Deficiency of L-carnitine may be responsible for the steatosis and steatohepatitis in patients receiving home parenteral nutrition [147]. Inclusion of carnitine in TPN based on plasma concentration seems reasonable, but there is no evidence that carnitine constitutes effective prophylaxis against PNALD [148].

THERAPEUTIC USES OF ENTERAL OR PARENTERAL PROTEIN

In certain conditions and situations, protein intake is modified for therapeutic indications. The modifications may involve the timing of protein intake as in the case of early amino acid infusions in preterm infants. In other situations, the total protein intake may be increased or decreased, or the source of the protein may be modified.

Early Amino Acid Infusion in Neonates: Rationale

The reluctance to start amino acid administration soon after birth is attributed to increased risk of hyperammonemia, metabolic acidosis and uremia. These side effects, caused by immaturity of metabolic pathways of preterm infants, have been reported in association with the use of protein hydrolysates [117, 118]. Although protein hydrolysates have been replaced by solutions of crystalline amino acids, reluctance to introduce these solutions during the first day persists. Preterm infants who receive only supplemental glucose lose 1% of protein store each day [149]. Early introduction of amino acids with glucose in preterm infant during neonatal period decreases the protein breakdown as compared to intravenous glucose alone, and does not appear to be associated with toxicity [150]. The intravenous administration of amino acids (1.5 g/.kg/day) to ill preterm infants beginning on the 1st day of life improves whole-body protein balance as a result of increased protein synthesis [150]. Early amino acid intake was associated with better growth and neurodevelopment outcome.

Poindexter et al [151] reported in a secondary analysis from a study of glutamine supplementation, that early amino acid intakes was associated with significantly better growth outcomes at 36 weeks postmenstrual age, and fewer infants who received early amino acid were found to have suboptimal head growth at 18 months corrected age. Although elevated blood urea nitrogen is often cited as a reason to decrease amino acid administration, a recent study found no correlation between blood urea nitrogen and amino acid intake in parenterally fed preterm infants [152]. Elevated blood urea nitrogen may be a reflection of higher amino acid oxidation; which more closely resembles the in-utero situation whereby the fetus uses amino acids as a significant source of energy. Many studies have shown no adverse effects of introducing intravenously administered amino acids as early as first day of life [125, 150, 153].

Early aggressive nutrition, defined as introduction of TPN in the first hours after birth in conjunction with minimal enteral feeding, is recommended as a means to avoid early malnutrition and achieve rapid maximal nutrition [16]. A randomized open label trial [153] evaluating safety and efficacy of amino acid administration to VLBW infants (n=135) initiated immediately after birth, randomized infants to receive either glucose and 2.4 g/kg/day of amino acids within 2 hours of birth or glucose alone on the first day of life, with step wise increase in amino acid intake to 2.4 g/kg/day on day 3. The investigators demonstrated positive nitrogen balance at 2 and 4 days of life, without any major adverse effects, in the group of infants who received early amino acids, whereas infants in the control group were in negative nitrogen balance on day 2 [153]. Studies in children and preterm infants suggest that the effect of amino acid on protein accretion occurs by increased protein synthesis rather than inhibition of protein breakdown [150, 154-156].

Controversy remains regarding the impact of energy source, both with respect to composition and amount, on protein metabolism. Glucose and lipid intake ratios that maximize protein accretion have not been determined and will most likely vary with gestational age, prior nutrition status, and underlying disease state [16].

High versus Low Protein Intakes at Commencement of Total Parenteral Nutrition

Several studies have shown the beneficial effect of early parenteral amino acids on decreasing protein loss in preterm infant. Parenteral intakes of 1.5-2 g/kg/day of amino acids is sufficient to avoid negative protein balance in all neonatal populations if concomitant catabolic conditions are not present, and this value may be as little as 1.0-1.5 g/kg/day in ELBW infants [119, 157-159]. This modest intake will not promote growth and as a result, deficits will continue to increase. Amino acids are introduced gradually due to concerns that preterm infants or very ill neonates may be intolerant to protein. Then again, higher concentrations of blood urea nitrogen may be indicative of effective use of amino acids rather than protein intolerance [160]. To achieve intrauterine protein accretion rates, preterm infants particularly ELBW infants may require up to 3.8 g/kg/day of protein [161].

Higher parenteral amino acid intake for only 24 hours produced increase protein deposition in ELBW/VLBW infants. Thureen et al [125] reported a randomized control trial of early parenteral administration of low (1 gm/kg/day) versus high (3 gm/kg/day) amino acid in 28 preterm infants (mean weight ~ 950 g). Metabolic studies showed that protein balance, measured by nitrogen balance and leucine stable isotope methodologies, was significantly higher with a higher amino acid intake. There was no evidence of toxicity with higher amino acids [125]. Plasma concentration of essential and non essential amino acids (except threonine and lysine) in the 3 gm/kg/day group were similar to those of normally growing 2nd and 3rd trimester human fetuses who were sampled by cordocentesis [162], whereas in the 1 gm/kg/day group, concentrations of most amino acids were at least 50% lower than fetal concentrations.

The common practice of starting start amino acids at lower doses and stepwise increase in amino acid intake seems unnecessary. A Cochrane Systematic Review has been proposed to determine whether commencement of TPN nutrition at higher amounts of amino acid is associated with improved growth and development [163].

Restriction of Protein Intake

Restriction of protein intake has been utilized in the conditions which include impaired hepatic or renal function. One of the hallmarks of dietary therapy of renal disease is protein restriction. Protein metabolism breakdown products account for at least some of the adverse effects seen in uremic infants and children [164, 165]. Similarly, infants and children with hepatocellular disease may be unable to adequately utilize and clear endogenously produced or exogenously produced protein from catabolic pathways or dietary protein, leading to hyperammonemia and eventually encephalopathy [166, 167]. In other situations such as inborn errors of metabolism, specific amino acids are eliminated from the diet to avoid the build-up of neurotoxic breakdown products. Among the most common of these is Phenylketonuria in which an inborn error of phenylalanine hydroxylase limits the utilization of ingested phenylalanine. The provision of phenylalanine-free diets prevents or decreases damage to the developing brain [168].

Increased Protein Intake

Provision of large amounts of protein may be required in infants or children with exaggerated ongoing protein losses (through renal or bowel pathologic conditions) or those at risk of relative protein deficiency due to very high growth rates. Poor total protein intake is a component of all types of starvation including the most extreme forms, marasmus and protein-calorie malnutrition (kwashiorkor) [169].

Special Amino Acids in Parenteral and Enteral Nutrition

Glutamine and arginine, the branched chain amino acid have received considerable attention as conditionally indispensable amino acids for stressed individuals (e.g. infants) receiving either parenteral or enteral nutrition. These amino acids, along with cysteine and tyrosine, which are insoluble in aqueous solution and therefore not present in any TPN amino acid mixture are discussed below in greater detail.

Glutamine: Glutamine is a branched chained amino acid, traditionally considered as a nonessential amino acid in healthy individuals, but it is now known to be "conditionally essential" in states of serious illness or injury [170]. It is one of the most abundant amino acids in both plasma and breast milk, however it is not a component of standard amino acid solutions because of its instability in solution. Glutamine is the preferred respiratory fuel for rapidly proliferating cells such as enterocytes of the small bowel [171] and lymphocytes and macrophages in the immune system [172, 173]. Glutamine provides nitrogen for the synthesis of purine and pyrimidine nucleotides, which are necessary for the synthesis of new DNA and RNA during proliferation of lymphocytes as well as for mRNA synthesis and DNA repair in macrophages. Glutamine is also a regulator of acid-base balance via ammonium, as well as an important precursor of nucleic acids, nucleotides, amino sugars and protein [173]. Glutamine is involved in the production of the body's most important antioxidants, glutathione and amino acid taurine [174].

A Cochrane Systematic Review evaluating the effects of glutamine supplementation compared to no supplementation on mortality and morbidity included 6 randomized trials involving 2365 preterm, VLBW infants [175]. Of the six trials, three trials assessed enteral glutamine supplementation [176-178], and three trials assessed parenteral glutamine supplementation [179-181]. Glutamine supplementation, enteral or parenteral, in preterm infants did not have significant effect on the incidence of systemic infection (relative risk (RR) 1.02, 95% CI 0.92 to 1.13), NEC (RR 1.02, 95% CI 0.79 to 1.33), days to full enteral nutrition (WMD -1.1 days, 95% CI -3.4 to 1.2 days), or duration of hospital stay (WMD 0.65 days, 95% CI -2.9 to 4.2 days). Glutamine supplementation did not alter mortality rates for VLBW infants (typical RR 0.98, 95% CI 0.80 to 1.21). In one study the authors found no statistically significant difference in various assessments of neurodevelopment (including Bayley scales) at 18 month corrected age [151]. There is no evidence from good quality randomized trials to suggest that the routine use of parenteral or enteral glutamine supplementation alters mortality or other important clinical outcomes for VLBW infants.

Arginine: Arginine is a precursor of nitric oxide and proteins [182]. Recent evidence has confirmed that de novo arginine synthesis in the neonate is dependent upon small intestinal metabolism and, dietary arginine requirements could be higher in parenterally fed infants [183]. This would explain the greater parenteral versus enteral requirement for this amino acid. Preterm infants have low arginine levels. Preterm infants fed TPN commonly demonstrate elevated plasma ammonia concentration [184]. Inadequate arginine availability has been implicated in many neonatal diseases including persistent pulmonary hypertension [185, 186] and NEC [187, 188].

L-arginine supplementation has been shown to reduce the incidence of NEC in a randomized trial [189]. In this trial 152 preterm neonates $\leq$32 weeks and $\leq$ 1250g were randomized either to receive L-arginine (1.5 mmol/kg/day) or placebo (normal saline) in equivalent volume. The primary outcome was incidence of NEC. L-arginine was added to TPN solution until infants were able to tolerate >40% of feeds, after which L-arginine or placebo supplementation was provided via enteral route. There was a significant reduction in the risk of developing NEC (any stage) in the arginine group as compared with placebo group (RR 0.24, 95% CI 0.10, 0.61, Risk Difference -0.21, 95% CI -0.32, -0.09). The number of patients needed to treat with arginine to prevent the development of NEC among preterm infants $\leq$ 32 weeks according to this study was 5 (95% CI 3, 11). This is a single centre study and the primary outcome was combining all stages of NEC. A larger multicentre trial looking at advanced NEC (Bell Stage II or greater) would be important. There was no difference in the long-term neurodevelopment outcome at 3 years between the L-arginine supplemented groups compared to the placebo group [190].

Phenylalanine and Tyrosine: Tyrosine is synthesized by the hydroxylation of phenylalanine, via phenylalanine hydroxylase in the liver. Tyrosine is insoluble in aqueous solution; hence the concentration of tyrosine in commercially available parenteral solutions is limited to less than 1 % [124]. When compared with neonatal reference proteins such as human milk or human fetal tissue, which contain approximately equal quantities of phenylalanine and tyrosine (3.7-4.1 and 2.9-4.6 g/100 g protein, respectively) [191, 192], the levels provided by these synthetic nutrient solutions appear grossly inadequate. To circumvent this problem, increased phenylalanine has been included in some solutions in the hope that the neonates will endogenously hydroxylate sufficient phenylalanine to tyrosine to meet total aromatic amino acid needs.

Roberts et al [193] studied phenylalanine and tyrosine metabolism in 16 preterm neonates receiving TPN containing either high (Vamin, 7.9%) or moderate (Primene, 4.2%) phenylalanine and low tyrosine concentration (0.7 to 0.9%). Using a 24 hour continuous stable isotope infusion of phenylalanine and tyrosine, the authors demonstrated that the hydroxylation of phenylalanine to tyrosine was twofold greater in infants fed high phenylalanine (i.e.Vamin). Plasma tyrosine and phenylalanine were not measured in this study. Urinary phenylalanine and tyrosine concentrations were greater in infants receiving Vamin as compared with those receiving Primene. Phenylalanine oxidation was two fold greater in infants receiving Vamin as compared to the Primene group [193]. However, oxidation of phenylalanine was decreased as the energy intake increased in the Vamin group. This suggests that energy intake maybe a crucial factor in the utilization of phenylalanine for protein synthesis and that energy intake was inadequate in some of these infants. No such relationship was observed in infants receiving Primene. There was greater urinary excretion of catabolites of phenylalanine and tyrosine in infants receiving Vamin compared to those receiving Primene. This supports the conclusion that the level of phenylalanine in Vamin is in excess of requirement. Therefore neither of the parenteral solutions had either an adequate intake or ideal pattern of aromatic amino acids. Phenylalanine metabolism in neonates receiving TPN containing high phenylalanine is of critical concern due to documented development of hyperphenylalanaemia [194-196] and production of alternate metabolites of phenylalanine [197]. This suggests that phenylalanine hydroxylation is limited and thus tyrosine may be a conditionally indispensable amino acid in the neonate.

Alternative measures for the provision of tyrosine to infants receiving TPN include the use of soluble tyrosine-containing compounds or dipeptides which, upon administration, yield tyrosine in vivo. For example, N-acetyl tyrosine (NAT) has been utilized as a water soluble form of tyrosine for TPN. It has been shown to be an efficiently used source of tyrosine for rats when incorporated into TPN solutions [198, 199]. N-acetyl tyrosine has been included in commercially available pediatric amino acid solution TrophAmine. However, plasma tyrosine concentrations of infants receiving these mixtures, although higher than those of infants receiving tyrosine free regimens, remain low [120, 200, 201]. In addition, plasma concentrations of NAT are as high as or higher than plasma tyrosine concentrations, particularly in preterm neonates [120].

Alternatively, tyrosine dipeptides, such as L-alanyl-L-tyrosine and glycyl L-tyrosine (GT), have been shown to be biologically effective as parenteral tyrosine sources in both laboratory animals and humans [198, 202, 203]. Using graded intakes of tyrosine in the form of glycyl tyrosine, House et al [204] determined that the mean requirement and safe tyrosine intake in the piglet receiving a fixed phenylalanine intake of 4.1% of total amino acids to be 2.7 and 3.2% of amino acids, respectively. Roberts et al [205] estimated the tyrosine requirement of parenterally fed 13 term human neonates receiving 4.2% phenylalanine and graded intakes of glycyl-L-tyrosine as a source of tyrosine. The tyrosine requirement of the parenteral fed neonate lies between 66-82 mg/kg/day, representing 2.8 to 3.4% of total amino acids. The corresponding safe level of intake was 3.8 to 4.0% of total amino acids. The ideal balance of phenylalanine to tyrosine was 56:44 [205]. This study had a small sample size and most of the infants were term (mean gestational age of 37 weeks). The extent, to which the parenterally fed preterm neonates can adapt to a high phenylalanine intake, by increasing phenylalanine hydroxylation, remains to be determined. We need larger studies in preterm infants before recommending the dosage.

Cysteine: Cysteine is considered a conditionally essential amino acid in newborn. Cysteine is synthesized from methionine in the liver via the enzyme cystathionase. However the hepatic cystathionase, the enzyme that normally converts methionine to cysteine, was absent or in low concentration in the fetus and newborn infants [206, 207]. This report has led to a general acceptance that cysteine must be exogenously supplied and therefore is an essential amino acid in neonates. Cystathionase activity increases with postnatal and possibly gestational age [208].

Cysteine is unstable in solution and is easily oxidized to the insoluble form, cystine [209]. Commercially available parenteral amino acid solution contains little or no cysteine [124] because its solubility imposes a short shelf life on the TPN solution. The plasma concentration of cysteine is low in preterm infants receiving cysteine free TPN despite availability of excess methionine [210]. Another potential advantage of addition of cysteine to TPN solution is enhanced calcium and phosphorus solubility. Cysteine hydrochloride must be added to TPN solutions at the time of preparation. Such infusate results in metabolic acidosis particularly in low birth weight (LBW) infants [120, 211] but are effective in normalizing plasma cysteine concentration.

The potential importance of cysteine lies in its role as a precursor for both glutathione and taurine. Glutathione is an important antioxidant particularly for the preterm infants. Newborn infants receiving oxygen therapy for respiratory distress have low plasma cysteine concentration compared with those infants who do not have respiratory distress and who are either receiving IV dextrose infusion or oral feedings [212], perhaps because of use of glutathione as an antioxidant. In a study of 28 LBW infants, plasma total (both free and bound) cysteine concentration and plasma cysteine concentration are approximately 50% higher in LBW infants than term infants [120]. This may suggest that LBW infants require less cysteine than term infants. Until this is clarified by further studies, it appears wise to limit parenteral cysteine intake in LBW infants especially those with birth weight ≤ 1250 gm. Zlotkin et al [209] reported that cysteine supplementation in TPN did not improve nitrogen retention and short term growth in term and preterm infants but it did normalize plasma concentration of cystein. However, muscle protein catabolism was affected by cysteine supplementation. Similarly, Malloy et al [213] reported that cysteine supplementation increased plasma and urine free cysteine concentrations and enhanced total sulfur retention, but did not enhance nitrogen retention.

Helms et al [214] reported that in six infants receiving TPN at home, cysteine supplementation increased serum taurine concentration to normal levels. Cysteine supplementation was required even though there was some taurine in the amino acid preparation. Addition of cysteine in TPN reduced methionine requirement for optimal protein synthesis by about 40% in both enterally and parenterally fed piglets [215]. There is association between cholestatic liver disease and intravenous methionine intake [140]. N-acetyl cystein has been used as a precursor for cysteine to support growth and protein (nitrogen) accretion in piglets administered a parenteral solution [216]. However it did not improve the nitrogen retention or weight gain.

The rate of maturation of the enzyme system (cystathionase) and the age at which methionine conversion to cysteine becomes adequate are not known. Future research should investigate the ideal concentration of methionine and cysteine in neonates. The effect of replacing a portion of methionine with cysteine in intravenous solutions for human neonates needs further investigation. Early nutritional support with an appropriate balance of methionine and cysteine may be of critical importance in infants requiring oxygen and to reduce cholestasis.

Carnitine: Carnitine, a quaternary amino acid, is formed in the liver from the hydroxylation of gamma butyrobetaine, which is synthesized in the kidney from lysine and methionine. Carnitine plays an important role in long chain fatty acid oxidation. It is essential for the transport of very long chain fatty acids across the inner mitochondrial membrane into the mitochondrial matrix where the enzymes for beta oxidation are located. This is achieved via a three-enzyme cycle with carnitine palmityl transferases on the outer and inner aspects of the inner mitochondrial membranes and carnitine translocase as a transport protein in between. Both breast milk and infant formulas contain carnitine. However, it is not provided routinely in TPN formula. Carnitine deficiency developed among VLBW infants who did not receive exogenous supplementation for 2 to 4 weeks [217]. This may be related to minimal body store at birth, limited carnitine production, high carnitine loss in urine and lack of L-carnitine supplements in the standard TPN. Carnitine deficiency would limit the metabolism of lipid leading to a rise in plasma triglycerides and free fatty acids, and impaired ketogenesis. The available non protein energy would be impaired, amino acids would be increasingly used to meet endogenous energy needs and thus tissue growth would be compromised. A Cochrane systematic review evaluating the role of carnitine supplementation in parenterally fed neonates was reported [218]. This review included 6 randomized trials of carnitine supplementation of parenterally fed neonates compared with placebo [219-224]. Among infants supplemented with carnitine there was no evidence of effect on weight gain, lipid utilization or ketogenesis. The results of the review are limited by the fact that the studies were generally short term and studied different outcomes. One study examined short term and long term weight gain, three reported only short term weight gain, three reported biochemical results in response to a short lipid challenge, and two reported results obtained during normal TPN. Subsequently three more randomized trials were reported [225-227]. Meta-analysis of these studies also failed to show any benefits of long term carnitine supplementation in preterm neonates. Hence there is no evidence to support the routine supplementation of parenterally fed neonates with carnitine.

Conclusion

In clinical terms it is evident that the provision of adequate protein intake is challenging given the heterogeneity in the preterm infant population with respect to gestational age, weight, and acuity of illness; the lack of clear guidelines about nutritional goals; and difficulties associated with measuring body composition in the clinical setting. The overall gestalt of the literature reviewed for this chapter suggests that early protein management of preterm infants can be improved considerably without increasing adverse consequences such as metabolic sequelae. Studies performed to date which guide our understanding and clinical strategies have major biological significance with important implications for future health and wellness of preterm infants. At present there is uncertainty with regards to the impact of the proposed strategies in this chapter on long-term growth and development, as well as clinical disease (e.g., obesity, hypertension, diabetes, and bone health).

Acknowledgements

Tanis Fenton for her contribution to the Cochrane Systematic Review on "Higher versus lower protein intake in formula-fed low birth weight infants" which is an integral part of this chapter, and Dr. Majeeda Kamaluddeen for reading the chapter and providing feedback.

References

[1] Raiha NC, Fazzolari A, Cayozzo C, Puccio G, Minoli I., Moro G, et al. Protein nutrition during infancy: Effects on growth and metabolism. In: Martorell R, Haschke F, editors. *Nutrition and growth.* Philadelphia: Lippincott Williams and Wilkins; 2001. p. 73-83.

[2] Kashyap S, Schulze KF, Forsyth M, Zucker C, Dell RB, Ramakrishnan R, et al. Growth, nutrient retention, and metabolic response in low birth weight infants fed varying intakes of protein and energy. *J. Pediatr.* 1988 Oct; 113(4):713-21.

[3] Denne SC. Protein and energy requirements in preterm infants. Semin Neonatol 2001; 6:377-82.

[4] Nayak KC, Sethi AS, Aggarwal TD, Chadda VS, Kumar KK. Bactericidal power of neutrophils in protein calorie malnutrition. *Indian J. Pediatr.* 1989; 56:371-71.

[5] Pencharz PB, Masson M, Desgranges F, Papageorgiou A. Total-body protein turnover in human premature neonates: Effects of birth weight, intra-uterine nutritional status and diet. *Clin. Sci.* 1981; 61:207-15.

[6] Hay WWJ. Assessing the effect of disease on nutrition of the preterm infant. *Clin. Biochem.* 1996; 29:399-417.

[7] Kalhan SC, Iben, S. Protein metabolism in the extremely low-birth weight infant. *Clin. Perinatol.* 2000; 27:35-56.

[8] Kashyap S, Schulze KF, Ramarkrishnan R, Dell RB, Heird WC. Evaluation of a mathematical model for predicting the relationship between protein and energy intakes of low-birth-weight infants and the rate and composition of weight gain. *Pediatr. Res.* 1994; 35:704-12.

[9] Micheli J-L, Fawer C-L, Schutz Y. Protein requirement of the extremely low-birthweight preterm infant. In: Ziegler EE, Lucas A, Moro GE, editors. *Nutrition of the very low birthweight infant.* Philadelphia: Lippincott Williams and Wilkins; 1999. p. 155-77.

[10] Putet G, Rigo J, Salle B, Senterre J. Supplementation of pooled human milk with casein hydrolysate: energy and nitrogen balance and weight gain composition in very low birth weight infants. *Pediatr. Res.* 1987 May; 21(5):458-61.

[11] Polberger SK, Fex GA, Axelsson IE, Raiha NC. Eleven plasma proteins as indicators of protein nutritional status in very low birth weight infants. *Pediatr.* 1990 Dec; 86(6):916-21.

[12] Polberger SK, Axelsson IE, Raiha NC. Amino acid concentrations in plasma and urine in very low birth weight infants fed protein-unenriched or human milk protein-enriched human milk. *Pediatr.* 1990 Dec; 86(6):909-15.

[13] Polberger SK, Axelsson IE, Raiha NC. Urinary and serum urea as indicators of protein metabolism in very low birthweight infants fed varying human milk protein intakes. *Acta Paediatr. Scand.* 1990 Aug-Sep; 79(8-9):737-42.

[14] Ronnholm KA, Siimes MA. Haemoglobin concentration depends on protein intake in small preterm infants fed human milk. *Arch. Dis. Child* 1985 Feb; 60(2):99-104.

[15] Embleton NE, Pang N, Cooke RJ. Postnatal malnutrition and growth retardation: An inevitable consequence of current recommendations in preterm infants? *Pediatr.* 2001; 107:270-3.

[16] Thureen PJ, Hay WH Jr. Early aggressive nutrition in preterm infants. *Semin. Neonatol.* 2001; 6:403-15.

[17] Denne S. Protein and energy requirements in preterm infants. *Semin. Neonatol.* 2001; 6:377-82.

[18] Adamkin DH. Pragmatic approach to in-hospital nutrition in high-risk neonates. *J. Perinatol.* 2005; 25:S7-S11.

[19] Plagemann A. 'Fetal programming' and 'functional teratogenesis': on epigenetic mechanisms and prevention of perinatally acquired lasting health risks. *J. Perinat Med.* 2004; 32(4):297-305.

[20] Armitage JA, Taylor PD, Poston L. Experimental models of developmental programming: consequences of exposure to an energy rich diet during development.[see comment]. *J. Physiol.* 2005 May 15; 565(Pt 1):3-8.

[21] Lau C, Rogers JM. Embryonic and fetal programming of physiological disorders in adulthood. Birth Defects Research Part C, *Embryo Today: Reviews.* 2004 Dec; 72(4):300-12.

[22] Gale CR, Martyn CN, Kellingray S, Eastell R, Cooper C. Intrauterine programming of adult body composition. *J. Clin. Endocrinol. Metab.* 2001 Jan; 86(1):267-72.

[23] Barness LA. The nutrition of premature infants. In: Smith GF, Vidyasagar D, editors. *Historical review and recent advances in neonatal and perinatal medicine.* Mead Johnson Nutritional Division; 1980.

[24] Thureen P, Heird WC. Protein and energy requirements of the preterm/low birthweight (LBW) infant. *Pediatr. Res.* 2005 May; 57(5 Pt 2):95R-8R.

[25] Fomon SJ. Requirements and recommended dietary intakes of protein during infancy. *Pediatr. Res.* 1991; 30:391-5.

[26] Ziegler EE, Biga RL, Fomon SJ. Nutritional requirements of the premature infant. In: Suskind RM, editor. *Textbook of Pediatric Nutrition.* New York: Raven Press; 1981. p. 29-39.

[27] Ziegler EE, O'Donnell AM, Nelson SE, Fomon SJ. Body composition of the reference fetus. *Growth* 1976 Dec; 40(4):329-41.

[28] American Academy of Pediatrics. Committee on N. Nutritional needs of preterm infants. In: Kleinman RE, editor. *Pediatric Nutrition Handbook.* 3rd ed. Elk Grove Village; 1998.

[29] Kramer MS, Platt RW, Wen SW, Joseph KS, Allen A, Abrahamowicz M, et al. A new and improved population-based Canadian reference for birth weight for gestational age. *Pediatr* 2001 Aug; 108(2):E35.

[30] Kashyap S, Heird WC. Protein requirements of low birthweight, very low birthweight, and small for gestational age infants. In: Raiha NCR, editor. *Protein metabolism during infancy.* New York: Nestee Ltd., Vevey/Raven Press Ltd; 1994. p. 136-46.

[31] Kashyap S, Schulze KF, Forsyth M, Dell RB, Ramakrishnan R, Heird WC. Growth, nutrient retention, and metabolic response of low-birth-weight infants fed supplemented and unsupplemented preterm human milk. *Am. J. Clin. Nutr.* 1990 Aug; 52(2):254-62.

[32] Polberger SK, Axelsson IA, Raiha NC. Growth of very low birth weight infants on varying amounts of human milk protein. *Pediatr. Res.* 1989; 25:414-9.

[33] Embleton NE, Pang N, Cooke RJ. Postnatal malnutrition and growth retardation: An inevitable consequence of current recommendations in preterm infants? *Pediatr.* 2001; 107:270-3.

[34] Carlson SJ, Ziegler EE. Nutrient intakes and growth of very low birth weight infants. *J. Perinatol.* 1998; 18:252-8.

[35] Wilson DC, Cairns P, Halliday HL, Reid M, McClure G, Dodge JA. Randomised controlled trial of an aggressive nutritional regimen in sick very low birthweight infants. *Arch. Dis. Child Fetal Neonatal.* Ed. 1997 Jul; 77(1):F4-11.

[36] Wilson DC, McClure G, Halliday HL, Reid MM, Dodge JA. Nutrition and bronchopulmonary dysplasia [see comment]. *Arch. Dis. Child 1991 Jan*; 66(1 Spec No):37-8.

[37] Zello GA, Menendez CE, Rafii M, Clarke R, Wykes LJ, Ball RO, et al. Minimum protein intake for the preterm neonate determined by protein and amino acid kinetics. *Pediatr. Res*. 2003 Feb; 53(2):338-44.

[38] Ehrenkranz RA, Younes N, Lemons JA, Fanaroff AA, Donovan EF, Wright LL, et al. Longitudinal growth of hospitalized very low birth weight infants. *Pediatr*. 1999 Aug; 104(2 Pt 1):280-9.

[39] Zello GA, Wykes LJ, Ball RO, Pencharz PB. Recent advances in methods of assessing dietary amino acid requirements for adult humans. *J. Nutr*. 1995 Dec; 125(12):2907-15.

[40] Fanaro S, Cristofori G, Mosca F, Savino F, Vigi V. Considerations and approaches in determining the protein and energy composition of preterm infant formulas. Acta *Paediatr. Suppl.* 2005 Oct; 94(449):57-63.

[41] Schulze KF, Stefanski M, Masterson J, Spinnazola R, Ramakrishnan R, Dell RB, et al. Energy expenditure, energy balance, and composition of weight gain in low birth weight infants fed diets of different protein and energy content. *J. Pediatr.* 1987 May; 110(5):753-9.

[42] Romera G, Figueras J, Rodriguez-Miguelez JM, Ortega J, Jimenez R. Energy intake, metabolic balance and growth in preterm infants fed formulas with different nonprotein energy supplements. *J. Pediatr. Gastroenterol. Nutr*. 2004 Apr; 38(4):407-13.

[43] Kashyap S, Forsyth M, Zucker C, Ramakrishnan R, Dell RB, Heird WC. Effects of varying protein and energy intakes on growth and metabolic response in low birth weight infants. *J. Pediatr*. 1986 Jun; 108(6):955-63.

[44] Dietz W. Obesity in infants, children, and adolescents in the United States. *Nur. Res.* 1981; 1:117-37.

[45] Fairey AK, Butte NF, Mehta N, Thotathuchery M, Schanler RJ, Heird WC. Nutrient accretion in preterm infants fed formula with different protein:energy ratios. *J. Pediatr. Gastroenterol. Nutr*. 1997 Jul; 25(1):37-45.

[46] Klein CJ. Nutrient requirements for preterm infant formulas.[see comment]. *J. Nutr*. 2002 Jun; 132(6 Suppl 1):1395S-577S.

[47] Rigo J, Senterre, J. Significance of plasma amino acid pattern in preterm infants. *Biol. Neonate* 1987; 52(Suppl 1):41-9.

[48] Tikanoja T, Simell O, Jarvenpaa AL, Raiha NC. Plasma amino acids in preterm infants after a feed of human milk or formula. *J. Pediatr*. 1982 Aug; 101(2):248-52.

[49] Moro G, Minoli I, Boehm G, Georgi G, Jelinek J, Sawatzki G. Postprandial plasma amino acids in preterm infants: influence of the protein source. *Acta Paediatr*. 1999 Aug; 88(8):885-9.

[50] Nutrition Committee, Canadian Paediatric Society. Nutrient needs and feeding of premature infants. *CMAJ* 1995 June 1; 152(11):1765-85.

[51] Bayes R, Campoy C, Molina-Font JA. Some current controversies on nutritional requirements of full-term and pre-term newborn infants. *Early Hum. Devel*. 1998 Dec; 53 Suppl:S3-13.

[52] Atkinson SA, Bryan MH, Anderson GH. Human milk feeding in premature infants: protein, fat, and carbohydrate balances in the first two weeks of life. *J. Pediatr.* 1981 Oct; 99(4):617-24.

[53] Atkinson SA, Radde IC, Anderson GH. Macromineral balances in premature infants fed their own mothers' milk or formula. *J. Pediatr.* 1983 Jan; 102(1):99-106.

[54] Lucas A, Gore SM, Cole TJ, Bamford MF, Dossetor JF, Barr I, et al. Multicentre trial on feeding low birthweight infants: effects of diet on early growth. *Arch. Dis. Child* 1984 Aug; 59(8):722-30.

[55] Senterre J, Voyer, M., Putet, G., Rigo, J. Nitrogen, fat and mineral balance studies in preterm infants fed bank human milk, a human milk formula, or a low birth-weight infant formula. In: Baum D, editor. *Human milk processing, fractionation, and the nutrition of the low birth-weight baby.* New York: Raven Press; 1983. p. 102-11.

[56] Premji SS, Fenton TR, Sauve RS. Higher versus lower protein intake in formula-fed low birth weight infants. *Cochrane Database of Systematic Reviews.* 2006(1):CD003959.

[57] Bhatia J, Rassin DK, Cerreto MC, Bee DE. Effect of protein/energy ratio on growth and behavior of premature infants: preliminary findings. *J. Pediatr.* 1991 Jul; 119(1 (Pt 1)):103-10.

[58] Hillman LS, Salmons SS, Erickson MM, Hansen JW, Hillman RE, Chesney R. Calciuria and aminoaciduria in very low birth weight infants fed a high-mineral premature formula with varying levels of protein. *J. Pediatr.* 1994 Aug; 125(2):288-94.

[59] Svenningsen NW, Lindroth M, Lindquist B. A comparative study of varying protein intake in low birthweight infant feeding. *Acta Paediatr. Scand Suppl.* 1982; 296:28-31.

[60] Wauben I, Westerterp K, Gerver WJ, Blanco C. Effect of varying protein intake on energy balance, protein balance and estimated weight gain composition in premature infants. *Eur. J. Clin. Nutr.* 1995 Jan; 49(1):11-6.

[61] Goldman HI, Freudenthal R, Holland B, Karelitz S. Clinical effects of two different levels of protein intake on low-birth-weight infants. *J. Pediatr.* 1969 Jun; 74(6):881-9.

[62] Raiha NC, Heinonen K, Rassin DK, Gaull GE. Milk protein quantity and quality in low-birthweight infants: I. Metabolic responses and effects on growth. *Pediatr.* 1976 May; 57(5):659-84.

[63] Goldman HI, Goldman J, Kaufman I, Liebman OB. Late effects of early dietary protein intake on low-birth-weight infants. *J. Pediatr.* 1974 Dec; 85(6):764-9.

[64] Goldman HI, Liebman OB, Freudenthal R, Reuben R. Effects of early dietary protein intake on low-birth-weight infants: evaluation at 3 years of age. *J. Pediatr.* 1971 Jan; 78(1):126-9.

[65] Gaull GE, Rassin DK, Raiha NC, Heinonen K. Milk protein quantity and quality in low-birth-weight infants. III. Effects on sulfur amino acids in plasma and urine. *J. Pediatr.* 1977 Mar; 90(3):348-55.

[66] Rassin DK, Gaull GE, Heinonen K, Raiha NC. Milk protein quantity and quality in low-birth-weight infants: II. Effects on selected aliphatic amino acids in plasma and urine. P*ediatr.* 1977 Mar; 59(3):407-22.

[67] Rassin DK, Gaull GE, Raiha NC, Heinonen K. Milk protein quantity and quality in low-birth-weight infants. IV. Effects on tyrosine and phenylalanine in plasma and urine. *J. Pediatr.* 1977 Mar; 90(3):356-60.
[68] Embleton ND, Cooke RJ. Protein requirements in preterm infants: effect of different levels of protein intake on growth and body composition. *Pediatr. Res* 2005 Nov; 58(5):855-60.
[69] Cooke R, Embleton N, Rigo J, Carrie A, Haschke F, Ziegler E. High protein pre-term infant formula: effect on nutrient balance, metabolic status and growth. *Pediatr. Res.* 2006 Feb; 59(2):265-70.
[70] Rigo J, Putet G, Picaud JC, Pieltain C, DeCurtis M, Salle BL, et al. Nitrogen balance and plasma amino acids in the evaluation of protein sources for extremely low birthweight infants. In: Ziegler EE, Lucas A, Moro GE, editors. *Nutrition of the very low birthweight infant.* Philadelphia: Nestec, Ltd., and Lippincott Williams and Wilkins; 1999. p. 139-53.
[71] Gomella TL, Cunningham, MD, Eyal FG, Zenk KE. Neonatology: Management, procedures, on-call problems, diseases, and drugs. 4th ed. New York: *Lang Medical Books*/McGraw Hill; 1999.
[72] Putet G, Senterre J, Rigo J, Salle B. Nutrient balance, energy utilization, and composition of weight gain in very-low-birth-weight infants fed pooled human milk or a preterm formula. *J. Pediatr.* 1984 Jul; 105(1):79-85.
[73] Schanler RJ. The use of human milk for premature infants. *Pediatr. Clin. North Am.* 2001 Feb; 48(1):207-19.
[74] Gordon HH, Levine SZ, McNamara H. Feeding of premature infants: a comparison of human and cow's milk. A*m. J. Dis. Child* 1947; 73:442.
[75] Schanler RJ. Suitability of human milk for the low-birthweight infant. *Clin. Perinatol.* 1995 Mar; 22(1):207-22.
[76] Atkinson SA, Bryan MH, Anderson GH. Human milk: difference in nitrogen concentration in milk from mothers of term and premature infants. *J. Pediatr.* 1978 Jul; 93(1):67-9.
[77] Atkinson SA, Anderson GH, Bryan MH. Human milk: comparison of the nitrogen composition in milk from mothers of premature and full-term infants. *Am. J. Clin. Nutr.* 1980 Apr; 33(4):811-5.
[78] Chessex P, Reichman B, Verellen G, Putet G, Smith JM, Heim T, et al. Quality of growth in premature infants fed their own mothers' milk. *J. Pediatr.* 1983 Jan; 102(1):107-12.
[79] Schanler RJ, Oh W. Nitrogen and mineral balance in preterm infants fed human milks or formula. J. *Pediatr. Gastroenterol. Nutr.* 1985 Apr; 4(2):214-9.
[80] Thureen PJ, Hay WH Jr. *Neonatal nutrition and metabolism.* 2nd ed. Cambridge University Press; 2006.
[81] Kuschel CA, Harding JE. Protein supplementation of human milk for promoting growth in preterm infants. *Cochrane Database of Systematic Reviews.* 2004(3):CD000433.
[82] Boehm G, Muller DM, Beyreiss K, Raiha NC. Evidence for functional immaturity of the ornithine-urea cycle in very-low-birth-weight infants. *Biol. Neonate* 1988; 54(3):121-5.

[83] Ronnholm KA, Sipila I, Siimes MA. Human milk protein supplementation for the prevention of hypoproteinemia without metabolic imbalance in breast milk-fed, very low-birth-weight infants. *J. Pediatr.* 1982 Aug; 101(2):243-7.

[84] Ehrenkranz RA, Gettner PA, Nelli CM. Nutrient balance studies in premature infants fed premature formula or fortified preterm human milk. *J. Pediatr. Gastroenterol. Nutr.* 1989 Jan; 8(1):58-67.

[85] Schanler RJ, Garza C, Nichols BL. Fortified mothers' milk for very low birth weight infants: results of growth and nutrient balance studies. *J. Pediatr.* 1985 Sep; 107(3):437-45.

[86] Ronnholm KA, Simell O, Siimes MA. Human milk protein and medium-chain triglyceride oil supplementation of human milk: plasma amino acids in very low-birth-weight infants. *Pediatr* 1984 Nov; 74(5):792-9.

[87] Kuschel CA, Harding JE. *Multicomponent fortified human milk for promoting growth in preterm infants*. 2000(2):CD000343.

[88] Kuschel CA, Harding JE. *Carbohydrate supplementation of human milk to promote growth in preterm infants*. 2000(2):CD000280.

[89] Kuschel CA, Harding JE. *Calcium and phosphorus supplementation of human milk for preterm infants*. 2001(4):CD003310.

[90] Kuschel CA, Harding JE. *Fat supplementation of human milk for promoting growth in preterm infants*. 2000(2):CD000341.

[91] Pereira GR, Baumgart S, Bennett MJ, Stallings VA, Georgieff MK, Hamosh M, et al. Use of high-fat formula for premature infants with bronchopulmonary dysplasia: metabolic, pulmonary, and nutritional studies.[see comment]. *J. Pediatr.* 1994 Apr; 124(4):605-11.

[92] Boehm G, Muller DM, Senger H, Borte M, Moro G. Nitrogen and fat balances in very low birth weight infants fed human milk fortified with human milk or bovine milk protein. *Eur. J. Pediatr.* 1993 Mar; 152(3):236-9.

[93] Polberger S, Raiha NC, Juvonen P, Moro GE, Minoli I, Warm A. Individualized protein fortification of human milk for preterm infants: comparison of ultrafiltrated human milk protein and a bovine whey fortifier. *J. Pediatr. Gastroenterol. Nutr.* 1999 Sep; 29(3):332-8.

[94] Moro GE, Minoli I, Fulconis F, Clementi M, Raiha NC. Growth and metabolic responses in low-birth-weight infants fed human milk fortified with human milk protein or with a bovine milk protein preparation. *J. Pediatr. Gastroenterol. Nutr.* 1991 Aug; 13(2):150-4.

[95] Morley R. Nutrition and cognitive development.[see comment]. *Nutrition*. 1998 Oct; 14(10):752-4.

[96] Agostoni C, Axelsson I, Goulet O, Koletzko B, Michaelsen KF, Puntis J, Rieu D, et al. Soy protein infant formulae and follow-on formulae: A commentary by the ESPGHAN Committee on nutrition. *J. Pediatr. Gastroenterol. Nutr.* 2006; 42:352-61.

[97] Klish WJ, Baker SS, Cochran WJ, Flores CA, Georgieff MK, Jacobson MS, et al. Soy protein-based formulas: Recommendations for use in infant feeding. *Pediatr.* 1998; 101:148-53.

[98] Mauer AM, Dweck HS, Holmes F, Reynolds JW, Suskind RM, Walker WA, et al. Soy-protein formulas: Recommendations for use in infant feeding. *Pediatr.* 1983; 72:359-63.

[99] Fomon SJ, Ziegler EE, Filer LJ Jr., Nelson SE, Edwards BB. Methionine fortification of a soy protein formula fed to infants. *Am. J. Clin. Nutr.* 1979 Dec; 32(12):2460-71.

[100] Fomon SJ, Ziegler EE, Nelson SE, Edwards BB. Requirement for sulfur-containing amino acids in infancy. *J. Nutr.* 1986 Aug; 116(8):1405-22.

[101] Fomon SJ, Thomas LN, Filer LJ Jr., Anderson TA, Bergmann KE. Requirements for protein and essential amino acids in early infancy. Studies with a soy-isolate formula. *Acta Paediatr. Scan.* 1973 Jan; 62(1):33-45.

[102] Naude SP, Prinsloo JG, Haupt CE. Comparison between a humanized cow's milk and a soy product for premature infants. *SAMJ* 1979 Jun 9; 55(24):982-6.

[103] Hall RT, Callenbach JC, Sheehan MB, Hall FK, Thibeault DW, Kurth CG, et al. Comparison of calcium- and phosphorus-supplemented soy isolate formula with whey-predominant premature formula in very low birth weight infants. *J. Pediatr. Gastroenterol. Nutr.* 1984 Sep; 3(4):571-6.

[104] Shenai JP, Jhaveri BM, Reynolds JW, Huston RK, Babson SG. Nutritional balance studies in very low-birth-weight infants: role of soy formula. *Pediatr.* 1981 May; 67(5):631-7.

[105] Koo WW, Kaplan LA. Aluminum and bone disorders: with specific reference to aluminum contamination of infant nutrients. *J. Am. Coll Nutr.* 1988 Jun; 7(3):199-214.

[106] Koo WW, Kaplan LA, Krug-Wispe SK. Aluminum contamination of infant formulas. *JPEN J. Parenter Enteral. Nutr.* 1988 Mar-Apr; 12(2):170-3.

[107] Callenbach JC, Sheehan MB, Abramson SJ, Hall RT. Etiologic factors in rickets of very low-birth-weight infants. *J. Pediatr.* 1981 May; 98(5):800-5.

[108] Kulkarni PB, Hall RT, Rhodes PG, Sheehan MB, Callenbach JC, Germann DR, et al. Rickets in very low-birth-weight infants. *J. Pediatr.* 1980 Feb; 96(2):249-52.

[109] Bos C, Metges CC, Gaudichon C, Petzke KJ, Pueyo ME, Morens C, et al. Postprandial kinetics of dietary amino acids are the main determinant of their metabolism after soy or milk protein ingestion in humans. *J. Nutr.* 2003 May; 133(5):1308-15.

[110] Moro GE, Warm A, Arslanoglu S, Miniello V. Management of bovine protein allergy: new perspectives and nutritional aspects. *Ann. Allergy Asthma Immunol.* 2002 Dec; 89(6): Supplement 1: 91-6.

[111] American Academy of Pediatrics. Committee on N. American Academy of Pediatrics. Committee on Nutrition. Soy protein-based formulas: recommendations for use in infant feeding. *Pediatr.* 1998 Jan; 101(1 Pt 1):148-53.

[112] Lasekan JB, Koo WWK, Walters J, Neylan M, Luebbers S. Growth, tolerance and biochemical measures in healthy infants fed a partially hydrolyzed rice protein-based formula: a randomized, blinded, prospective trial. *J. Am. Coll. Nutr.* 2006 Feb; 25(1):12-9.

[113] Vinnars E, Wilmore D. Jonathan Roads Symposium Papers. History of parenteral nutrition. *JPEN J. Parenter Enteral. Nutr.* 2003 May-Jun; 27(3):225-31.

[114] Wretlind KAJ. Free amino acids in dialyzed casein digest. *Acta Physiol. Scand.* 1947; 13:45-54.

[115] Furst P, Albers S, Stehle P. Dipeptides in clinical nutrition. *Proc. Nutr. Soc.* 1990 Oct; 49(3):343-59.

[116] Wilmore DW, Dudrick SJ. Growth and development of an infant receiving all nutrients exclusively by vein. *JAMA* 1968 Mar 4; 203(10):860-4.

[117] Heird WC, Dell RB, Driscoll JM Jr., Grebin B, Winters RW. Metabolic acidosis resulting from intravenous alimentation mixtures containing synthetic amino acids. *New Engl. J. Med.* 1972 Nov 9; 287(19):943-8.

[118] Johnson JD, Albritton WL, Sunshine P. Hyperammonemia accompanying parenteral nutrition in newborn infants. *J. Pediatr.* 1972 Jul; 81(1):154-61.

[119] Duffy B, Gunn T, Collinge J, Pencharz P. The effect of varying protein quality and energy intake on the nitrogen metabolism of parenterally fed very low birthweight (less than 1600 g) infants. *Pediatr. Res.* 1981 Jul; 15(7):1040-4.

[120] Heird WC, Hay W, Helms RA, Storm MC, Kashyap S, Dell RB. Pediatric parenteral amino acid mixture in low birth weight infants. *Pediatr.* 1988 Jan; 81(1):41-50.

[121] Helms RA, Christensen ML, Mauer EC, Storm MC. Comparison of a pediatric versus standard amino acid formulation in preterm neonates requiring parenteral nutrition. *J. Pediat*r. 1987 Mar; 110(3):466-70.

[122] Wu PY, Edwards N, Storm MC. Plasma amino acid pattern in normal term breast-fed infants. *J. Pediatr.* 1986 Aug; 109(2):347-9.

[123] Rassin DK, Gaull GE, Jarvenpaa AL, Raiha NC. Feeding the low-birth-weight infant: II. Effects of taurine and cholesterol supplementation on amino acids and cholesterol. *Pediat*r. 1983 Feb; 71(2):179-86.

[124] Brunton JA, Ball RO, Pencharz PB. Current total parenteral nutrition solutions for the neonate are inadequate. *Curr. Opin. Clin. Nutr. Metab. Care* 2000 Jul; 3(4):299-304.

[125] Thureen PJ, Melara D, Fennessey PV, Hay WW Jr. Effect of low versus high intravenous amino acid intake on very low birth weight infants in the early neonatal period. *Pediatr. Res.* 2003 Jan; 53(1):24-32.

[126] Heird WC. Amino acids in pediatric and neonatal nutrition. *Curr. Opin. Clin. Nutr. Metab. Care* 1998 Jan; 1(1):73-8.

[127] Hay WW Jr. Intravenous nutrition of the very preterm neonate. *Acta Paediatr. Suppl.* 2005 Oct; 94(449):47-56.

[128] Kwan V, George J. Liver disease due to parenteral and enteral nutrition. *Clin. Liver Dis.* 2004 Nov; 8(4):893-913.

[129] Beale EF, Nelson RM, Bucciarelli RL, Donnelly WH, Eitzman DV. Intrahepatic cholestasis associated with parenteral nutrition in premature infants. *Pediatr.* 1979 Sep; 64(3):342-7.

[130] Merritt RJ. Cholestasis associated with total parenteral nutrition. *J. Pediatric. Gastroenterol. Nutr.* 1986 Jan; 5(1):9-22.

[131] Sondheimer JM, Asturias E, Cadnapaphornchai M. Infection and cholestasis in neonates with intestinal resection and long-term parenteral nutrition. *J. Pediatr. Gastroenterol. Nutr.* 1998 Aug; 27(2):131-7.

[132] Kubota A, Yonekura T, Hoki M, Oyanagi H, Kawahara H, Yagi M, et al. Total parenteral nutrition-associated intrahepatic cholestasis in infants: 25 years' experience. *J. Pediatr. Surg.* 2000 Jul; 35(7):1049-51.

[133] Peden VH, Witzleben CL, Skelton MA. Total parenteral nutrition. *J. Pediatr.* 1971 Jan; 78(1):180-1.

[134] Postuma R, Trevenen CL. Liver disease in infants receiving total parenteral nutrition. *Pediatr.* 1979 Jan; 63(1):110-5.

[135] Pereira GR, Sherman MS, DiGiacomo J, Ziegler M, Roth K, Jacobowski D. Hyperalimentation-induced cholestasis. Increased incidence and severity in premature infants. *Am. J. Dis. Child* 1981 Sep; 135(9):842-5.

[136] Sankaran K, Berscheid B, Verma V, Zakhary G, Tan L. An evaluation of total parenteral nutrition using Vamin and Aminosyn as protein base in critically ill preterm infants. *JPEN J. Parenter Enter Nutr.* 1985 Jul-Aug; 9(4):439-42.

[137] Vileisis RA, Inwood RJ, Hunt CE. Prospective controlled study of parenteral nutrition-associated cholestatic jaundice: effect of protein intake. *J. Pediatr.* 1980 May; 96(5):893-7.

[138] Graham MF, Tavill AS, Halpin TC, Louis LN. Inhibition of bile flow in the isolated perfused rat liver by a synthetic parenteral amino acid mixture: associated net amino acid fluxes. *Hepatology* 1984 Jan-Feb; 4(1):69-73.

[139] Zahavi I, Shaffer EA, Gall DG. Total parenteral nutrition-associated cholestasis: acute studies in infant and adult rabbits. *J. Pediatr. Gastroenterol. Nutr.* 1985 Aug; 4 (4): 622-7.

[140] Moss RL, Haynes AL, Pastuszyn A, Glew RH. Methionine infusion reproduces liver injury of parenteral nutrition cholestasis. *Pediatr. Res.* 1999 May; 45(5 Pt 1):664-8.

[141] Merritt RJ, Sinatra FR, Henton D, Neustein H. Cholestatic effect of intraperitoneal administration of tryptophan to suckling rat pups. *Pediatr. Res.* 1984 Sep; 18(9):904-7.

[142] Grant JP, Cox CE, Kleinman LM, Maher MM, Pittman MA, Tangrea JA, et al. Serum hepatic enzyme and bilirubin elevations during parenteral nutrition. *Surg. Gynecol. Obstet* 1977 Oct; 145(4):573-80.

[143] Bhatia J, Rassin DK. Photosensitized oxidation of tryptophan and hepatic dysfunction in neonatal gerbils. *JPEN J. Parenter Enter Nutr.* 1985 Jul-Aug; 9(4):491-5.

[144] Bhatia J, Moslen MT, Haque AK, McCleery R, Rassin DK. Total parenteral nutrition-associated alterations in hepatobiliary function and histology in rats: is light exposure a clue? *Pediatr. Res.* 1993 May; 33(5):487-92.

[145] Cooke RJ, Whitington PF, Kelts D. Effect of taurine supplementation on hepatic function during short-term parenteral nutrition in the premature infant. *J. Pediatr. Gastroenterol. Nutr.* 1984 Mar; 3(2):234-8.

[146] Spencer AU, Yu S, Tracy TF, Aouthmany MM, Llanos A, Brown MB, et al. Parenteral nutrition-associated cholestasis in neonates: multivariate analysis of the potential protective effect of taurine. *JPEN J. Parenter Enter Nutr.* 2005 Sep-Oct; 29(5):337-43; discussion 43-4.

[147] Bowyer BA, Fleming CR, Ilstrup D, Nelson J, Reek S, Burnes J. Plasma carnitine levels in patients receiving home parenteral nutrition. *Am. J. Clin. Nutr.* 1986 Jan; 43(1):85-91.

[148] Moukarzel AA, Dahlstrom KA, Buchman AL, Ament ME. Carnitine status of children receiving long-term total parenteral nutrition: a longitudinal prospective study.[see comment]. *J. Pediatr.* 1992 May; 120(5):759-62.

[149] Kayshap S, Heird WC. Protein requirements of low birthweight very low birthweight and small for gestational age infants. In: Raiha NCR, editor. *Protein metabolism during infancy*. New York: Raven Press; 1993. p. 133-51.

[150] Rivera A Jr., Bell EF, Bier DM. Effect of intravenous amino acids on protein metabolism of preterm infants during the first three days of life. *Pediatr. Res.* 1993 Feb; 33(2):106-11.

[151] Poindexter BB, Langer JC, Dusick AM, Ehrenkranz RA, National Institute of Child Health and Human Development Neonatal Research N. Early provision of parenteral amino acids in extremely low birth weight infants: relation to growth and neurodevelopmental outcome.[see comment]. *J. Pediatr.* 2006 Mar; 148(3):300-5.

[152] Ridout E, Melara D, Rottinghaus S, Thureen PJ. Blood urea nitrogen concentration as a marker of amino-acid intolerance in neonates with birthweight less than 1250 g. *J. Perinatol.* 2005 Feb; 25(2):130-3.

[153] te Braake FWJ, van den Akker CHP, Wattimena DJL, Huijmans JGM, van Goudoever JB. Amino acid administration to premature infants directly after birth.[see comment]. *J. Pediat*r. 2005 Oct; 147(4):457-61.

[154] Goulet O, DePotter S, Salas J, Robert JJ, Rongier M, Ben Hariz M, et al. Leucine metabolism at graded amino acid intakes in children receiving parenteral nutrition. Am *J. Physiol.* 1993 Oct; 265(4 Pt 1):E540-6.

[155] Denne SC, Karn CA, Ahlrichs JA, Dorotheo AR, Wang J, Liechty EA. Proteolysis and phenylalanine hydroxylation in response to parenteral nutrition in extremely premature and normal newborns. *J Clin. Invest* 1996 Feb 1; 97(3):746-54.

[156] van den Akker CHP, te Braake FWJ, Wattimena DJL, Voortman G, Schierbeek H, Vermes A, et al. Effects of early amino acid administration on leucine and glucose kinetics in premature infants. *Pediatr. Res.* 2006 May; 59(5):732-5.

[157] Thureen PJ, Anderson AH, Baron KA, Melara DL, Hay WW Jr., Fennessey PV. Protein balance in the first week of life in ventilated neonates receiving parenteral nutrition [erratum appears in Am J Clin Nutr 1999 Aug;70(2):303]. *Am. J. Clin. Nutr.* 1998 Nov; 68(5):1128-35.

[158] van Lingen RA, van Goudoever JB, Luijendijk IH, Wattimena JL, Sauer PJ. Effects of early amino acid administration during total parenteral nutrition on protein metabolism in pre-term infants. *Clin. Sci.* 1992 Feb; 82(2):199-203.

[159] Zlotkin SH, Bryan MH, Anderson GH. Intravenous nitrogen and energy intakes required to duplicate in utero nitrogen accretion in prematurely born human infants. J *Pediatr.* 1981 Jul; 99(1):115-20.

[160] Thureen PJ. Early aggressive nutrition in the neonate. *Pediatr. Rev.* 1999 Sep; 20(9):e45-55.

[161] Ziegler EE. Protein in premature feeding. *Nutrition.* 1994 Jan-Feb; 10(1):69-71.

[162] Cetin I, Corbetta C, Sereni LP, Marconi AM, Bozzetti P, Pardi G, et al. Umbilical amino acid concentrations in normal and growth-retarded fetuses sampled in utero by cordocentesis. *Am. J. Obstet Gynecol.* 1990 Jan; 162(1):253-61.

[163] Thomas N, Sinn JKH. *High versus low amino acid intake at the commencement of parenteral nutrition to improve growth in neonates receiving intensive care* [protocol]. 2006 [cited; Issue 3].

[164] Karlowicz MG, Adelman RD. Acute renal failure in the neonate. *Clin. Perinatol.* 1992 Mar; 19(1):139-58.

[165] Yiu VW, Harmon WE, Spinozzi N. High-calorie nutrition for infants with chronic renal disease. *J. Renal. Nutr.* 1966; 6:203.

[166] Gourley GR, Kreamer B, Cohnen M, Kosorok MR. Neonatal jaundice and diet.[see comment]. *Arch. Pediatr. Adolesc. Med* 1999 Feb; 153(2):184-8.

[167] Mezey E. Liver disease and nutrition. *Gastroenterology* 1978 Apr; 74(4):770-83.

[168] Scriver CR, Beaudet AL, Sly WS, Valle D. *The metabolic basis of inherited disease.* 6th ed. New York: McGraw Hill Co.; 1989.

[169] Gomez F, Ramos-Galvan R, Cravioto J, Frenk S. Prevention and treatment of chronic severe infantile malnutrition (kwashiorkor). *Ann. NY Acad Sci.* 1958; 69:969-81.

[170] Lacey JM, Wilmore DW. Is glutamine a conditionally essential amino acid? *Nutr. Rev.* 1990 Aug; 48(8):297-309.

[171] Windmueller HG. Glutamine utilization by the small intestine. *Adv. Enzymol. Relat. Areas of Mol. Biol.* 1982; 53:201-37.

[172] Newsholme P. Why is L-glutamine metabolism important to cells of the immune system in health, postinjury, surgery or infection? *J. Nutr.* 2001 Sep; 131(9 Suppl):2515S-22S; discussion 23S-4S.

[173] Krebs H. Glutamine metabolism in the animal body. In: Mora J, Palacios R, editors. *Glutamine: metabolism, enzymology and regulation.* New York: Academic Press; 1980. p. 319-29.

[174] Boelens PG, Houdijk APJ, de Thouars HN, Teerlink T, van Engeland MIA, Haarman HJTM, et al. Plasma taurine concentrations increase after enteral glutamine supplementation in trauma patients and stressed rats. *Am. J. Clin. Nutr.* 2003 Jan; 77(1):250-6.

[175] Tubman TRJ, Thompson SW, McGuire W. Glutamine supplementation to prevent morbidity and mortality in preterm infants.[update of Cochrane Database Syst Rev. 2001;(4):CD001457; PMID: 11687112]. *Cochrane Database of Systematic Reviews.* 2005(1):CD001457.

[176] Neu J, Roig JC, Meetze WH, Veerman M, Carter C, Millsaps M, et al. Enteral glutamine supplementation for very low birth weight infants decreases morbidity. *J. Pediatr.* 1997 Nov; 131(5):691-9.

[177] van den Berg A, van Elburg RM, Twisk JWR, Fetter WPF. Glutamine-enriched enteral nutrition in very low birth weight infants. Design of a double-blind randomised controlled trial [ISRCTN32545831]. *BMC Pediatr.* 2004 Sep 1; 4:17.

[178] Vaughn P, Thomas P, Clark R, Neu J. Enteral glutamine supplementation and morbidity in low birth weight infants. *J. Pediatr.* 2003 Jun; 142(6):662-8.

[179] Lacey JM, Crouch JB, Benfell K, Ringer SA, Wilmore CK, Maguire D, et al. The effects of glutamine-supplemented parenteral nutrition in premature infants. *JPEN J. Parenter Enter Nutr.* 1996 Jan-Feb; 20(1):74-80.

[180] Poindexter BB, Ehrenkranz RA, Stoll BJ, Wright LL, Poole WK, Oh W, et al. Parenteral glutamine supplementation does not reduce the risk of mortality or late-onset sepsis in extremely low birth weight infants. *Pediatr.* 2004 May; 113(5):1209-15.

[181] Thompson SW, McClure BG, Tubman TRJ. A randomized, controlled trial of parenteral glutamine in ill, very low birth-weight neonates.[see comment]. *J. Pediatr. Gastroenterol. Nutr.* 2003 Nov; 37(5):550-3.

[182] Wu G, Morris SM Jr. Arginine metabolism: nitric oxide and beyond. *Biochem. J.* 1998 Nov 15; 336(Pt 1):1-17.

[183] Brunton JA, Bertolo RF, Pencharz PB, Ball RO. Proline ameliorates arginine deficiency during enteral but not parenteral feeding in neonatal piglets. *Am. J. Physiol.* 1999 Aug; 277(2 Pt 1):E223-31.

[184] Batshaw ML, Wachtel RC, Thomas GH, Starrett A, Brusilow SW. Arginine-responsive asymptomatic hyperammonemia in the premature infant. *J. Pediatr.* 1984 Jul; 105(1):86-91.

[185] Castillo L, DeRojas-Walker T, Yu YM, Sanchez M, Chapman TE, Shannon D, et al. Whole body arginine metabolism and nitric oxide synthesis in newborns with persistent pulmonary hypertension. *Pediatr. Res.* 1995 Jul; 38(1):17-24.

[186] Vosatka RJ, Kashyap S, Trifiletti RR. Arginine deficiency accompanies persistent pulmonary hypertension of the newborn. *Biol. Neonate.* 1994; 66(2-3):65-70.

[187] Di Lorenzo M, Bass J, Krantis A. Use of L-arginine in the treatment of experimental necrotizing enterocolitis. *J. Pediatr. Surg.* 1995 Feb; 30(2):235-40; discussion 40-1.

[188] Zamora SA, Amin HJ, McMillan DD, Kubes P, Fick GH, Butzner JD, et al. Plasma L-arginine concentrations in premature infants with necrotizing enterocolitis. *J. Pediatr.* 1997 Aug; 131(2):226-32.

[189] Amin HJ, Zamora SA, McMillan DD, Fick GH, Butzner JD, Parsons HG, et al. Arginine supplementation prevents necrotizing enterocolitis in the premature infant.[see comment]. *J. Pediatr.* 2002 Apr; 140(4):425-31.

[190] Amin HJ, Soraisham AS, Sauve R. Does L-arginine supplementation used in the prevention of NEC in $\leq$ 1250gm birth weight infants adversely impact neurodevelopment? *Pediatr. Res.* 2006; 59:5570.385.

[191] Pencharz PB, House JD, Wykes LJ, Ball RO. What are the essential amino acids for the preterm and term infant? In: Bindels JG, Goedhart AC, Visser HKA, editors. *Recent developments in infant nutrition* (10th nutricia symposium). Dordrech: Kluwer Academic Publishers; 1996. p. 278-96.

[192] Rassin DK. Amino acid metabolism in total parenteral nutrition during development. In: Friedman M, editor. *Absorption and utilization of amino acids.* Boca Raton, FL: CRC Press; 1989. p. 71-85.

[193] Roberts SA, Ball RO, Filler RM, Moore AM, Pencharz PB. Phenylalanine and tyrosine metabolism in neonates receiving parenteral nutrition differing in pattern of amino acids. *Pediatr. Res.* 1998 Dec; 44(6):907-14.

[194] Lucas A, Baker BA, Morley RM. Hyperphenylalaninaemia and outcome in intravenously fed preterm neonates. *Arch. Dis. Child* 1993 May; 68(5 Spec No):579-83.

[195] Mitton SG, Burston D, Brueton MJ. Hyperphenylalanaemia in parenterally fed newborn infants. *Lancet* 1988 Dec 24-31; 2(8626-8627):1497-8.

[196] Walker V, Hall MA, Bulusu S, Allan A. Hyperphenylalaninaemia in parenterally fed newborn babies. *Lancet* 1986 Nov 29; 2(8518):1284.

[197] Walker V, Mills GA. Metabolism of intravenous phenylalanine by babies born before 33 weeks of gestation. *Biol. Neonate* 1990; 57(3-4):155-66.

[198] Neuhauser M, Wandira JA, Gottmann U, Bassler KH, Langer K. Utilization of N-acetyl-L-tyrosine and glycyl-L-tyrosine during long-term parenteral nutrition in the growing rat. *Am. J. Clin. Nutr.* 1985 Oct; 42(4):585-96.

[199] Im HA, Meyer PD, Stegink LD. N-acetyl-L-tyrosine as a tyrosine source during total parenteral nutrition in adult rats. *Pediatr. Res.* 1985 Jun; 19(6):514-8.

[200] Heird WC, Dell RB, Helms RA, Greene HL, Ament ME, Karna P, et al. Amino acid mixture designed to maintain normal plasma amino acid patterns in infants and children requiring parenteral nutrition. *Pediatr.* 1987 Sep; 80(3):401-8.

[201] Van Goudoever JB, Sulkers EJ, Timmerman M, Huijmans JG, Langer K, Carnielli VP, et al. Amino acid solutions for premature neonates during the first week of life: the role of N-acetyl-L-cysteine and N-acetyl-L-tyrosine. *JPEN J. Parenter Enter Nutr.* 1994 Sep-Oct; 18(5):404-8.

[202] Druml W, Lochs H, Roth E, Hubl W, Balcke P, Lenz K. Utilization of tyrosine dipeptides and acetyltyrosine in normal and uremic humans. *Am. J Physiol.* 1991 Feb; 260(2 Pt 1):E280-5.

[203] Wykes LJ, House JD, Ball RO, Pencharz PB. Aromatic amino acid metabolism of neonatal piglets receiving TPN: effect of tyrosine precursors. *Am. J. Physiol.* 1994 Nov; 267(5 Pt 1):E672-9.

[204] House JD, Pencharz PB, Ball RO. Tyrosine kinetics and requirements during total parenteral nutrition in the neonatal piglet: the effect of glycyl-L-tyrosine supplementation. *Pediatr. Res.* 1997 Apr; 41(4 Pt 1):575-83.

[205] Roberts SA, Ball RO, Moore AM, Filler RM, Pencharz PB. The effect of graded intake of glycyl-L-tyrosine on phenylalanine and tyrosine metabolism in parenterally fed neonates with an estimation of tyrosine requirement. *Pediatr. Res.* 2001 Jan; 49(1): 111-9.

[206] Sturman JA, Gaull G, Raiha NC. Absence of cystathionase in human fetal liver: is cystine essential? *Science* 1970 Jul 3; 169(940):74-6.

[207] Gaull G, Sturman JA, Raiha NC. Development of mammalian sulfur metabolism: absence of cystathionase in human fetal tissues. *Pediatr. Res.* 1972 Jun; 6(6):538-47.

[208] Gaull GE, Von Berg W, Raiha NC, Sturman JA. Development of methyltransferase activities of human fetal tissues. *Pediatr. Res.* 1973 May; 7(5):527-33.

[209] Zlotkin SH, Bryan MH, Anderson GH. Cysteine supplementation to cysteine-free intravenous feeding regimens in newborn infants. *Am. J. Clin. Nutr.* 1981 May; 34(5):914-23.

[210] Heird WC, Gomez MR. Parenteral nutrition in low-birth-weight infants. *Annu. Rev. Nutr.* 1996, 16:471-99.

[211] Laine L, Shulman RJ, Pitre D, Lifschitz CH, Adams J. Cysteine usage increases the need for acetate in neonates who receive total parenteral nutrition.[see comment]. *Am. J. Clin. Nutr.* 1991 Sep; 54(3):565-7.

[212] White CW, Stabler SP, Allen RH, Moreland S, Rosenberg AA. Plasma cysteine concentrations in infants with respiratory distress. *J. Pediatr.* 1994 Nov; 125(5 Pt 1):769-77.

[213] Malloy MH, Rassin DK, Richardson CJ. Total parenteral nutrition in sick preterm infants: effects of cysteine supplementation with nitrogen intakes of 240 and 400 mg/kg/day. *J. Pediatr. Gastroenterol. Nutr.* 1984 Mar; 3(2):239-44.

[214] Helms RA, Storm MC, Christensen ML, Hak EB, Chesney RW. Cysteine supplementation results in normalization of plasma taurine concentrations in children receiving home parenteral nutrition. J. *Pediatr.* 1999 Mar; 134(3):358-61.

[215] Shoveller AK, Brunton JA, House JD, Pencharz PB, Ball RO. Dietary cysteine reduces the methionine requirement by an equal proportion in both parenterally and enterally fed piglets. *J. Nutr.* 2003 Dec; 133(12):4215-24.

[216] Shoveller AK, Brunton JA, Brand O, Pencharz PB, Ball RO. N-acetylcysteine is a highly available precursor for cysteine in the neonatal piglet receiving parenteral nutrition. *JPEN J. Parenter Enter Nutr.* 2006 Mar-Apr; 30(2):133-42.

[217] Smith RB, Sachan DS, Plattsmier J, Feld N, Lorch V. Plasma carnitine alterations in premature infants receiving various nutritional regimes. *JPEN J. Parenter Enter Nutr.* 1988 Jan-Feb; 12(1):37-42.

[218] Cairns PA, Stalker DJ. Carnitine supplementation of parenterally fed neonates. *Cochrane Database of Systematic Reviews.* 2000(4):CD000950.

[219] Coran AG, Drongowski RA, Baker PJ. The metabolic effects of oral L-carnitine administration in infants receiving total parenteral nutrition with fat. *J. Pediatr. Surg.* 1985 Dec; 20(6):758-64.

[220] Helms RA, Mauer EC, Hay WW Jr., Christensen ML, Storm MC. Effect of intravenous L-carnitine on growth parameters and fat metabolism during parenteral nutrition in neonates. *JPEN J. Parenter Enter Nutr.* 1990 Sep-Oct; 14(5):448-53.

[221] Larsson LE, Olegard R, Ljung BM, Niklasson A, Rubensson A, Cederblad G. Parenteral nutrition in preterm neonates with and without carnitine supplementation. *Acta Anaesthesiol. Scand.* 1990 Aug; 34(6):501-5.

[222] Bonner CM, DeBrie KL, Hug G, Landrigan E, Taylor BJ. Effects of parenteral L-carnitine supplementation on fat metabolism and nutrition in premature neonates. *J. Pediatr.* 1995 Feb; 126(2):287-92.

[223] Shortland GJ, Walter JH, Stroud C, Fleming PJ, Speidel BD, Marlow N. Randomised controlled trial of L-carnitine as a nutritional supplement in preterm infants. *Arch. Dis. Child Fetal Neonatal.* Ed 1998 May; 78(3):F185-8.

[224] Schmidt-Sommerfeld E, Penn D, Wolf H. Carnitine deficiency in premature infants receiving total parenteral nutrition: effect of L-carnitine supplementation. *J. Pediatr.* 1983 Jun; 102(6):931-5.

[225] O'Donnell J, Finer NN, Rich W, Barshop BA, Barrington KJ. Role of L-carnitine in apnea of prematurity: a randomized, controlled trial. *Pediatr.* 2002 Apr; 109(4):622-6.

[226] Whitfield J, Smith T, Sollohub H, Sweetman L, Roe CR. Clinical effects of L-carnitine supplementation on apnea and growth in very low birth weight infants. *Pediatr.* 2003 Mar; 111(3):477-82.

[227] Pande S, Brion LP, Campbell DE, Gayle Y, Esteban-Cruciani NV. Lack of effect of L-carnitine supplementation on weight gain in very preterm infants. *J. Perinatol.* 2005 Jul; 25(7):470-7.

In: Protein Research Progress ISBN 978-1-60456-663-3
Editors: A.B. Boscoe and C. R. Listow, pp. 99-115

Chapter 3

Supervised Learning Under Sample Selection Bias From Protein Structure Databases

K. Peng, Z. Obradovic, and S. Vucetic*
Information Science and Technology Center,
Temple University - Philadelphia, PA 19122, U.S.A.

Abstract

In supervised learning it is commonly assumed that labeled data available for learning a prediction model is a random sample from the underlying distribution. In real-life scenarios, however, this assumption is often violated due to sample selection bias. Consequently, a prediction model learned from such labeled data could be suboptimal and may not generalize well on unseen examples from the same population. In this chapter, we provide a brief review of the sample selection bias problem and existing methods for detecting and correcting such bias. We then describe our contrast classifier framework for addressing this problem and illustrate its effectiveness in several bioinformatics applications related to learning from biased protein structure databases.

Key Words: sample selection bias, contrast classifier, supervised learning.

AMS Subject Classification: Primary 43A60.

1 Introduction

In supervised learning [1, 2], the objective is to learn a prediction model using certain *learning algorithm* from a labeled dataset D_L with N examples $\{(\mathbf{x}_i, y_i)\}_{i=1,2,...,N}$ sampled independently from an unknown distribution $p(\mathbf{x}, y)$, where $\mathbf{x}_i = [x_{i1}, x_{i2}, ..., x_{iK}]$ is the feature vector from K-dimensional domain $\mathbb{X}$ and y_i is the label from domain $\mathbb{Y}$. In statistical terminology, $x_{i1}, x_{i2}, ..., x_{iK}$ are also called independent variables and y_i is called dependent variable or outcome. The resulting model should optimally approximate

*E-mail address: vucetic@ist.temple.edu; This study was supported in part by NSF grant IIS-0219736 to Z. Obradovic and S. Vucetic and NSF grant CSE-IIS-0196237 to Z. Obradovic and A.K. Dunker.

the unknown relationship between $\mathbf{x}$ and y, such that it can predict y for $\mathbf{x}$ unseen in D_L as accurately as possible.

A supervised learning task can either be a *regression* or *classification* task depending on whether label y is continuous or categorical. For regression tasks, the goal is to approximate the conditional expectation $E[y|\mathbf{x}]$ as the predicted y for given $\mathbf{x}$. For classification tasks, the goal is to predict the exact class label y for given $\mathbf{x}$. This can be achieved by first estimating the class conditional probability density $p(y = c|\mathbf{x})$, where class label c = 1, 2, ..., C (number of classes), and then predicting y as the c that has maximal $p(y = c|\mathbf{x})$.

A basic assumption in supervised learning is that labeled dataset D_L should be a true random sample from the underlying distribution $p(\mathbf{x}, y)$ such that the model learned from D_L is applicable to unseen examples from the same distribution. However, this assumption is often violated in real-life scenarios due to the problem of *sample selection bias*, or *selection bias*, *sampling bias* [3], where D_L is not representative of the population. Consequently, the resulting prediction model could be suboptimal and may not perform well on unseen examples.

Sample selection bias may be caused by many reasons. The labeled data may be *truncated* or *censored* according to values of $\mathbf{x}$ and y. In social study surveys, the participants are often *self-selected* instead of chosen randomly from the general population. A similar situation in direct marketing is the *non-response* problem in which people who do not respond to mailings may have significantly different characteristics from those that do. *Attrition bias* is due to the loss of participants in clinical trial evaluations and other studies. In supervised classification, one type of bias is that labeled data have significantly different class distribution from the population, possibly due to the class-dependent costs to obtain class labels.

Since sample selection bias might significantly deteriorate the learning performance, it is important to detect and correct such bias. However, the underlying bias mechanism is often unknown in real life. Fortunately, quite often an unlabeled dataset $D_U = \{\mathbf{x}_i\}_{i=1,2,...,M}$, $M \gg N$, $\mathbf{x}_i \in \mathbb{X}$, is also available and can be assumed as representative of the underlying distribution. Under certain conditions, it is possible to model the bias mechanism by comparing the labeled dataset D_L and the unlabeled dataset D_U, and to use this knowledge to remedy the problem of bias.

Research on sample selection bias originated from econometrics and related areas in the late 1970s. Since the seminal work by James J. Heckman [3], various statistical methods to address the sample selection bias problem have been proposed in different areas such as econometrics [3, 4], credit scoring [5, 6, 7], social sciences [8], and clinical trials [9]. In machine learning community, specific forms of the sample selection bias problem have been studied for years, such as learning from skewed class distribution [10, 11, 12], learning from labeled and unlabeled data [13], and outlier detection [14]. It was not until recently that studies began to emerge which systematically investigate the bias problem in machine learning context.

Zadrozny [15] formalized the sample selection bias in machine learning terms and classified several popular supervised learning algorithms into two categories, "local" or "global", depending on their sensitivities to a common type of bias. Fan et al. [16] further improved Zadrozny's classification and provided more insights into the effects of sample selection bias in machine learning. Smith and Elkan [17] proposed a Bayesian network framework as another attempt to formalize sample selection bias in the context of machine learning. In an empirical study by Chawla and Karakoulas [18], the influence of sample selection bias (along with other factors) on performance of semi-supervised learning algorithms was systematically evaluated; they also examined the effectiveness of several bias correction methods in semi-supervised learning scenarios.

The rest of this chapter is organized into two parts. In the first part we provide an overview of the sample selection bias problem as well as existing methods for bias detection and correction. In the second part we present our contrast classifier framework [19] for addressing this problem. After a brief introduction to the contrast classifier, we illustrate its effectiveness on several bioinformatics applications related to protein structure prediction.

2 Sample Selection Bias

2.1 Formal Definition

Sample selection bias can be formally modeled [15] by associating a binary random variable s with each possible $(\mathbf{x}, y) \in X \times Y$; an example $(\mathbf{x}, y)$ may be selected into (or observed in) D_L if and only if $s = 1$. The *sample selection probability* $p(s = 1|\mathbf{x}, y)$ quantifies the bias toward a given example $(\mathbf{x}, y)$ and determines how likely it can be sampled besides its original probability density $p(\mathbf{x}, y)$. From this perspective, the labeled dataset D_L is in effect sampled from distribution $p(\mathbf{x}, y, s = 1) = p(s = 1|\mathbf{x}, y) \cdot p(\mathbf{x}, y)$ instead of the original distribution.

If indicator s is independent of both $\mathbf{x}$ and y, the sample is unbiased. If this is not the case, three types of bias are possible: (I) s depends on $\mathbf{x}$ but is independent of y given $\mathbf{x}$, i.e. $p(s = 1|\mathbf{x}, y) = p(s = 1|\mathbf{x})$ and $p(y|\mathbf{x}, s = 1) = p(y|\mathbf{x})$; (II) s depends on y but is independent of $\mathbf{x}$ given y, i.e. $p(s = 1|\mathbf{x}, y) = p(s = 1|y)$ and $p(\mathbf{x}|y, s = 1) = p(\mathbf{x}|y)$; (III) s depends on both $\mathbf{x}$ and y. In supervised learning, Type I or Type II bias can often be assumed. Type III bias is not tractable unless some additional restrictive assumptions can be made.

Note that it is also possible that indicator s depends on features $\mathbf{x}_s$ not included in $\mathbf{x}$, such that the sample selection probability is specified as $p(s = 1|\mathbf{x}, \mathbf{x}_s, y)$. In such case, the sample selection bias cannot be modeled in general. In machine learning, however, the emphasis is on optimizing learning performance from given data rather than on understanding the exact data generating mechanism. Therefore, we do not consider this scenario further.

2.2 Effects of Sample Selection Bias on Popular Learning Algorithms

Two recent studies [15, 16] were conducted to investigate the effect of Type I bias on several popular supervised classification algorithms. As claimed by Zadrozny [15], "local" learning algorithms whose outputs depend only on estimated $p(y|\mathbf{x})$ should not be sensitive to Type I bias, since the actual probability density $p(y|\mathbf{x}, s = 1)$ estimated from the biased labeled dataset D_L would be similar to $p(y|\mathbf{x})$ from the whole population. On the other hand, "global" learning algorithms that rely on both $p(y|\mathbf{x})$ and $p(\mathbf{x})$, or equivalently $p(\mathbf{x}, y)$, are more likely to be affected by the bias since $p(\mathbf{x}, s = 1) \neq p(\mathbf{x})$. Fan et al. [16] argued that these are not always true. They illustrated that "local" learning algorithms on particular biased datasets might still be affected by the bias if the estimated $p(y|\mathbf{x}, s = 1)$ is inaccurate due to incorrect model assumption, while "global" learning algorithms would not be sensitive to the bias if y and s depend on two disjoint feature subsets of **x**.

For Type II bias, the sample selection depends only on label y, i.e. $p(s = 1|\mathbf{x}, y) = p(s = 1|y)$. In supervised classification, this implies that D_L has different class distribution from that of the population but the class conditional distribution $p(\mathbf{x}|y)$ remains unaltered. One common cause of this type of bias is the class-dependent costs for obtaining class labels (y), or equivalently, obtaining examples for certain classes are more difficult than obtaining those for other classes. As illustrated by Weiss and Provost [11], predictor models learned from D_L with biased class distribution are likely to have, although not always, deteriorated performance on unseen examples from the same population. However, such bias may also be created intentionally to improve the learning performance under certain circumstances, e.g. when the class distribution is extremely imbalanced [10]. In the case of binary classification, this means that one class (minority class) is rare while the other (majority class) is abundant in the population, and the labeled data is assumed to be unbiased. Since many traditional learning algorithms perform poorly under such circumstances, a common strategy is to make a balanced training set by either down-sampling the majority class or up-sampling the minority class [10], and thus creating the bias.

2.3 Estimating Sample Selection Probability

It is possible to estimate $p(s = 1|\mathbf{x}, y)$ if we are given an unlabeled dataset $D_U = \{\mathbf{x}_i\}_{i=1..M}$, $M \gg N$, that is sampled from the underlying distribution $p(\mathbf{x}, y)$. It is worth noting that, unlike labeled data, unlabeled data is typically easier to obtain in large amount. As we will show later, the sample selection probability $p(s = 1|\mathbf{x})$ for Type I bias can be estimated indirectly using the contrast classifier approach [19]. For Type II bias, the sample selection probability $p(s = 1|y)$ can also be obtained indirectly by estimating the class proportion in D_U using a bootstrapping methodology [12]. For Type III bias, the sample selection probability $p(s = 1|\mathbf{x}, y)$ cannot be modeled in similar manner as for Type I and II bias due to its dependence on both **x** and y.

Recently [20], a method for estimating $p(s = 1|\mathbf{x}, y)$ under Type III bias was proposed which utilizes the equation $E_D[g(\mathbf{x})] = E_D[s \cdot g(\mathbf{x})/p(s = 1|\mathbf{x}, y)]$, where $g(\mathbf{x})$ is a real-valued function, $E_D[.]$ denotes expectation over D representing the whole population containing both labeled ($s = 1$) and unlabeled ($s = 0$) examples. Assuming that $p(s = 1|\mathbf{x}, y)$ is from a parametric family with k parameters, k equations can be established using empirical estimates of $E_D[.]$ for k different functions $g_1(\mathbf{x})$, $g_2(\mathbf{x})$, ..., $g_k(\mathbf{x})$. By solving the k equations, the parameters in $p(s = 1|\mathbf{x}, y)$ can be determined. However, its practical use might be limited due to the inherit difficulties in choosing appropriate functional forms for both $p(s = 1|\mathbf{x}, y)$ and $g(\mathbf{x})$ and solving non-linear equations. Fortunately, under certain assumptions, explicit estimation of $p(s = 1|\mathbf{x}, y)$ may not be necessary for bias detection and correction as we will see in the next section.

2.4 Methods for Sample Selection Bias Detection and Correction

In this section we will review some existing methods that the address sample selection bias problem. All of them utilize unlabeled data and model the sample selection probability explicitly or implicitly.

The famous Heckman's two-step procedure [3] was originally developed for regression analysis in studies of labor force supply. In this procedure, two regression models are built: the *outcome model* is an ordinary linear squares (OLS) model for estimating the outcome (y) of interest, while the *selection model* is a probit linear model for modeling the unknown sample selection mechanism. In the first step, the selection model is built using both labeled ($s = 1$) and unlabeled ($s = 0$) data. Using the estimated model parameters, the *inverse Mill's ratio* λ , which is related to the sample selection probability, can be calculated for each $\mathbf{x}$. In the second step, λ is included as an additional feature (or independent variable) to build an unbiased outcome model using labeled examples ($s = 1$) only.

An important assumption of the Heckman's procedure, under which λ can be calculated, is that error terms of the two initial models (not including λ) follow a bivariate normal distribution. If there is no correlation between the two error terms (i.e. a diagonal covariance matrix), the sample selection is independent of y given $\mathbf{x}$, i.e. we have Type I bias; otherwise, we have Type III bias. In the later case, the selection model output may not be regarded as a good estimate for the sample selection probability $p(s = 1|\mathbf{x}, y)$ since its dependence on y given $\mathbf{x}$ is reflected in the error term. In this sense, the Heckman's procedure avoids direct estimation of $p(s = 1|\mathbf{x}, y)$ but is still able to correct the bias.

The bivariate probit model [6, 7] can be viewed as a variant of the Heckman's procedure that is specific for classification tasks, where the OLS outcome model is replaced by another probit model for binary classification. Like the Heckman's procedure, the bivariate probit model also assumes that error terms of the two models follow a bivariate normal distribution and is applicable to scenarios where Type III bias is present.

In a procedure [21] similar to the Heckman's, Elkan used machine learning algorithms such as Naive Bayes classifiers or decision trees to estimate the selection model. Instead of the inverse Mill's ratio λ, which is no longer applicable, he used the selection model output, or estimated sample selection probability, as the additional feature to build the outcome model. Since no assumption is made about the correlation between the error terms of two models, this procedure implicitly assumes that sample selection depends only on $\mathbf{x}$ but not on y and is therefore only applicable to Type I bias. In their empirical study [18], Chawla and Karakoulas further adapted this procedure to solve semi-supervised classification problems.

3 Learning from Biased Data with Contrast Classifier

The contrast classifier [19] is a binary classifier which models the distributional difference between the labeled dataset D_L and the unlabeled dataset D_U. Its name *contrast* comes from the meaning of its output, which represents a measure of difference, or contrast, in probability density of a given data point $\mathbf{x}$ between labeled and unlabeled data. Given this property, the contrast classifier could be used in a wide range of important machine learning or data mining applications such as outlier detection, one-class classification, density estimation, and learning from biased data [19]. In the following sections we will show that contrast classifier can be used to estimate the sample selection probability under the Type I bias.

3.1 Contrast Classifier

To build a contrast classifier, a new dataset is constructed by assigning class labels 0 and 1 to examples from D_L and D_U, respectively. Note that the original label y is not used and the new label is not related to y in labeled examples. The contrast classifier is trained as a binary classifier on the constructed dataset using classification algorithms (e.g. feedforward neural networks [1, 2]) which estimate the posterior conditional class probability. An optimally trained contrast classifier would output

$$cc(\mathbf{x}) = \frac{r \cdot u(\mathbf{x})}{(1-r) \cdot l(\mathbf{x}) + r \cdot u(\mathbf{x})}$$

where $l(\mathbf{x})$ and $u(\mathbf{x})$ are the probability density functions of D_L and D_U respectively, and r is the fraction of unlabeled examples in training data.

In practice D_L and D_U are often *imbalanced* because unlabeled data could be obtained easier and in large amount. Thus, instead of the original D_L and D_U, balanced training set with $r = 0.5$ should be used. For $r = 0.5$ it follows that

$$\frac{l(\mathbf{x})}{u(\mathbf{x})} = \frac{1 - cc(\mathbf{x})}{cc(\mathbf{x})}$$

Thus, the contrast classifier output $cc(\mathbf{x})$ is a monotonically decreasing function of the ratio $l(\mathbf{x})/u(\mathbf{x})$, which reveals the distributional difference between the two datasets.

If Type I bias can be assumed, i.e. $p(s = 1|\mathbf{x}, y) = p(s = 1|\mathbf{x})$, the contrast classifier output $cc(\mathbf{x})$ is related to the sample selection probability $p(s = 1|\mathbf{x})$. While $l(\mathbf{x})$ approximates $p(\mathbf{x}, s = 1)$, $u(\mathbf{x})$ may approximate either $p(\mathbf{x}, s = 0)$ or $p(\mathbf{x})$ depending on the actual application. For example, in credit scoring problem [6] D_L contains all accepted applicants ($s = 1$) while D_U contains all rejected applicants ($s = 0$); thus, it is appropriate to assume $u(\mathbf{x}) = p(\mathbf{x}, s = 0)$. In this case

$$cc(\mathbf{x}) = 1 - p(s = 1|\mathbf{x})$$

In some other scenarios it is more reasonable to assume that unlabeled dataset D_U is representative of the whole population, i.e. $u(\mathbf{x}) = p(\mathbf{x})$. In this case

$$cc(\mathbf{x}) = \frac{1}{1 + p(s = 1|\mathbf{x})}$$

In either case, $cc(\mathbf{x})$ is a decreasing function of $p(s = 1|\mathbf{x})$, and could therefore be a useful measure of sample selection bias. In the following discussion, it is assumed that $u(\mathbf{x}) = p(\mathbf{x})$.

3.2 Selecting Underrepresented Examples

According to the interpretation above, contrast classifier output $cc(\mathbf{x})$ could be a useful measure of sample selection bias. An example $\mathbf{x}$ is *underrepresented* if $cc(\mathbf{x}) > 0.5$, or *overrepresented* if $cc(\mathbf{x}) < 0.5$. Using the $cc(\mathbf{x})$ distribution for D_L as a *null distribution*, we can determine if an unlabeled example $\mathbf{x}$ is likely to come from the same distribution as

D_L, i.e. if $\mathbf{x}$ is well represented by D_L. A threshold θ^p is first determined such that only $100p\%$ examples in D_L satisfy $cc(\mathbf{x}) > \theta^p$. If an unlabeled example $\mathbf{x}$ has $cc(\mathbf{x}) > \theta^p$, we conclude that it is underrepresented in D_L.

In a classification problem of $C(\geq 2)$ classes, the labeled dataset D_L is inherently heterogeneous, and consequently a single contrast classifier built using the whole D_L versus D_U might not detect the sample selection bias effectively. A better approach might be building C *class-specific* contrast classifiers $cc_c(\mathbf{x}), c = 1, 2, ..., C$. For an example $\mathbf{x}$ to be underrepresented in D_L, we require that $cc_c(\mathbf{x}) > \theta_c^p$ holds for all c = 1,2,...,C. Similar to the discussion above, the threshold θ_c^p is chosen for the c-th contrast classifier such that only $100p\%$ class c labeled examples in D_L satisfy the inequality.

3.3 Assessing Overall Level of Bias

Contrast classifier output $cc(\mathbf{x})$ can be used to measure the overall level of bias. One can compare the difference between the two $cc(\mathbf{x})$ distributions (histograms) of D_L and D_U: the bias is negligible if the two distributions largely overlap; it is significant if the two distributions are well separated and/or have different shapes. The two-sample *Kolmogorov-Smirnov* (KS) goodness-of-fit test [22] could be applied to compare the two $cc(\mathbf{x})$ distributions and the resulting KS statistic could serve as a quantitative measure of the overall bias.

Alternatively, the $cc(\mathbf{x})$ distribution of D_U alone could reveal the level of overall bias if D_U is representative of the population. If this distribution is centered on 0.5, $l(\mathbf{x})$ should be similar to $u(\mathbf{x})$ almost everywhere, which indicates that the overall level of bias is negligible. On the other hand, if this distribution is scattered and/or its mean deviates considerably from 0.5, the overall level of bias is significant. We therefore define a quantitative measure Δ as

$$\Delta = \sqrt{\sum_{i=1}^{|D_U|} (cc(\mathbf{x}_i) - 0.5)^2 / \, |D_U|}$$

where $\mathbf{x}_i$ is i-th example in D_U. The value of Δ will approach 0 when bias is negligible since all $cc(\mathbf{x}_i)$ should be close to 0.5. It will be large if the bias is significant and many $cc(\mathbf{x}_i)$ are far away from 0.5. In the extreme case when all $cc(\mathbf{x}_i) = 1$, Δ will be 0.5.

3.4 Correcting Sample Selection Bias

It has been shown that resampling the labeled dataset D_L with weight proportional to $1/p(s = 1|\mathbf{x})$ could be a useful way to correct sample selection bias [15, 23], given that $p(s = 1|\mathbf{x})$ is nonzero at any $\mathbf{x} \in \mathbb{X}$ and can be properly estimated. However, in practice this approach is not always applicable. For example, if D_L is not large enough and/or highly biased, some regions in the feature space $\mathbb{X}$ will be highly underrepresented and very few examples in D_L will come from these regions. Consequently, resampling D_L alone would be of little use in correcting the bias. Furthermore, the presence of duplicate examples as a result of resampling might lead to overfitting by some learning algorithms. In such a case, it becomes necessary to enlarge D_L by labeling additional examples.

In general, one can always randomly select unlabeled examples for labeling, and as more examples are labeled the sample selection bias should be eventually corrected. Obviously, this would be inefficient if the labeling cost is significant. Therefore, the unlabeled examples should be selected in a cost-effective way. The contrast classifier output $cc(\mathbf{x})$ can serve this purpose since it directly measures the bias toward individual examples. We can first rank all unlabeled examples from D_U based on their $cc(\mathbf{x})$ values, and then select those with high values for labeling. Intuitively, this approach would be more effective in bias reduction.

Based on the discussion above, a one-step approach would consist of building a single contrast classifier from the initial D_L and D_U and selecting G underrepresented examples from D_U based on $cc(\mathbf{x})$. In practice it is often difficult to determine an appropriate value for G such that the labeling cost is minimized. We therefore propose a procedure that iteratively builds contrast classifiers and incrementally selects underrepresented examples [24]. At each iteration, a contrast classifier is built from current D_L and D_U and then applied to D_U. If the $cc(\mathbf{x})$ distribution for D_U indicates significant overall bias, i.e. high Δ value (Section 3.3), a set of $B(B < G)$ underrepresented examples will be selected and added into D_L for building the contrast classifier at the next iteration. The whole procedure iterates until the required G examples have been selected or the sample selection bias becomes negligible.

It should be noted that the actual labeling is not necessary during the procedure since the label information is not used for selection. The labeling for all selected examples can be done after the procedure stops, which is the set difference between the final and initial D_L. This property is a major difference between the proposed procedure and active learning methods [25, 26]. Another difference is that active sampling methods are primarily designed for supervised classification tasks; they tend to select more examples close to the decision boundary and therefore might introduce more bias. On the other hand, the proposed procedure aims at reducing bias over the entire feature space given a sufficiently large and representative unlabeled dataset. In this sense, it is also applicable to other learning tasks such as nearest-neighbor classification and density estimation [2].

3.5 Implementation Issues

Since the unlabeled dataset D_U is often much larger than D_L, learning from such imbalanced data would result in a low-quality contrast classifier, while learning time could be prohibitively long. We addressed this by training an ensemble of neural networks on balanced training sets consisting of equal number of labeled (class 0) and unlabeled (class 1) examples randomly sampled from the available data. Similar to bagging [27], we constructed a contrast classifier by aggregating the predictions of these neural networks through averaging. An additional benefit of using an ensemble of neural networks is that the averaging is known as a successful technique for increasing their accuracy by reducing variance while retaining low prediction bias.

4 Bioinformatics Applications

In this section, we illustrate how contrast classifiers could be used to detect and correct sample selection bias in protein structure databases and to improve accuracy of protein structure prediction. Before this, we first give a brief introduction to protein structure and objectives of protein structure prediction.

4.1 Protein Structure and Protein Structure Prediction

As the "machinery of life", protein is one of the most important macromolecules in living organisms that are essential for the structure and function of all living cells. Proteins constitute more than half of the dry weight of most cell matter and are involved in diverse biological functions such as enzyme action, regulation of cellular functions, signaling, structural support, and transportation and storage of matter A protein may consist of one or more amino acid sequences (or polypeptide chains), each of which is a linear polymer of amino acids connected by covalent peptide bonds. It is commonly believed that under physiological conditions the amino acid sequence(s) of a protein spontaneously folds into a stable three dimensional (3D) structure, which in turn determines the protein function [28]. Therefore it is of great importance to know the structure of a protein for understanding its functions and the underlying biological mechanisms it is part of.

Protein 3D structure can be determined experimentally mainly by two types of techniques, X-ray diffraction and nuclear magnetic resonance (NMR). For both techniques, the target protein has to be expressed and purified to produce protein solutions of sufficient quantity, purity and concentration. Furthermore, X-ray diffraction requires growing high quality crystals from protein solutions. Although progress has been made in experimental technologies, structure determination is still an expensive and time-consuming process. Consequently, there are only about 30,000 protein structures that have been experimentally determined and deposited in Protein Data Bank (PDB) [29], the main protein structure database. This is only a small subset of the more than 2 million protein sequences deposited in UniProt (Release 6.0) protein database [30]. As more genomes are being completely sequenced, this sequence-structure gap is likely to become even larger. Therefore, computational methods that predict protein structures from amino acid sequence [31] are becoming increasingly attractive.

In a very simplified view, protein structure prediction can be regarded as a supervised learning task where amino acid sequence is represented as feature $\mathbf{x}$ and the corresponding structure is label y. Therefore, proteins with known structures form the labeled dataset D_L, and the goal of learning is to model the sequence-structure relationship. While it is commonly accepted that protein structure is encoded by amino acid sequence [28], predicting 3D structure from sequence is still an open problem since the actual sequence-structure relationship is very complicated and is affected by many factors. Apart from the inherent difficulties, protein structure prediction may also suffer from the sample selection bias problem, i.e. the current protein structure database (PDB) is not a representative sample of all natural proteins.

4.2 Characterizing Bias in Database of Disordered Proteins

We first illustrate the effectiveness of the contrast classifier on a typical protein structure prediction problem - prediction of protein disordered regions. Protein disorder [32] refers to an important structural property of some protein regions that do not have a stable 3D structure under physiological conditions. Protein regions that have this property are called *disordered regions*. Proteins that have disordered regions are called *disordered proteins*; otherwise, they are called *ordered proteins*. To develop predictors of protein disorder, a non-redundant database of disordered proteins has been developed [33]. Due to the historical overlooking of this property, our database is relatively small and is not a good representative of disordered proteins in nature. Thus, in a recent study [19] we applied contrast classifiers to characterize the sample selection bias and identify proteins from large sequence repositories that are statistically underrepresented in our database. Understanding biological properties of these outliers could lead to more effective studies of protein disorder.

In this study, the labeled dataset consisted of 442 proteins, including 152 disordered proteins with disordered regions longer than 30 consecutive residues, and 290 ordered proteins. The unlabeled dataset of 17,676 proteins was constructed as a non-redundant representative subset of the Swiss-Prot (Release 40) database [34]. This unlabeled dataset was assumed to be a representative of all natural proteins. Instead of a single contrast classifier built with all labeled proteins, two class-specific contrast classifiers were trained - $cc_{disorder}(\mathbf{x})$ as disordered *versus* unlabeled, and $cc_{order}(\mathbf{x})$ as ordered *versus* unlabeled. For both contrast classifiers, the same knowledge representation as for the VL3 disorder predictor [35, 36] was used.

To select underrepresented proteins, we first filtered the unlabeled dataset and kept only a subset of 6,964 proteins which were 200 to 500 residues long and were not homologous to any labeled proteins. After applying the two contrast classifiers to these proteins, we summarized each protein with two numerical values, cc_avg_{order} and $cc\_avg_{disorder}$, representing the average predictions of the contrast classifiers over its sequence. Following the method described in Section 3.2, we selected a total of 1,259 proteins that had both high cc_avg_{order} and high $cc\_avg_{disorder}$ values, and were therefore statistically underrepresented by the labeled proteins.

To understand the biological properties of these proteins, we compared the frequencies of 840 protein function keywords listed in the Swiss-Prot database between the selected proteins and 895 Swiss-Prot homologues of the labeled proteins. As the results revealed, keywords Inner Membrane, Membrane, and Transmembrane appear more frequently in the selected proteins. This indicates that many selected proteins belong to a large family of membrane proteins that are known to be structurally and functionally diverse and are highly underrepresented among our labeled order and disorder sequences. It is likely that sample selection bias in our labeled data comes mainly from these membrane proteins. This result is also consistent with the results of our related study [37] where outliers were identified by learning class-conditional distribution models of ordered and disordered protein sequences.

4.3 Exploring Bias in Protein Structure Database

As mentioned above, the content of Protein Data Bank (PDB) [29] - the major database of experimental protein structures - is biased in the sense that it does not adequately cover the protein space. As illustrated in [38] the PDB sequences have significantly different amino acid compositions and predicted secondary structures from those in eight complete genomes. It was estimated that current structural information in PDB could cover only about 1/3 of all possible proteins [39, 40]. On the other hand, PDB structures may not be representative of the structure space either. Brenner et al. [41] analyzed the SCOP [42] structural classification of PDB proteins and reported high skewness at all classification levels. Furthermore, the estimated numbers of folds and superfamilies (high-level structural classes) in nature [43] suggest that there are more distinct structures that have not been represented by known PDB structures.

The sample selection bias in PDB occurs due to several reasons. In general, PDB is positively biased towards proteins that are more amenable to expression, purification and crystallization. Another source of bias is the fact that different research groups usually have different objectives when selecting the target proteins. Understanding the sample selection bias in PDB is important to protein structure prediction since it is the ultimate source of labeled data for building predictors. This is especially true for structure prediction methods based on supervised learning, e.g. secondary structure prediction, transmembrane helix prediction, and protein disorder prediction. Other structure prediction methods could also be affected by the bias. For example, comparative modeling or homology modeling can be viewed as a nearest-neighbor classification method, where the evolutionary distance or sequence similarity measures serve as the distance measure [44]. For these methods, the key to success is a database of representative fold structures that could serve as templates for as many sequences as possible.

In a recent study [45] we applied contrast classifiers to explore the sample selection bias in the Protein Data Bank (PDB). In this application, PDB was considered as labeled data while Swiss-Prot was used as unlabeled data. Swiss-Prot was assumed to be representative of all natural proteins. A contrast classifier was trained on non-redundant subsets of PDB and Swiss-Prot and then applied to the Swiss-Prot sequences to select those underrepresented by PDB. By analyzing the Swiss-Prot keywords and other functional annotations associated with these outlying sequences, the bias in PDB towards different functional protein properties was characterized. The results confirmed some well-known biases, e.g. non-globular regions such as transmembrane, disordered and low complexity regions are significantly underrepresented, while disulfide bonds, metal binding sites, and sites involved in enzyme activity are overrepresented. In addition, this analysis also revealed some less-known bias, e.g. hydroxylation and phosphorylation posttranslational modification sites were found to be underrepresented while acetylation sites were significantly overrepresented [45].

4.4 Prioritizing Protein Targets for Structural Genomics

The ongoing structural genomics projects [46] aim at experimentally determining structures for a carefully selected set of protein targets. Along with those already in PDB, these structures can serve as structural templates to computationally model or predict structures for other proteins with reasonable accuracy. Thus, an important problem, or the *target selection* problem, is how to select protein targets such that the resulting structures can cover as many proteins as possible. Many existing target selection strategies [47] rely on sequence comparison methods [44] to group protein sequences into homologous families, and then select one or several representatives from each family that has no structurally characterized members.

In a recent study (unpublished), we applied contrast classifiers to the target registration database (TargetDB) [48], which contains protein targets selected by different structural genomics projects along with information about their current progress. We used protein targets at different structural determination stages [48], e.g. *selected*, *expressed*, *purified*, *crystallized*, and *structure_determined*, as the labeled datasets and built a series of contrast classifiers. As the results revealed, the targets at *selected* stage seem to be indeed unbiased (with respect to our unlabeled dataset from Swiss-Prot), indicating the effectiveness of current target selection strategies. On the other hand, the targets at *structure_determined* stage exhibit similar bias as in Protein Data Bank [45]. In addition, we observed a clear trend of increasing bias from *selected* stage to *structure_determined* stage. Further analysis of these results might help understand the underlying mechanism of sample selection bias in current protein structure databases.

While existing target selection strategies might be effective in selecting representative protein targets, the target lists can still be very long due to the vast number of known proteins. Since structural determination is still costly and time-consuming, it is highly desirable to prioritize protein targets in order to *rapidly* achieve a good coverage of the protein space. This target prioritization problem can be cast as a problem of bias correction. As in previous studies, labeled examples are proteins with experimentally determined structures and their homologues, while unlabeled examples are all natural proteins. The goal is to select the most informative unlabeled proteins for labeling (i.e., experimental structure characterization) to rapidly reduce the bias existing in the labeled proteins. As illustrated in another study [24], the iterative procedure for correcting sample selection bias (Section 3.4) could be a suitable choice for this purpose.

We applied the iterative procedure to 28,334 protein sequences (40 amino acids or longer) from a model organism called *Arabidopsis thaliana* (ATH) that is extensively studied in plant biology [49]. ATH is also a major source of structural genomics targets for the Center for Eukaryotic Structural Genomics (CESG). We first clustered the ATH sequences using the CD-HIT program [50] and obtained 14,988 clusters (families). By selecting the longest sequence from each cluster, we obtained a non-redundant set of 14,988 sequences, such that no two sequences have pairwise identity higher than 40%. Out of these sequences 838 were identified to have known structures and thus formed the labeled dataset D_L, while all of the 14,988 proteins were assigned to the unlabeled dataset D_U. Note that this process can be viewed as a simplified target selection process and D_U can be regarded as our target lists.

It was further assumed that at most 500 proteins, i.e. $G = 500$, could be selected from D_U for labeling due to available resources for experimental structure determination. We examined different values of $B = \{50, 100, 250, 500\}$ and also compared the proposed procedure to simple random selection. As the results showed, the 500 proteins selected by the proposed iterative procedure ($B = 50$) were more effective in correcting sample selection bias than those selected randomly. Furthermore, the iterative procedure with small B value, e.g. $B = 50$, performed considerably better than the one-step approach ($B = 500$) in the sense that it could achieve much lower Δ value, or lower level of overall bias, after selected 500 proteins.

At the time of the study, only a small fraction of the 500 proteins were selected as structural genomics targets by the Center for Eukaryotic Structural Genomics and none of them had been solved. We argue that the remaining proteins should also be selected as structural genomics targets and should be given higher priorities. Along with proteins with known structures, these proteins should be very helpful in achieving a rapid and good coverage of the protein space for *Arabidopsis thaliana* genome [49]. We believe that revealing structural properties of such proteins is likely to produce highly significant biological results.

5 Discussions and Conclusion

Sample selection bias is a common problem in almost all areas that rely on statistical methods to infer or learn population characteristics from samples. It is therefore of great importance to understand the underlying bias mechanism such that the learning process can be adjusted accordingly to obtain meaningful results. Many statistical methods have been developed for bias detection and correction in various disciplines, although they typically rely on certain restrictive assumptions which may limit their practical use. In machine learning context, systematic studies on sample selection bias are receiving increasing attention. It is expected that more robust learning algorithms would be possible if the sample selection bias is taken into account.

As illustrated in Section 4, the contrast classifier framework provides a simple yet effective way of addressing sample selection bias. However, a major challenge in applying contrast classifiers to learning from biased protein structure data is how to select relevant features. Since the protein sequences have variable length, it is always necessary to map the sequences into certain space ($\mathbb{X}$) with fixed number of features before a contrast classifier can be learned. Only if the features extracted from amino acid sequence are relevant to the underlying bias mechanism can the resulting contrast classifier be effective in bias detection and correction. For example, amino acid compositions and sequence complexity could readily reveal the well-known bias toward non-globular regions in PDB, but are less sensitive to bias toward different structural fold classes [45].

To improve the resulting contrast classifiers, it might be desirable to perform feature selection to remove features that might be irrelevant to the bias. However, existing feature selection techniques [51] may not work well for contrast classifiers, since the labeled dataset could be part of or largely overlapped with the unlabeled dataset. In addition, standard measures of feature relevance for normal binary classifiers may not be suitable for contrast classifiers. It is therefore desirable to modify existing or to develop new feature selection techniques that are more effective for contrast classifier applications.

In current approaches, contrast classifiers are usually implemented as ensembles of neural networks. However, the proper ensemble size and the complexity of component networks (e.g. number of hidden neurons) have to be chosen manually by trial-and-error procedures, which can be inefficient and ineffective especially when the unlabeled dataset is large. Thus, a recently proposed procedure [52] for cost-effective learning of predictors as ensembles of neural networks from arbitrary large datasets will be applied to improve contrast classifier learning. It is expected that the resulting contrast classifiers would utilize the data diversity more efficiently and be adaptive to the inherent complexity of given data.

Methods for incorporating contrast classifiers into the process of protein structure prediction, e.g. protein disorder prediction, will also be investigated to improve the prediction accuracy. A straightforward approach is to assign prediction confidence based on the contrast classifier output $cc(\mathbf{x})$ to the structure predictors. If a protein produces high $cc(\mathbf{x})$, it is very likely that it is underrepresented in the labeled dataset and that the resulting predictor may not generalize well on it. Therefore it might be necessary to report that the prediction confidence is low or even refuse to provide prediction. In this way, the prediction accuracy might be improved, but at the expense of lower coverage.

6 Acknowledgements

This study was supported in part by NSF grant IIS-0219736 to Z. Obradovic and S. Vucetic and NSF grant CSE-IIS-0196237 to Z. Obradovic and A.K. Dunker.

References

[1] T. Mitchell, *Machine Learning*, McGraw Hill, New York, 1997.

[2] R. Duda, P. Hart, and D. Stork, *Pattern Classification*, John Wiley & Sons, New York, 2000.

[3] J. J. Heckman, *Sample Selection Bias as a Specification Error*, Econometrica, Vol. **47**(1979) 153-161.

[4] W. H. Greene, *Sample Selection Bias as a Specification Error: Comment*, Econometrica, Vol. **49**(1981) 795-798.

[5] A. J. Feelders, *An Overview of Model Based Reject Inference for Credit Scoring*, Technical Report, Institute of Information and Computing Sciences, Utrecht University, 2003.

[6] J. Crook, J. Banasik, and L. Thomas, *Sample Selection Bias in Credit Scoring Models*, Journal of the Operational Research Society, Vol. **54**(2003) 822-832.

[7] W. H. Greene, *Sample Selection in Credit-Scoring Models*, Japan and the World Economy, Vol. **10**(1998) 299-316.

[8] G. Cuddeback, E. Wilson, J. G. Orme, and T. Combs-Orme, *Detecting and Statistically Correcting Sample Selection Bias*, Journal of Social Service Research, Vol. **30**(2004) 19-33.

[9] K. D. Miller, Z. U. Rahman, and G. W. J. Sledge, *Selection Bias in Clinical Trials*, Breast Disease, Vol. **14**(2001) 31-40.

[10] N. Japkowicz and S. Stephen, *The Class Imbalance Problem: a Systematic Study*, Intelligent Data Analysis Journal, Vol. **6**(2002) 429-450.

[11] G. M. Weiss and F. Provost, *Learning When Training Data Are Costly: The Effect of Class Distribution on Tree Induction*, Journal of Artificial Intelligence Research, Vol. **19**(2003) 315-354.

[12] S. Vucetic and Z. Obradovic, *Classification on Data with Biased Class Distribution*, Proc. 12th European Conference on Machine Learning (2001), Freiburg, Germany, pp. 527-538.

[13] M. Seeger, *Learning with Labeled and Unlabeled Data*, Technical Report, Institute for Adaptive and Neural Computation, University of Edinburgh, 2001.

[14] V. J. Hodge, and J. Austin, *A Survey of Outlier Detection Methodologies*, Artificial Intelligence Review, Vol. **22**(2004) 85-126.

[15] B. Zadrozny, *Learning and Evaluating Classifiers under Sample Selection Bias* , Proc. 21st International Conference on Machine learning (2004), Banff, Alberta, Canada, pp. 903-910.

[16] W. Fan, I. Davidson, B. Zadrozny, and P. S. Yu, *An Improved Categorization of Classifier's Sensitivity on Sample Selection Bias* , Proc. 5th IEEE International Conference on Data Mining (2005), Houston, TX.

[17] A. Smith and C. Elkan, *A Bayesian Network Framework for Reject Inference*, Proc. 10th ACM SIGKDD International Conference on Knowledge Discovery and Data Mining (2004), Seattle, WA, pp. 286-295.

[18] N. V. Chawla and G. Karakoulas, *Learning From Labeled and Unlabeled Data: an Empirical Study Across Techniques and Domains*, Journal of Artificial Intelligence Research, Vol. **23**(2005) 331-366.

[19] K. Peng, S. Vucetic, B. Han, H. Xie, and Z. Obradovic, *Exploiting Unlabeled Data for Improving Accuracy of Predictive Data Mining*, Proc. 3rd IEEE International Conference on Data Mining (2003), Melbourne, FL, pp. 267-274.

[20] S. Rosset, J. Zhu, H. Zou, and T. Hastie, *A Method for Inferring Label Sampling Mechanisms in Semi-supervised Learning*, Advances in Neural Information Processing Systems, eds. L. K. Saul, Y. Weiss, and L. Bottou. Vol. **17**, MIT Press, Cambridge, MA, 2005.

[21] C. Elkan, *Cost-Sensitive Learning and Decision-Making When Costs Are Unknown* , Proc. 7th International Conference on Machine Learning (2000), Stanford University, CA, pp. 204-213.

[22] F. J. J. Massey, *The Kolmogorov-Smirnov Test of Goodness of Fit*, Journal of The American Statistical Association, Vol. **46**(1951) 68-78.

[23] Y. Vardi, *Empirical Distributions in Selection Bias Models* , Annals of Statistics, Vol. **13**(1985) 178-203.

[24] K. Peng, S. Vucetic, and Z. Obradovic, *Correcting Sampling Bias in Structural Genomics through Iterative Selection of Underrepresented Targets* , Proc. 5th SIAM International Conference on Data Mining (2005), Newport Beach, CA, pp. 621-625

[25] D. Cohn, L. Atlas, and R. Ladner, *Improved Generalization with Active Learning*, Machine Learning, Vol. **15**(1994) 201-221.

[26] H. S. Seung, M. Opper, and H. Sompolinsky, *Query by Committee*, Proc. 5th Annual ACM Workshop on Computational Learning Theory (1992), Pittsburgh, PA, pp. 287-294

[27] L. Breiman, *Bagging Predictors*, Machine Learning, Vol. **24**(1996) 123-140.

[28] C. B. Anfinsen, *Principles that Govern the Folding of Protein Chains* , Science, Vol. **181**(1973) 223-230.

[29] H. M. Berman, J. Westbrook, Z. Feng, G. Gilliland, T. N. Bhat, H. Weissig, I. N. Shindyalov, and P. E. Bourne, *The Protein Data Bank*, Nucleic Acids Research, Vol. **28**(2000) 235-242.

[30] A. Bairoch, R. Apweiler, C. H. C. H. Wu, W. C. Barker, B. Boeckmann, S. Ferro, E. Gasteiger, H. Huang, R. Lopez, M. Magrane, M. J. Martin, D. A. Natale, C. O'Donovan, N. Redaschi, and L. S. Yeh, *The Universal Protein Resource (UniProt)*, Nucleic Acids Research, Vol. **33**(2005) 154-159.

[31] B. Rost. *Protein Structure Prediction in 1D, 2D, and 3D* , Encyclopedia of Computational Chemistry, eds. P. von Rague-Schleyer, N. L. Allinger, T. CClark, J. Gasteiger, P. A. Kollman and H. F. Schaefer, John Wiley, Sussex, UK, 1998.

[32] A. K. Dunker and Z. Obradovic, *The Protein Trinity - Linking Function and Disorder* , Nature Biotechnology, Vol. **19**(2001) 805-806.

[33] S. Vucetic, Z. Obradovic, V. Vacic, P. Radivojac, K. Peng, L. M. Iakoucheva, M. S. Cortese, J. D. Lawson, C. J. Brown, J. G. Sikes, C. D. Newton, and A. K. Dunker, *DisProt: a Database of Protein Disorder* , Bioinformatics, Vol. **21**(2005) 137-140.

[34] B. Boeckmann, A. Bairoch, B. Apweiler, M. C. Blatter, A. Estreicher, E. Gasteiger, M. J. Martin, K. Michoud, C. O'Donovan, I. Phan, S. Pilbout, and M. Schneider, *The SWISS-PROT Protein Knowledgebase and Its Supplement TrEMBL in 2003* , Nucleic Acids Research, Vol. **31**(2003), 365-370.

[35] K. Peng, S. Vucetic, P. Radivojac, C. J. Brown, A. K. Dunker, and Z. Obradovic, *Optimizing Long Intrinsic Disorder Predictors with Protein Evolutionary Information* , Journal of Bioinformatics and Computational Biology, Vol. **3**(2005) 35-60.

[36] Z. Obradovic, K. Peng, S. Vucetic, P. Radivojac, C. J. Brown, and A. K. Dunker, *Predicting Intrinsic Disorder from Amino Acid Sequence* , Proteins, Vol. **53(S6)**(2003) 566-572.

[37] S. Vucetic, D. Pokrajac, H. Xie, and Z. Obradovic, *Detection of Underrepresented Biological Sequences Using Class-Conditional Distribution Models* , Proc. 3rd SIAM International Conference on Data Mining (2003), San Francisco, CA, pp. 179-283.

[38] M. Gerstein, *How Representative Are the Known Structures of the Proteins in a Complete Genome? A Comprehensive Structural Census* , Fold Design, Vol. **3**(1998) 497-512.

[39] J. Liu and B. Rost, *Target Space for Structural Genomics Revisited*, Bioinformatics, Vol. **18**(2002) 922-933.

[40] D. Vitkup, E. Melamud, J. Moult, and C. Sander, *Completeness in Structural Genomics*, Nature Structural Biology, Vol. **8**(2001) 559-566.

[41] S. E. Brenner, C. Chothia, and T. J. Hubbard, *Population Statistics of Protein Structures: Lessons from Structural Classifications*, Current Opinion in Structural Biology, Vol. **7**(1997) 369-376.

[42] A. G. Murzin, S. E. Brenner, T. J. Hubbard, and C. Chothia, *SCOP: a Structural Classification of Proteins Database for the Investigation of Sequences and Structures*, Journal of Molecular Biology, Vol. **247**(1995) 536-540.

[43] E. V. Koonin, Y. I. Wolf, and G. P. Karev, *The Structure of the Protein Universe and Genome Evolution*, Nature, Vol. **420**(2002) 218-223.

[44] S. E. Brenner, C. Chothia, and T. J. Hubbard, *Assessing Sequence Comparison Methods with Reliable Structurally Identified Distant Evolutionary Relationships*, Proceedings of the National Academy of Sciences USA, Vol. **95**(1998) 6073-6078.

[45] K. Peng, Z. Obradovic, and S. Vucetic, *Exploring Bias in the Protein Data Bank Using Contrast Classifiers*, Proc. Pacific Symposium on Biocomputing (2004), Hawii, pp. 435-446.

[46] S. E. Brenner, *A Tour of Structural Genomics*, Nature Reviews Genetics, Vol. **2**(2001) 801-809

[47] M. Linial and G. Yona, *Methodologies for Target Selection in Structural Genomics*, Progress in Biophysics & Molecular Biology, Vol. **73**(2000) 297-320.

[48] L. Chen, R. Oughtred, H. M. Berman, and J. Westbrook, *TargetDB: a Target Registration Database for Structural Genomics Projects*, Bioinformatics, Vol. **20**(2004) 2860-2862.

[49] M. Garcia-Hernandez, T. Z. Berardini, G. Chen, D. Crist, A. Doyle, E. Huala, E. Knee, M. Lambrecht, N. Miller, L. A. Mueller, S. Mundodi, L. Reiser, S. Y. Rhee, R. Scholl, J. Tacklind, D. C. Weems, Y. Wu, I. Xu, D. Yoo, J. Yoon, and P. Zhang, *TAIR: a Resource for Integrated Arabidopsis Data*, Functional and Integrative Genomics, Vol. **2**(2002) 239-253.

[50] W. Li, L. Jaroszewski, and A. Godzik, *Clustering of Highly Homologous Sequences to Reduce the Size of Large Protein Database*, Bioinformatics, Vol. **17**(2001), 282-283

[51] A. L. Blum and P. Langley, *Selection of Relevant Features and Examples in Machine Learning*, Artificial Intelligence, Vol. **97**(1997) 245-271.

[52] K. Peng, Z. Obradovic, and S. Vucetic, *Towards Efficient Learning of Neural Network Ensembles from Arbitrarily Large Datasets*, Proc. 16th European Conference on Artificial Intelligence (2004), Valencia, Spain, pp. 623-627.

In: Protein Research Progress
Editor: A.B. Boscoe and C. R. Listow, pp.117-168
ISBN 978-1-60456-663-3

Chapter 4

NEW TECHNOLOGIES AND PHYSIOLOGICAL FUNCTIONAL PROPERTIES OF MILK PROTEINS

***Maria Teresa Bertoldo Pacheco*[1], *Adriane E. Costa Antunes*[2] *and Valdemiro C. Sgarbieri*[3]**

[1]Food Chemistry and Applied Nutrition Center-Institute of Food Technology (ITAL). Av. Brasil, 2880, Campinas, SP, Brazil, Box 139, Zip Code 13070-178

[2]Pos-doc at Dairy Research and Development Center – Institute of Food Technology (ITAL)., Av. Brasil, 2880, Campinas, SP, Brazil, Box 139, Zip Code 13070-178

[3]Food and Nutrition Department (UNICAMP), Campinas/SP, Brazil – Box 6121 CEP 13083-970

ABSTRACT

The present chapter emphasizes the multifunctional properties of milk proteins and the new technologies to obtain biological active proteins and peptides. New approaches have recently been applied in dietary protein science and technology due to improved knowledge regarding the performance of peptides in the modulation of biological systems. Milk proteins have been the object of special attention not only because milk is the first and the most complete food for newborns, but also because it continues to be part of the diet of many adults. Besides the nutrients that are present in milk, this product contains bioactive or biofunctional peptides with effects and activities that resemble those produced by hormones or pharmacological preparations. These peptides result from the hydrolysis of casein and whey proteins during digestion, cheese-making and milk fermentation processes. Reports on immunomodulatory, hypocholesterolemic, hypotensive, antimicrobial, sleep-inducing, epidermal growth factor, transforming growth factor, anticancer, opioid, mineral-binding and bone-formation activities and properties associated with milk proteins have been published in recent literature. Phosphopeptides were found to exert positive effects on the solubility and absorbability of minerals, in addition to being capable of improving bone mineralization. The immunomodulating potential of whey proteins is unquestionable and is in part attributed to their capacity to stimulate glutathione synthesis that promotes the synthesis of immunoglobulins in lymphocytes. Glutathione is necessary for lymphocyte proliferation. Some milk protein hydrolysates carry information concerning the organism from which they were originated. These milk hydrolysates contain opioid peptides having pharmacological similarities to opium (morphine), which, when introduced into the bloodstream, induce an analgesic and sedative effect due to their action on the nervous system. Secondarily, they act on the circulatory system, the passage rate through the digestive tract and the rate of intestinal absorption of foods. Milk whey contains lactoferrin and lactoperoxidase, both of which are known to have antimicrobial and antiviral activity, inhibiting the proliferation of Gram-positive and Gram-negative bacteria, moulds and viruses. Several scientific communications report the inhibition of the hepatitis C virus and *H. pylori* (bacterium that causes gastritis and peptic ulcers). As a nutritive and tasty food, milk remains a constituent of the human diet after childhood, whether in its original form or as dairy products. Besides, this paper discusses the functional properties of whey proteins, the advantages of their use as food supplements in sports nutrition, as well as the possible benefits of the several growth factors present in whey.

INTRODUCTION

In 1838, a Swedish scientist called Jöns Jacob Berzelius suggested the term “protein” after the Greek word “proteios,” which means “primary” or "of the first rank". Since then, many scientific discoveries have been made about these large molecules indispensable for the good functioning of the body's cells, tissues, and organs.

In the past it was believed that the function of proteins was restricted to tissue-formation, while carbohydrates and lipids provided energy to the body. The microelements (vitamins and minerals) were thought of playing a pivotal role in the regulatory function. Another out-of-date concept was the belief that dietary proteins were completely hydrolyzed in the gastro-intestinal tract and that only free amino acids could be absorbed from the gut. The new concept is that macro- and microelements may interact to perform numerous functions in the body. Amino acids and peptides formed in the digestion of natural proteins are absorbed and incorporated (anabolism) into various tissues and organs as body proteins. In addition, dietary proteins and peptides may perform numerous metabolism-modulating functions in various body systems, such as the circulatory and immune systems, transport and absorption mechanisms, in addition to antioxidant and antibiotic effects, thereby contributing to decrease the risk of non-transmittable chronic diseases and improve longevity and quality of life.

Several studies carried out over the past decade have shown that proteins are no longer considered only as nutritional components. Nutritionally, they are a source of energy and amino acids, which are essential for growth and maintenance. Moreover, many dietary proteins possess specific biological properties that make these components potential ingredients of functional or heath-promoting foods (Korhonen *et al*, 1998).

There has been a growing interest in products that naturally contain functional components or are fortified with them. A functional food can be defined as a set of dietary components that affect its host in a target manner exerting effects that may eventually justify certain health claims (Costa *et al.*, 2003).

Recently, much scientific interest has focused on physiologically active peptides derived from food proteins. Bioactive peptides have been defined as specific protein fragments that have a positive impact on body functions or conditions and may ultimately influence health (Kitts and Weiler, 2003). Such peptides may occur naturally in raw food materials exerting their physiological action upon enzymatic hydrolysis *in vitro* or *in vivo*. The *in vivo* production of peptides from casein stems from the action of either the enzymes secreted by the gastrointestinal tract and those released by the intestinal microflora (Sabikhi and Mathur, 2001). For example, it has become clear that dietary proteins are a source of biologically active peptides that act as regulators of physiological processes. Scientific works show that these bioactive peptides can translocate through the intestinal epithelium and reach peripheral tissues through systemic circulation with the ability of performing specific functions at a local (gastrointestinal tract) and systemic level (Baró *et al.*, 2001).

To exert physiological effects *in vivo*, bioactive peptides must be released during intestinal digestion and must then reach their target sites at the luminal side of the intestinal tract or after absorption in the peripheral organs (Pihlanto-Leppälä, 2001). It should be emphasized that one of the first requirements for a physiological effect is that bioactive peptides, after being released from milk proteins, they need to be resistant to further attack by digestive proteinases (Mullally *et al.*, 1997a). According to Hernández-Ledesma *et al.* (2004), resistance of theses peptides to gastrointestinal enzymes is poorly documented. Di- and tripeptides can be easily absorbed in the intestine, however, it is not clear whether larger bioactive peptides containing more than three amino acids are absorbed from the intestine and reach the target organ (Shah, 2000).

Individual amino acids considered to have special physiological activity are arginine, glutamine, histidine, lysine, taurine, tyrosine and tryptophan (Marshall, 1994). In conventional industrial manufacture of various foodstuffs, the proteins contained in raw materials are readily subjected to alterations with regards to their functional or biological properties. pH changes and certain chemical treatments affect functional properties by modifying one or more amino acids. For example, acidic treatments destroy glutamine and asparagine, whereas alkaline treatments partially destroy cystine, serine and threonine (Finot, 1997). The heat treatment can affect the activity of bioactive proteins. At a standard pasteurization temperature of 72°C for 15 s, the bioactive whey proteins retain most of their activity (Finot, 1997).

TECHNOLOGY TO OBTAIN BIOACTIVE PROTEINS AND PEPTIDES

The knowledge of bioactive peptides has steadily increased since 1979 (Pihlanto-Leppälä, 2001). Industrial-scale technologies suitable for the commercial production of bioactive milk peptides have been developed and launched recently (Kornen and Pihlanto, 2006).

Achievements in separation techniques in the dairy industry and enzyme technology offer opportunities to isolate, concentrate or modify milk compounds, so that their application in functional foods, dietary supplements, nutraceuticals and medical foods has become possible (Steijns, 2001).

Although often considered a "perfect" food, milk hypersensitivity may develop. The immune-mediated hypersensitivity includes skin, gastrointestinal and/or respiratory symptoms (Pelto *et al.*, 1999). If the allergic reaction occurs against caseins, albumins or β-lactoblobulin, it is necessary to avoid the consumption of milk (Riccio, 2004).

Milk protein can be hydrolized to obtain bioactive peptides or to lower the allergenicity of dietary proteins, particularly for children fed on formula diets and who are sensitive to milk proteins.

Pathologies that impair the protein intake are: inborn errors of metabolism such as phenylketonuria and cystic fibrosis, patients with impaired gastrointestinal function, and specific stomach, renal or hepatic failures.

The heat treatment of the milk or whey is a key factor. High temperatures (such as 121°C for 20 min) are better than pasteurization at 75°C for 15s to reduce the allergenicity of milk proteins (Riccio, 2004), however, such treatments affect the properties of the bioactive whey proteins (Finot, 1997). So, the manufacture processes of dairy products should be specifically designed to achieve the desired properties.

The production of extensive protein hydrolysates by sequencial action of endopeptidases and exoproteases coupled with the development of post-hydrolysis procedures is considered the most effective way to obtain protein hydrolysates with defined characteristics (Clemente, 2000).

Hydrolysis of proteins can be accomplished by enzymes, acids or alkalis. According to Clemente (2000), acid and alkaline hydrolysis tend to be processes difficult to control, in addition to yielding products with reduced nutritional qualities and forming potentially toxic substances. On the other hand, enzymatic hydrolysis is developed under mild conditions of pH (6-8), and temperature (40-60°C), avoiding the extremes usually required for chemical and physical treatments thus minimizing side reaction.

According to Costa *et al.*, (2003), the most common way to produce bioactive peptides is to submit the protein to limited hydrolysis using pancreatic enzymes, mainly trypsin. However, other enzymes and combinations of proteases have also been used. Another example is alcalase, an industrial alkaline protease that is very stable in organic solvents and serves as a catalyst for the resolution of N-protected amino acids in aqueous solution and organic solvents, and is used to prepare optically pure peptides (Costa *et al.*, 2003).

Commercial production of bioactive peptides from milk proteins has been limited by a lack of suitable large-scale technologies. There is a need to develop technologies that retain or even enhance the activity of bioactive peptides in food systems. Also it is essential to study the optimum utilization of such peptides during passage through the gastrointestinal tract (Korhonen and Pihlanto, 2003).

According to Korhonen and Pihlanto (2006), ultrafiltration is routinely employed to enrich bioactive peptides from protein hydrolysates. Enzymatic hydrolysis can be performed through conventional batch hydrolysis or continuous hydrolysis using ultrafiltration membranes. The traditional batch method has several disadvantages, such as the relatively high cost of the enzymes and their inefficiency as compared with a continuous process (Korhonen and Pihlanto, 2006).

According to Ellegard *et al.* (1999), it is possible to establish a robust process scale production of tryptic casein phophopeptides from caseinate with an excellent yield of 20% (w/w) product and a process output of 40Kg sodium enriched CPP (Na-CCP) or calcium enriched (Ca-CPP). The unit operation process scale chromatography permitted eminent control over the process in terms of producing exclusively phosphorylated peptides in order to ensure the high quality of the product with the molar serine/phosphate ratio of 1. The unique complete solubility of the isolated CPP in the tested pH range pH 0.5 to pH 9.0, the slight sodium taste of Na-CPP and the neutral taste of Ca-CPP gives the product great potential in various applications (Ellegard *et al.*, 1999).

Commercial interest is increasing in lactoferrin and other components with bactericidal activity as a result of a rising consumer demand for "natural" foods and cosmetics that contain no man-made preservatives (Tomita *et al.*, 1991). Industrial-scale processes are already established for the recovery of bovine lactorrefin from cheese whey in a non-denatured functional form (Tomita *et al.*, 1991).

NEW TECHNOLOGIES

The functionality of foods is based on bioactive proteins and peptides, which may be contained naturally in the foods but usually require appropriate technologies in order to concentrate or isolate them and optimize the desired beneficial properties. To this end, it is often necessary to develop and apply novel technologies. The processing technologies applied have focus specifically on two aspects: (a) maximum retention of activity during processing and storage and (b) delivery of the desired bioactive components to target sites in the body (Korhonen, 2002).

New nutritional concepts require the development and application of technologies that can preserve or improve the overall quality of the raw materials and their physicochemical functionality; maintain or enhance the nutritional and physiological performance of the end-product, and increase the safety of the product.

The advancement and diversification of new concepts of protein utilization in human food was possible due to technological advancements in the techniques of protein fractionation and purification of bioactive materials.

In the case of milk proteins, the first step for the concentration of the proteins was to devise official methods to eliminate low molecular mass components, such as water, lactose, fat, minerals and vitamins. For this purpose, techniques like microfiltration, ultrafiltration, nanofiltration and reverse osmosis have been applied. For protein fractionation, the most used commercial processes are (a) selective precipitation, (b), membrane techniques, (c) selective adsorption, (d) selective elution (Etzel, 2004).

Selective precipitation involves adjusting the solution's physical properties to promote insolubity. Lowest solubility of the proteins is typically found at low ionic strength. Aggregates are normally formed.

Membrane filtration has been based solely on differences in molecular mass. Using membranes for fractionation of proteins is possible when proteins present great differences in molecular mass. Sometimes combined differences in charge and size are required. Membrane development has gone through three phases. The first generation of membranes consisted of highly hydrophobic cellulose derivatives. The second generation of microfiltration membranes were constructed from polypropylene, polysulfone and polyvinylidene fluoride. The third generation consisted of very thin and selective membranes made from aluminum oxide or zirconium and supported by coarse graphite or ceramic material (Shuryo and Modler, 2000). Mixtures of peptides can be fractioned with nanofiltration membranes (NF) and the separation mechanisms are based on a molecular sieve effect and\or charge effect depending on the membrane type and the composition of the feed phase. This kind of separation is strongly influenced by physicochemical parameters such as the pH and ionic strength of the feed phase and the concentration of amino acids (Pouliot *et al,* 1999).

Selective adsorption: a single purified protein is produced in conjunction with the use of immobilized selective compounds. For the adsorption process of α-lactoalbumin, Gurgel *et al.* (2000) used immobilized hexapeptide, while Noppe *et al.* (1998) used immobilized retinal. The cost of manufacture must be borne by revenues generated from that single protein product.

Selective elution: all the proteins in a mixture are trapped simultaneously onto the adsorbent, rinsed free of contaminants, and then eluted one-by-one to manufacture many different purified proteins. The process is operated at a high flow rate and uses an adsorvent and buffers that are inexpensive and food grade. With selective elution, once the proteins are bound to the adsorbent, simply changing the elution buffers causes separation of protein fractions. Upon completion of elution, the column is regenerated and the cycle can be started all over again.

Today, several whey protein concentrates and whey protein isolates are standard commercial products obtained by various membrane filtration techniques. Further fractionation of whey into individual proteins fractions, such as α–lactalbumin, β-lactoglobulin, immunoglobulins, lactoperoxidase, lactoferrin and glycomacropeptide, has been made possible by the development of industrial chromatographic and gel or membrane filtration methods. Improved technologies have been introduced using combined techniques based on enzymatic hydrolysis and ultrafiltration processes, in addition to new technologies, such as radial flow chromatography columns, i.e. the Sepralac TM-process, ion-charged chromatographic membranes and beads in columns and expanded bed absorption (EBA) chromatography techniques (Olander *et al.,* 2001).

On the other hand, there has been a growing interest over recent years in bioactive peptides derived from both caseins and whey proteins. Similarly to intact proteins, such peptides have been found to exert various bioactivities both *in vitro* and *in vivo.* However bioactive peptides are inactive within the sequence of the parent proteins and can be released in three ways: (a) by enzymatic hydrolysys using digestive enzymes; (b) by fermentation of milk using proteolytic starter cultures; and (c) through the action of enzymes derived from proteolytic microorganisms (Sannier *et al.,* 2000).

A more complex problem is the separation and purification of peptides. Since the differences in physicochemical properties of these peptides are often small, a separation technology which can discriminate small differences in charge, size and hydrophobicity of species is needed (Ahmed *et al.* 1998).

The most common way of producing bioactive peptides has been a partial enzymatic hydrolysis by trypsin or pepsin. The peptides in the hydrolysate can be fractioned and enriched using different methods, such as selective precipitation, ultrafiltration or ion exchange chromatography, but the yield and purity of the desired peptide fractions have proven fairly low in many studies (Korhonen *et al.*, 1998).

Chromatography processes are not widely used at industrial scale due to their low productivity and high costs. Nanofiltration is a promising technique that separates solutes based on solute charge and size. The ultrafiltration technique, using membranes with different cut-off values, has been applied successfully for the enrichment of antihypertensive and antibacterial peptides from whey protein hydrolysates. Ultrafiltration membrane reactors have been used for continuous extraction of permeates enriched with antithrombotic peptides (Pillanto-Leppälä *et al.,* 2000; Bouhallab *et al.*, 1992; Wijers *et al.*, 1998).

MULTIFUNCTIONAL PROPERTIES OF MILK PROTEINS

Milk has become a common component of the diet of many people after childhood probably because milk is a product that is at the same time palatable, well absorbed by the body, inexpensive, easily found, well tested over time, and finally because it is a food which reminds us of our first life experiences (Riccio, 2004).

Milk proteins appear to be an exciting link between nutrition, dietetics, and therapy (Fiat *et al.*, 1993). The bioactive peptides that result from the hydrolysis of casein and whey proteins, or that are generated during digestion, cheese-making and milk fermentation processes, can act in several ways, including promoting immunomodulation, stimulating bone-formation, and improving an array of other body functions as a result of their opioid, antimicrobial and antiviral properties.

Today, consumers can expect more than just nutrition from the intake of milk proteins. Milk contains a variety of bioactive compounds with special properties associated with the development, growth, and survival of infants beyond those provided by nutrition alone (Schanbacher *et al.*, 1998).

The studies reported in this chapter indicate the possibilities for the future development of special foods and food supplements capable of helping people to prevent diseases.

In addition to proteins and peptides, milk provides immunoglobulins, binding or carrier proteins, growth factors and antimicrobial agents (Fox and McSweeney, 1998). The caseins are known to exhibit biological activity, such as carrying of calcium, zinc, copper, iron and phosphate ions in the body (Korhonen *et al.,* 1998). Whey proteins in the form of concentrates or isolates, and their hydrolysates act as immunostimulants, inhibitors of pathogenic microorganisms and some viruses, including HIV and hepatitis C (Sgarbieri, 2004). Moreover, whey proteins play an important role in the protection against the development of tumors (particularly colon cancer and protection of the gastric mucosa against ulceration caused by aggressive ulcerogenic agents) and act on mechanisms involved in cardiovascular protection against degenerative disease (Sgarbieri, 2004). Some of the bioactive peptides obtained from whey proteins include α-lactorphin, β-lactorphin, albutensin A, and β-lactotensin (Shah, 2000).

A variety of naturally formed bioactive peptides have been found in fermented dairy products, such as yoghurt, sour milk and cheese (Korhonen and Pihlanto, 2006). It has been suggested by Meisel (2004) that the proteolytic system of lactic acid bacteria can contribute to the release of bioactive peptides. Many industrially utilized dairy starter cultures are highly proteolytic. Bioactive peptides can, thus, be generated by the starter and non-starter bacteria used in the manufacture of fermented dairy products (Korhonen and Pihlanto, 2006). *In vitro*, the purified cell wall proteinase of *Lactococcus lactis* was shown to liberate oligopeptides from β- and α-caseins which contain amino acid sequences present in casomorphins, casokinines, and immunopeptides. The sequences of practically all known biologically active peptides can also be cleaved by peptidases from lactic acid bacteria (Meisel and Bockelmann, 2004).

Bioactivities have been found in zein peptides, glycinin peptides, myofibril peptides, etc. However, milk proteins are the most important source of bioactive peptides (Marshall, 1994).

The principal physiological functionality of milk-derived bioactive peptides is summarized in Figure 1 and examples of bioactive peptides are shown in Table 1.

Table 1. Bioactive peptides obtained from milk proteins hydrolysis

Protein	Bioactive peptide	Described bioactivity	Reference
α and β-casein	caseinfosfopeptide	Mineral bioavailability	FitzGerald (1998)
α and β-casein	caseinfosfopeptide	Inmunomodulation	Otani *et al.* (2000)
α and β-casein	caseinfosfopeptide	Antioxidant property	Kitts (2005)
κ-casein	glycomacropeptide	Modulation of intestinal microbiota	Fox and McSweeney (1998)
κ-casein	casopiastrin	Antithrombotic activity	Fiat *et al.* (1993)
β-casein	β-casomorphin	Opioid agonist	Meisel and FitzGerald (2000)
α_{S1}-casein	α_{S1}-casein-exorphin	Opioid agonist	Meisel and FitzGerald (2000)
α_{S1}-casein	α_{S1}-casomorphin	Anti-cancer	Kampa *et al.* (1996)
α-lactalbumin	α-lactorphin	Opioid agonist	Meisel and FitzGerald (2000)
α-lactalbumin	α-lactorphin	Antihypertensive	FitzGerald *et al.* (2004)
β-lactoglobulin	β-lactorphin	Opioid agonist	Meisel and FitzGerald (2000)
Serum albumin	serorphin	Opioid agonist	Meisel and FitzGerald (2000)
κ-casein	casoxin	Opioid antagonist	Meisel and FitzGerald (2000)
Lactoferrin	lactoferricin	Antimicrobial	Jones *et al.* (1994)

EFFECTS ON GASTROINTESTINAL SYSTEM

The gastrointestinal tract is the first system to come in contact with the bioactive peptides introduced into the body by oral intake of certain proteins. The stomach and gut are target organs for some of those peptides, more specifically glycomacropeptide and casein phosphopeptides.

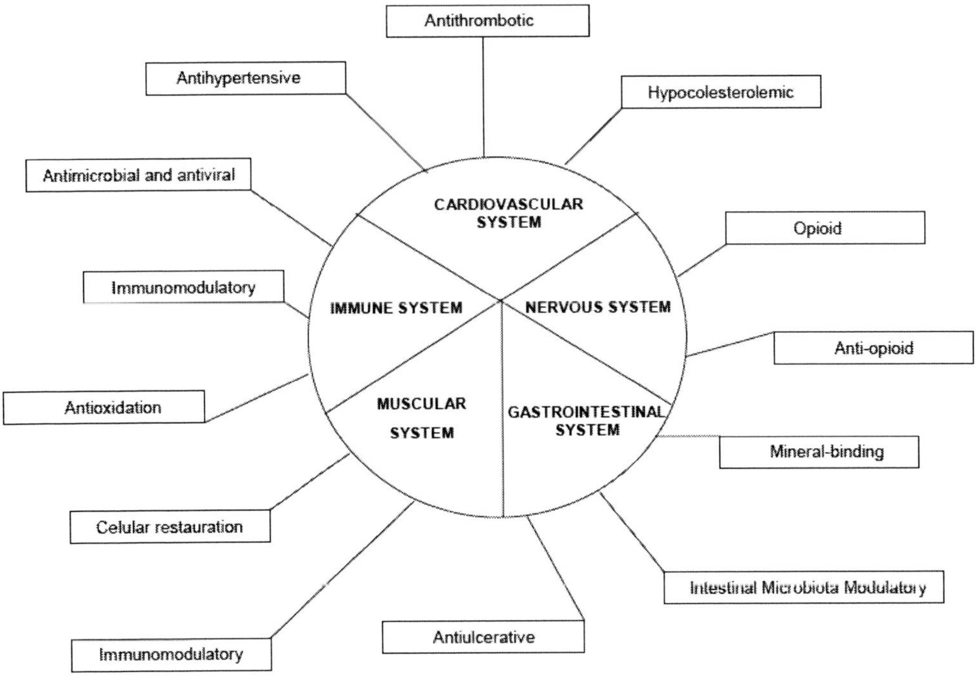

Figure 1. Physiological functionality of milk-derived peptides.

Mineral Binding Activity

The term caseinophosphopeptide appears to have been first introduced in 1950 to describe casein-derived phosphorylated peptides which enhanced vitamin D independent bone calcification in rachitic infants (FitzGerald, 1998, Scholz-Ahrens and Schrezenmeir, 2000).

The negatively charged side chain, in particular the phosphate groups, of those amino acids represent the binding sites for minerals (Meisel, 1998). Caseinophosphopeptides (CPPs) have been shown to bind to such macroelements as calcium, magnesium and iron as well as to oligoelements such as zinc, barium, chromium, nickel, cobalt and selenium (FitzGerald, 2003, Silva and Malcata, 2005). Numerous claims have been made with respect to the application of CPPs to enhance mineral bioavailability. Adequate supply of Ca^{2+} is required for bone development, bone recalcification following fracture, in the prevention and treatment of osteoporosis and during treatment of rickets. The bioavailability of dietary Fe^{2+}, Zn^{2+} and Mg^{2+} increased by the presence of CPPs, combat anemia, acrodermatitis enteropathica, hypogonadism, stress and so on (FitzGerald, 1998).

Consumption of high concentrations of Ca^{2+} in early life contributes to the development of maximal bone density, which in turn can prevent osteoporosis in later life. The high bioavailability of Ca^{2+} from milk and dairy products has, in part, been attributed to the production of caseinophosphopeptides (CPPs) which have different levels of phosphorylation and are produced *in vivo* following digestion of α_{s1}-, α_{s2}- and β-casein by the action of gastrointestinal proteinases (FitzGerald, 1998).

Bovine milk has a unique property of solvating calcium (1.2 $g.L^{-1}$) and inorganic phosphate (2.0 $g.L^{-1}$) in concentrations above the solubility of calcium phosphate (Ellegard *et al.*, 1999). This is possibly due to the presence of colloidal particles called micelles, where α_{s1}-, α_{s2}- and β-casein are linked by phosphorilated serine residues.

Milk fermentation increases calcium absorption through increased production of CPPs during food processing, which increases the production of more soluble calcium compounds (Meisel and Bockelmann, 1999).

Iron deficiency, a major worldwide nutritional problem, can be reduced by CPPs; in fact, binding of Fe to phosphopeptides prevents formation of high-molecular-weight ferric hydroxides, which are poorly absorbed (Silva and Malcata, 2005). Table 2 shows examples of caseinophosphopeptides derived form caseins.

CPPs produced during the digestion of caseins enhance the bioavailability of Ca^{2+} in the diet specifically by increasing the solubility of Ca^{2+} in the alkaline environment of the small intestine where maximal passive absorption of dietary Ca^{2+} is thought to take place (FitzGerald, 1998).

There are three modes of Ca^{2+} absorption in man, i.e. a saturable, vitamin D-dependent transport system located in the duodenum and jejunum and a passive, vitamin D-independent system located in the ileum and the distal small intestine (Bronner, 1999). According to FitzGerald (1998), several factors affect the absorption of Ca^{2+} from the diet including phytate, oxalate, phenol compounds, lactose, phosphorus and obviously vitamin D.

Divergent results have been reported as to the improvement of mineral absorption in the presence of CPP.

Caseinophosphopetides formed by proteolytic digestion of casein have been shown to enhance Ca absortion in studies with rats (Tsuchita *et al.*, 1993). The effect of dietary Ca-bound casein phosphopeptides (CaCPP) on the bones of aged ovariectomized rats was studied as a model for post-menopausal bone loss.

Table 2. Caseinophosphopeptides derived from αs_1-casein and β-casein

Protein	Amino acid sequence*	Activity
α_{s1} casein		
43-58	DIGPESTEDQAMEDIK	mineral transport
59-79	QMEAQPISSSQQIVPNSVEQK	mineral transport
106-119	VPQLQIVPNSAEER	mineral transport
66-74	SSSQQIVPN	mineral transport
β casein		
1-28[a]	RELEELRVPGEIVESLSSSEESITRINK	mineral transport
1-25	RELEELRVPGEIVESLSSSEESITR	mineral transport
1-28[b]	RELEELRVPGEIVESLSSSEESITRINK	mineral transport

* Amino acids: A – alanine; D – aspartic acid; E – glutamic acid; G – glycine; I – isoleucine; K – lysine; L – leucine; M – methionine; N – asparagine; P – proline; Q – glutamine; R – arginine; S – serine; T – threonine; V – valine.

[a] casein proteolysis; [b] isolated from intestine. Adapted from Sgarbieri (1999).

The relative bone mineral density (BMD) decreased approximatly 18% in the group of nonovariectomized rats fed on a diet without CPP, but with recommended levels of Ca. However, rats fed on Ca-CPP complex or $CaCO_3$ and CPP did not show this BMD decrease. On the other hand, Bennett *et al.* (2000), observed that Ca absorption from a meal was unaffected by inclusion of 100g CPP/Kg meal and was significantly reduced by CPP added at levels of 200, 350 and 500g/Kg in rats.

Human studies also presented controversial results. Hansen *et al.* (1995), investigated the effect of CPP on calcium and zinc absorption from infant food. Twenty-two men and women participated in the study. Absorption was calculated from measurements of whole-body retention of radioisotopes. It was observed that CPP addition improved both Ca and Zn absorption from rice-based cereal, while no effect was seen from whole-grain cereal. Heaney *et al.* (2006), studied the effect of CPP on absorbability of co-ingested Ca in normal postmenopausal women. The mean quotient of absorption of the Ca in meal with/without CPP addition was greater than 1.0, but statistically insignificant. However, when analysis focused on women with low absorption values, CPP was associated with significantly better absorption. Narva *et al.* (2003), examined the effect of added CPP in milk and fermented milk on acute calcium metabolism in postmenopausal women. Neither control milk nor fermented milk enriched with 1g of CPPs influenced Ca metabolism in healthy postmenopausal women within the first six hours following ingestion. This result does not, however, exclude the possible effect of CPPs in vitamin D deficient people, who have an increased need of Ca, or in those with poor Ca absorption levels.

The use of CPPs in hypertension treatment has also been proposed by FitzGerald (1998). A considerable body of evidence is available to demonstrate an inverse relationship between dietary Ca^{2+} intake and blood pressure. Ca^{2+} supplementation can lower blood pressure in both normotensive and hypertensive humans.

Milk and dairy products are recognized as low cariogenic potential foods, but they are also claimed to be cariostatic. The prevention of dental caries occurs due to the reduction in the adhesion of cariogenic mutans streptococci *in situ* and seems to reduce colonization (Johansson, 2002).

Tooth decay or dental caries arises due to demineralization of teeth brought about by organic acids produced by dental plaque odontopathogenic bacteria during the fermentation of dietary sugars (FitzGerald, 1998). The proposed anticariogenic mechanism of CPPs relates to their ability to localize high levels of amorphous Ca^{2+} phosphate at the tooth surface and as a consequence depress demineralization and enhance remineralisation of tooth enamel. By sequestering antibacterial metals such as Zn, Cu, Sn, Ag, Al, CPPs may also have an antimicrobial effect on odontopathogenic bacteria (FitzGerald, 1998).

The ability of CPP to reduce caries activity was investigated by use of specific-pathogen-free rats inoculated with *Streptococcus sobrinus* (Reynolds *et al.*, 1995). The animals consumed a defined cariogenic diet free of dairy products. Solutions (100μL) of CPP (0.1, 0.2, 0.5 1.0% w/v) were applied to the animals' molar teeth twice daily. The CPP significantly reduced dental caries activity in a dose-dependent fashion, with 1.0% CPP producing 55% and 46% reductions in smooth surface and fissure caries activity, respectively, being similar to that of 500 ppm fluoride.

In general, anticariogenic milk-derived bioactive peptides prevent dental lesions in multiple ways, including bacterial inhibition, competitive exclusion on enamel binding of bacteria, improved buffering capacity in the pellicle surrounding teeth, reduced enamel demineralization, and enamel remineralization (Aimutis, 2004). CPP and GMP (glycomacropeptide), have been patented for use in common personal hygiene products to prevent dental caries. Both compounds act as growth inhibitors of cariogenic bacteria, such as *Streptococcus mutans* and other species and, additionally, CPP forms nanoclusters with amorphous calcium phosphate (AMP) at the tooth surface to provide a reservoir of calcium and phosphate ions to maintain a state of super saturation with respect to tooth enamel (Aimutis, 2004).

Modulation of Intestinal Microbiota

One of the most intriguing functions of milk is its ability to modulate the intestinal flora by stimulating the growth of *Bifidobacteria* ssp. – a health promoting species - in the gut and, at the same time, inhibit the proliferation of pathogenic bacteria. It is critical that a beneficial and protective microbiota be established quickly and colonize the sterile gut of the newborn infant (Schanbacher *et al.,* 1998).

Human milk is said to have a "bifidus factor". Breast-feeding encourages the growth of bifidobacteria in the gut, whereas formula-fed infants have a more complex microflora that contains bifidobacteria, bacteroides, clostridia and streptococci (Isolauri, 2001). As a consequence, breast-fed infants are more resistant to gastroenteritis than bottle-fed babies. This is undoubtedly a multifactorial phenomenon, including better hygiene, more appropriate milk composition, several antibacterial systems (especially immunoglobulins, lysozyme, lactoferrin, vitamin-binding proteins and lactoperoxidase) (Fox and McSweeney, 1998). Babies fed exclusively with breast milk have lower intestinal pH - a consequence of both the lower buffering capacity of this milk and its lower content of protein and phosphate (Collingnon *et al.*, 2004; Fox and McSweeney, 1998).

Milk, and especially colostrum, contains a family of oligosaccharides comprised of lactose or *N*-acetyllactosamine oligomers with side glycosylations containing fucose or sialic acid that are potent promoters of bifidobacteria growth (Newburg, 1996, Brody, 2000).

Moreover, the kappa-casein glycomacropeptide is an important biological building block used in the construction of bacterial cell walls, stimulating the growth of *Bifidobacteria* ssp.

With age, the population of bifidobacteria has a strong tendency to decrease. In infants bifidobacteria represent 80% of the viable faecal bacteria versus 20% in adults (Marteau *et al.*, 2002). The reduction of *Lactobacilli* and *Bifidobacteria* species in the gut can facilitate the growth of pathogenic microorganisms. As a consequence, gastrointestinal disorders may occur, including constipation, diarrhea, urogenital tract diseases, Crohn's disease, irritable bowel syndrome, ulcerative colitis, pouchitis, colon cancer and other.

The human intestinal tract is no longer seen as a mere digestive tube, but it has become increasingly recognized as one of the main targets of immune responses and the inherent microbiota plays an essential role in promoting health or protecting against disease (Antunes et al., 2007).

Today, health promoting strains are called probiotics and many new dairy products have been developed with addition of theses cultures.

Antiulcer Activity

It is accepted that peptic ulcerogenesis (gastric and duodenal) results from an imbalance between infections agents like *Helicobacter pylori* or aggressive chemical agents versus protective substances such as mucus, bicarbonate, prostaglandins (PGE_2 and PGI_2), sulfhydryl compounds such as proteins, glutathione and others, as well as the blood flow to the mucosa (Abdel-Salam *et al.*, 2001, Allen *et al.*, 1993).

A new biological function of milk proteins has been proposed by Matsumoto *et al.* (2001). This group verified antiulcer activity as an active component of cow's milk protein, and suggested that α-lactalbumin may serve to protect against gastric mucosal injury, in part through endogenous prostaglandin synthesis.

Sulfhydryl substances such as cysteine, glutathione, and others can protect the mucosa by acting as antioxidants, free-radical scavengers and as regulators of the integrity and functioning of the cell membranes. Glutathione in conjunction with some prostaglandins seem to be the main substances providing protection against free radicals and reduce the influence of other aggressive agents on the gastric mucosa (Szabo *et al.*, 1992; Pacheco et al., 2006).

Mezzaroba *et al.* (2006), investigated the capacity of α-lactalbumin to protect the rat stomach against ulcerative lesions induced by administering indomethacin and absolute ethanol. They confirmed the protection of the stomach mucosa by whey protein and whey protein hydrolysates. Investigations on the mode of action of whey protein in gastric mucosa protection suggest that stimulation of prostaglandin synthesis and mucus production, and the presence of naturally occurring sulphydryl substances, as well as sulphydryl groups in α-lactoalbumin, are all factors that may actively participate in protecting the gastric mucosa against ulcer formation.

Effects on Cardiovascular System

Bioactive peptides originated from milk proteins may act synergistically on the health of the cardiovascular system, lowering the arterial blood pressure, reducing blood cholesterol levels and the risk of thrombosis. These factors, alone or combined, act to reduce the risk of clinical cardiovascular disease (CCD) and enhance the health of the cardiocirculatory system.

Peptides with Angiotensin I - Converting Enzyme Inhibition Activity

Hypertension is the major controllable risk factor associated with cardiovascular disease (CVD), events such as myocardial infarction, stroke, heart failure, and end-stage diabetes (Seppo *et al.* 2002, FitzGerald *et al.*, 2004).

Inadequate calcium intake may be a greater risk factor for hypertension than excess sodium intake. Epidemiological evidence implies that the consumption of milk and dairy products is inversely related to blood pressure (BP) and risk for hypertension. This effect is related not only to calcium, but also to other minerals, e. g. potassium and magnesium, and to the presence of peptides which have angiotensin converting enzyme (ACE) inhibitory activity (Nurminen *et al*, 2000).

Blood pressure regulation and hypertension (when it cannot be reversed), is partially dependent on the renin-angiotensin system; renin acts on angiotensinogen to form angiotensin I, which is further converted into the active peptide hormone angiotensin II - a potent vasoconstrictor - by the angiotensin-converting enzyme (Silva and Malcata, 2005).

Angiotensin I-converting enzyme (ACE - petidyl dipeptide hydrolase, EC 3.4.15.1), is one of the main molecules responsible for controlling the blood pressure due to its action on the formation of angiotensin II and its role in the degradation of bradykinin, a vasodilator (Ariyoshi, 1993, Quirós *et al.*,2005). Thus, inhibition of ACE results in a lowering of blood pressure. ACE is present in a large number of tissues including plasma, kidney, lungs and brain (Pihlanto-Leppälä *et al.*, 2000). This enzyme is present in the plasma and most of the organs of the body including the brain, heart, lungs, kidney, and reproductive systems (FitzGerald *et al.*, 2004). It exists under two isoforms: a somatic form (a chain of 1277 amino acids in mature human ACE) found in endothelial and neuronal cells and a smaller testicular form (Tauzin *et al.*, 2002). The somatic form contains an amino acid sequence and a carboxyl domain that exhibit 60% homology, suggesting gene duplication; these two domains have their own specific catalytic site and act independently of each other (Tauzin *et al.*, 2002).

Inhibitors of ACE (Captopril, Benazepril, Enalapril, Lisinopril, etc.) are used in hypertension therapy (Tauzin *et al.*, 2002). Various side effects are associated with the use of ACE inhibitory drugs in the control of blood pressure including hypotension, increased potassium levels, reduced renal function, cough, angioedema, skin rashes, and fetal abnormalities (FitzGerald *et al.*, 2004). Food-derived peptides are considered to be milder and safer and to have fewer side effects than the drugs currently used in the treatment of hypertension (Pihlanto-Leppälä *et al.*, 2000). As a consequence, whey protein-derived ACE inhibitory peptides/hydrolysates may find application as a nutraceutical in various "physiological functional foods" (Mullaly *et al.*, 1997b). Casein hydrolisates also may originate ACE inhibitory peptides.

The inhibition of ACE is one of the most comprehensively studied physiological effects of biologically active peptides (Hernández-Ledesma *et al.*, 2004). However, once the ACE-inhibitory peptides are released by food processing they have to be able to survive in the gastrointestinal system, be absorbed and reach the circulatory system in an active form (Hernández-Ledesma *et al*, 2004).

Some of the food-derived tripeptides with ACE inhibitory activity have been reported to be hypotensive when intravenously injected or even orally administered (Satake *et al.*, 2002). Angiotensin-converting enzyme inhibitory peptides have been identified and isolated from a fairly large number of food proteins including: gelatin, bovine casein, sardines, tuna, porcine glyceraldehydes-3-phosphate dehydrogenase (GAPDH), bonito, fig tree latex, α-zein, porcine plasma, soy bean, and antarctic krill muscle (Ariyoshi, 1993).

Hydrolysates of whole milk protein, caseinates, whey proteins, and enriched individual milk protein fractions are potentially good sources of ACE inhibitory peptides (FitzGerald *et al.*, 2004). Peptides showing ACE inhibitory activity were found in the milk proteins α-lactoalbumin, β-lactoglobulin and serum albumin. Whey hydrolysates show ACE-inhibitory activity after proteolysis with different digestive enzymes and several active peptides were identified. For example, α-lactorphin released by gastric or pancreatic hydrolysis from α-lactoalbumin possess ACE-inhibitory activity.

Some peptides with high inhibitory activity on ACE *in vitro* have no blood pressure lowering effects, whereas some peptides with low inhibitory activity on this enzyme *in vitro* have such effect (Li *et al.*, 2004). For example: peptides with potent *in vitro* ACE inhibitory activity such as α_{s1}-casein f(23-27) and α_{s1}-casein f(104-109) were subsequently shown to have no hypotensive effects *in vivo* (Maruyama *et al.*, 1987; Maeno *et al*, 1996). This was presumably due to degradation of inactive fragments during oral ingestion (FitzGerald *et al.*, 2004).

Satake *et al.* (2002), investigated the mechanism for the intestinal transport of tripeptides with ACE inhibitory activity, using monolayer-cultured human intestinal Caco-2 cells which express many enterocyte-like functions including the peptide transporter (PepT1)-mediated transport system. Their findings indicated that the ACE-inhibitory tripeptide Val-Pro-Pro (VPP) could be transported across the Caco-2 cell monolayer in an intact form. The mechanism for this transport was mainly *via* paracellular diffusion, and transcytosis and transporter-mediated transport are not likely to play a major role in the transepithelial transport of the "intact" tripeptide, although a small portion of the intact tripeptide could show transcellular transport *via* PepT1 (Satake *et al.*, 2002).

Masuda *et al.* (1996) demonstrated that two antihypertensive tripeptides Ile-Pro-Pro (IPP) and Val-Pro-Pro (VPP) were absorbed directly without being decomposed by digestive enzymes, reached the abdominal aorta, inhibited the ACE and finally showed antihypertensive effect in naturally hypertensive rats.

ACE inhibition is measured by the concentration of substance needed to inhibit 50% of original ACE activity (IC_{50}). As a conclusion, lower IC_{50} value indicates higher inhibitory power of the substance.

A novel angiotensin-I-converting enzyme inhibitory peptide from tryptic hydrolysis of bovine β-lactoglobulin was identified by Mullally *et al.* (1997a). The fraction of β-lactoglobulin which had a degree of hydrolysis value of 6.52%, gave an ACE inhibition index of 84.3%. Unhydrolysed β-lg gave an ACE inhibition index value of less than 10%. The peptide material eluting between 20 and 25% acetonitrile from the solid-phase extraction of hydrolysed β-lg had an ACE inhibition index of 93.6% and an IC_{50} value of 159.8mg.L^{-1}. This peptide f(142-148) contains a proline residue, a common characteristic of several food protein-derived ACE inhibitory peptides.

Nurminen *et al.* (2000), investigated the effect of subcutaneous administration of α-lactorphin on blood pressure and heart rate. They concluded that α-lactorphin dose-dependently decreased BP in spontaneously hypertensive rats (SHR). The lowest effective dose was 10 μg/kg. The maximal reductions in systolic blood pressure (was 23±4 mm Hg, $P<0.05$ vs saline) and in diastolic blood pressure (17±4 mm Hg, $P<0.05$) were obtained after α-lactorphin 100 μg/kg, and no further reductions were observed at a dose of 1 μg.kg^{-1}. The mechanism of the BP-lowering effect of α-lactorphin seems to be due to an interaction with opioid receptors and was not likely to be attributable to its individual constituent aminoacids tyrosine, glycine, leucine or phenylalanine, because the single amino acids failed to influence BP in SHR. The time-course of the blood pressure response after α-lactorphin 10 μg/kg was similar to 100 $\mu g.kg^{-1}$, whereas the maximal decrease in BP was obtained somewhat earlier (within 30-50 min) after the largest dose of α-lactorphin ($1mg.kg^{-1}$). According to these authors, the heart rate was not significantly influenced by α-lactorphin.

Ijäs *et al.* (2004), investigated whether peripherally administered α- lactorphin, at doses that lower blood pressure in rats, influences nociception, locomotor activity, motor coordination, rectal temperature or duration of pentobarbital anaesthesia in mice. They observed that peripherally administrated α-lactorphin at doses that lower blood pressure had no effect on nociception or locomotor activity in mice. Furthermore, neither motor coordination, rectal temperature nor the duration of pentobarbital anaesthesia was affected by α- lactorphin. In conclusion of that study, lack of central opioid receptor-mediated adverse effects of α-lactorphin was found. Further investigation of the antihypertensive effect of this tetrapeptide is encouraged and its possible use in humans pointed out.

The ACE-inhibitory activity of several infant formulas was evaluated by Hernández-Ledesma *et al.* (2004), in a two-stage *in vitro* enzymatic procedure, which simulates physiological digestion. The ACE-inhibitory activity of nine of the eleven products investigated did not reach one inhibitory unit (i.e., ACE-inhibitory indexes lower than 50%) and their peptide content ranged from 50 to 120mg nitrogen $100g^{-1}$. Only two of the commercial infant preparations exhibited potent activity with ACE-inhibitory indexes of 93 and 88% and IC_{50} values of 101.3μg peptide nitrogen mL^{-1} (1609 ACE-inhibitory U mL^{-1}) and 81.4μg peptide mL^{-1} (1978 ACE-inhibitory U mL^{-1}), respectively. The ACE-inhibitory activity of a non-hydrolyzed milk protein-based formula increased after simulated gastrointestinal digestion, but the activity of a hydrolyzed-whey formula did not change after hydrolysis with gastrointestinal enzymes (Hernández-Ledesma *et al.,* 2004).

Pihlanto-Leppälä *et al.* (2000), hydrolyzed α-lactalbumin and β-lactoglobulin with pepsin, trypsin, chymotrypsin, pancreatin, elastase or carboxipeptidase, alone and in combination, to obtain peptides with ACE inhibitory activity. The total hydrolysates were fractionated in a two-step ultrafiltration process, first with a 30 kDa membrane and then with a 1 kDa membrane. α-lactalbumin hydrolyzed with pepsin had no inhibitory activity, whereas α-la hydrolyzed with tripsin or enzyme combinations did display inhibitory activity. β-lactoglobulin hydrolyzed with pepsin, trypsin and different enzyme combinations also had ACE-inhibitory activity. According to the authors, with both proteins, the most active hydrolysates were obtained after hydrolyzis with trypsin, and the permeate fraction obtained after trypsin hydrolysis gave the highest ACE-inhibitory activity.

The best known ACE-inhibitory peptides, Val-Pro-Pro (VPP) and Ile-Pro-Pro (IPP), have been identified in milk fermented by *Lactobacillus helveticus* strains (Sipola *et al.,* 2002, Nakamura *et al.,* 1995).

ACE appears to prefer substrates or competitive inhibitors containing hydrophobic amino acids at the three C-terminal positions. Among the most favorable C-terminal amino acids are aromatic amino acids, as well as the amino acid proline (Hernández-Ledesma *et al.* 2004). The same research team (Hernández-Ledesma *et al.*, 2005) identified several peptides sharing the presence of hydrophobic residues, such as leucine, phenylalanine or valine, at C-terminal position, which favor binding to ACE.

Most of the studies on the production of ACE-inhibitory peptides by fermentation were performed with selected strains of lactic acid bacteria because the type of starter culture is one of the main factors influencing the formation of these peptides in fermented milk (Hernández-Ledesma *et al.*, 2004). Examples of pure or mixed strains used for milk fermentation, which have been shown to release ACE-inhibitory peptides are: *Lactobacillus delbrueckii* ssp *bulgaricus* and *Lactococcus lactis* ssp *cremoris* (Gobbetti *et al.*, 2000); *Lactobacillus helveticus* (Leclerc *et al.*, 2002; Fuglsand *et al.*, 2002, Seppo *et al.*, 2002, Pan *et al.*, 2005), *Lactobacillus delbrueckii* ssp. *bulgaricus, Streptococcus salivarius* ssp. *thermophilus* and *Lactococcus lactis* biovar. *diacetylactis* (Ashar and Chand, 2004), *Lactobacillus casei* ssp. *rhamnosus* (Rokka *et al.*, 1997).

Lactobacillus helveticus is the principal antihypertensive peptide producer during milk fermentation. According to Yamamoto *et al.* (1994), the biological activity of milk fermented by this microorganism has been attributed to the released bioactive tripeptide isoleucyl-prolyl-proline and valyl-prolyl-proline (VPP and TPP). Seppo *et al.* (2002), evaluated the effects of *Lactobacillus helveticus* LBK-16 H fermented milk on hypertension with a group of seventeen mildly hypertensive volunteers (mean blood pressure 148/94) who received 150 mL per day of sour milk peptide. The control group (mean blood pressure 148/93) received 150 mL of control sour milk. Systolic blood pressure decreased more in the study group than in the control group (-10.1% SEM 2.0 *vs.* -2.8% SEM 2.9) after 8 weeks of treatment. Diastolic blood pressure was also reduced in both study groups: there was a stronger reduction in the study group than in the control group.

Potent ACE-inhibitory activity was found in a commercial kefir made from goat's milk (Quirós *et al*, 2005). The low molecular mass peptides released from caseins during fermentation were mainly responsible for this activity. It was observed that the ACE-inhibitory activity after simulated digestion (incubation with tripsin and Corolase PP - a mixture of pancreatic proteases) was similar to or slightly lower than unhydrolyzed peptides, except for peptide β-casein f(47-52), which exhibited an activity 8 times greater after hydrolysis.

Nakamura *et al.* (1995), reported two potent ACE-inhibitory peptides purified from "Calpis", a Japanese soft drink made from bovine milk fermented by *Lactobacillus helveticus* and *Saccharomyces cerevisiae*. Most of the inhibitory activity in sour milk was attributed to the amino acid sequences VPP and IPP. The concentrations of peptides providing 50% inhibition of ACE were 9 and 5 μ*M*, respectively. Hata *et al.* (1996), conducted a double blind placebo controlled study and demonstrated that the oral ingestion of 95mL of the fermented milk "Calpis" per day reduced the SBP (systolic blood pressure) significantly 4 and 8 weeks after ingestion, by 9.4 ± 3.6 mm Hg and 14.1 ± 3.1 mmHg.

In a randomized placebo-controlled study, 39 hypertensive patients received 150mL of either *L. helveticus* LBK-16H fermented milk or a control product for 21 weeks (Seppo *et al.*, 2003). Mean differences of 6.7± 3.0 mmHg in SBP and of 3.6 ± 1.9 mmHg in DBP (diastolic blood pressure) were observed between the test product and control groups. Bioactive peptides with ACE-inhibitory activity have been found in cheese. Cheese ripening influences antihypertensive activity. According to Addeo *et al.* (1992), an α_{s1}-casein – derived antihypertensive peptide isolated from Parmesan cheese after 6 months ripening could not be found after 15 months. The antihypertensive activity found in long-ripened Gouda cheese was half that found in its medium-aged counterpart (Meisel *et al.* 1997).

Costa *et al.* (2003), studied the effect of intraperitoneally (*ip*) administered hydrolyzed whey protein on blood pressure and renal sodium handling in spontaneously hypertensive rats. This group was the pioneer in studying the effect of protein hydrolysates on renal function. They demonstrated that the *ip* administration of whey protein hydrolysate (WPH) caused a marked decrease in systolic arterial blood pressure (SBP) and glomerular filtration rate. This decrease in SBP was associated with a rise in proximal and post-proximal sodium reabsorption despite a supposed fall of *in vivo* ACE activity. According to these authors, WPH obtained by digestion with alcalase may be a potentially useful source of functional peptides because of its ability to inhibit *in vitro* ACE and lower SBP.

Another group of researchers (Tauzin *et al.*, 2002), obtained angiotensin-I-converting enzyme inhibitory peptides from tryptic hydrolysate of bovine α_{S2}-casein. The entire hydrolysate was first separated in 10 HPLC fractions, which were tested for ACE inhibitory activity. IC_{50} values were determined with corresponding purified synthetic peptides and the values were in agreement with the percentage of inhibition obtained at 50 μM primary amines with natural peptides.

In spite of the numerous studies on antihypertensive activity of milk proteins *in vitro*, there are few studies *in vivo,* and most of the *in vivo* studies evaluated the antihypertensive activity of fermented milk and not the consumption of the isolated biopeptides obtained by hydrolysis. Probably, the lack of *in vivo* studies with isolated biopeptides is due to the bitter taste of some peptides. According to Clemente (2000), a major hindrance to the use of protein hydrolysates is the unpleasant bitter taste of some peptides, which can result in refusal by the patients or consumers.

A whey protein hydrolysate has been evaluated *in vivo* by Pins and Keenan (2002). The hydrolysate 20g/d was orally ingested daily by 30 hypertensive male and female volunteers. Whey protein concentrate was used as control. Significant reduction in SBP and DBP was observed after 1week ingestion of the hydrolysate. It was also observed that this blood pressure reduction persisted for the remaining 5 weeks of the study (Pins and Keenan, 2002).

Table 3 shows the ACE inhibitory activity of some casein derived peptides.

In spite of the large number of milk peptides identified as antihypertensive under *in vitro* conditions, only a few have so far proven to be clinically effective in animal and human studies. As a conclusion, *in vivo* studies must confirm the *in vitro* results, preferably through the evaluation of oral consumption of isolated biopeptides or semi-purified fractions obtained by hydrolysis.

Table 3. ACE inhibitory activity of casein and whey protein derived peptides

Peptide	Amino acid sequence*	IC50 (μM)	Reference
Caseins			
α_{s1}-casein (f23-34)	FFVAPFPEVFGK	18	Tauzin *et al.* (2002)
α_{s1}-casein (f104-109)	YKVPGL	22	Maeno *et al.* (1996)
α_{s2}-casein (f25-32)	NMAINPSK	60	Tauzin *et al.* (2002)
α_{s2}-casein (f81-89)	ALNEINQFY	219	Tauzin *et al.* (2002)
α_{s2}-casein (f81-91)	ALNEINQFYQK	264	Tauzin *et al.* (2002)
α_{s2}-casein (f92-98)	FPQYLQY	14	Tauzin *et al.* (2002)
α_{s2}-casein (f174-181)	FALPQYLK	4.3	Tauzin *et al.* (2002)
α_{s2}-casein (f182-184)	TVY	15	Tauzin *et al.* (2002)
α_{s2}-casein (f198-202)	TKVIP	400	Maeno *et al.* (1996)
β-casein (f140-143)	LGSW	500	Maeno *et al.* (1996)
β-casein (f169-175)	KVLPVPG	1000	Maeno *et al.* (1996)
Whey proteins			
α-lactoalbumin (f50-52)	YGL	409	Philanto-Leppälä *et al.* (2000)
α-lactoalbumin (f99-108)	VGINYWLAHK	327	Philanto-Leppälä *et al.* (2000)
α-lactoalbumin (f105-110)	LAHKAL	621	Philanto-Leppälä *et al.* (1998)
β-lactoglobulin (f9-14)	GLDIQK	580	Philanto-Leppälä *et al.* (1998)
β-lactoglobulin (f15-20)	VAGTWY	1682	Philanto-Leppälä *et al.* (1998)
β-lactoglobulin (f81-83)	VFK	1029	Philanto-Leppälä *et al.* (2000)
β-lactoglobulin (f94-100)	VLDTDYK	946	Philanto-Leppälä *et al.* (2000)
β-lactoglobulin (f142-148)	ALPMHIR	946	Mullaly *et al.* (1997a)

* Amino acids: A – alanine; E – glutamic acid; D – aspartic acid; F – phenylalanine; G – glycine; H – histidine; I – Isoleucine; K – lysine; L – leucine; M – methionine; N – asparagine; P – proline; Q – glutamine; R – arginine; S – serine; T – threonine; V – valine; W – tryptophan; Y – tyrosine.

ANTITHROMBOTIC PEPTIDES

Thrombus is a fibrinous clot formed in a blood vessel or in a chamber of the heart. Some peptides have shown antithrombotic activity. According to Shah (2000), there are a number of similarities between clotting of blood and of milk. There are structural and functional homologies between the dodecapeptide of fibrinogen and the 106-110 sequence of κ-casein (called casopiastrin).

It was reported that κ-caseinoglycopeptides from several animal species are a source of antithrombotic peptides (Silva and Malcata, 2005). The sequence of amino acids in f106-171 of sheep's κ-casein, was shown to decrease thrombin- and collagen-induced platelet aggregation in a dose dependent manner (Qian *et al.*, 1995).

The κ-casein fragment named casopiastrin obtained from tryptic hydrolysates, shows antithrombotic activity by inhibiting fribrinogen binding to platelets (Fiat *et al.*, 1993). Before clotting fibrinogen has to be transformed into fibrin peptides (Sgarbieri, 1996).

According to Fiat *et al.* (1993), Lys-Arg-Asp-Ser (KRDS), a peptide with a sequence corresponding to residues 39-42 of human lactotransferrin, was tested *in vitro* on human platelet function. The IC_{50} of KRDS was 350μ*M*.

In the plasma of 5-day-old newborns, after breast feeding or ingestion of cow's milk-based formula, antithrombotic peptides derived from both human and bovine κ-caseinoglycopeptides were detected (Chabance *et al.*, 1995).

Hypocholesterolemic

Among the functional properties of milk whey proteins the hypocholesterolemic effect must be cited.

Proteins of animal origin, such as casein, are generally recognized as hypercholesterolemic and atherogenic, when compared to vegetable proteins, such as soybean protein, in both experimental animals and in humans (Carrol, 1991; Carrol and Kurowska, 1995). On the other hand, Sautier *et al.* (1983), and Nagaoka *et al.* (2001), reported on the hypocolesterolemic effect of milk whey protein comparatively to casein and soybean protein.

A study conducted by Sautier *et al.* (1983) on rats fed diets containing 23% of one of the following proteins: whey protein concentrate, casein, soy or sunflower protein isolates. The blood cholesterol level was significantly higher for the rats on casein diet compared with the rats fed the other protein diets. Liver cholesterol was significantly lower for the rats on whey protein diet, compared to the other protein sources.

Nagaoka *et al.* (2001) designed a study to clarify the mechanisms of hypolesterolemic action of β-lg. A tryptic peptide from β-lg (f71-75; IIAEK) exhibited a greater hypocholesterolemic activity in comparison with the drug β-sitosterol in animal studies. The mechanism of action would appear to be related to inhibition of micellar solubility of the cholesterol, which in turn inhibits cholesterol reabsorption.

Another comparative study with whey protein concentrate (WPC), soy protein isolate (SPI) and casein diets was reported by Jacobucci *et al.* (2001). Growth and lipid concentration in rat liver and blood serum were compared in a 45 days feeding experiment. Hypercholesterolemic diets containing 6% coconut fat plus1% cholesterol and 20% protein from whey protein concentrate (WPC), soy protein isolate (SPI) and commercial casein (CC) were prepared. Commercial casein and soy protein isolate promoted higher food intake and better rat growth than whey protein concentrate. Both SPI and WPC were effective in preventing blood serum cholesterol to rise when the rats were fed hypercholesterolemic diets. Soy protein isolate exhibited a stronger hypocholesterolemic effect than whey protein concentrate. Casein raised liver cholesterol to a level significantly higher than both SPI and WPC, after 45 days on the diets. With regard to total blood serum cholesterol, this result showed some differences with the data reported by Sautier *et al.* (1983) and Nagaoka *et al.* (2001), as described above.

The hypocholesterolemic effect of dairy products can be potentialized with the use of probiotic microorganisms such as: *Lactobacillus acidophilus* RP32 (Gilliland *et al.*, 1985), *Lactobacillus paracasei* subps. *paracasei* NTU 101, *Lactobacillus plantarum* NTU 102 and

L. acidophilus BRCR 17010 (Chiu *et al*., 2005), *Lactobacillus gasseri* SBT 0270 (Usmam and Hosono 1999, Usmam and Hosono 2000). Microorganisms with hypocholesterolemic effect may act through deconjugation, releasing into the medium amino acids and bile acids which reduce cholesterol reabsortion by increasing cholesterol faecal excretion (Lovegrove and Jackson, 2003).

The hypocholesterolaemic properties of milk and fermented milk products have been investigated by Beena and Prasad (1997). Yogurts were fortified with skim milk powder, condensed whey or lactose-hydrolyzed condensed whey. This study has demonstrated in a rat model that ordinary yogurt had no hypocholesterolaemic effect, but, bifidus yogurts and yogurts fortified with whey proteins can reduce total and LDL-cholesterol, and suggests that if they have the same effect in human subjects they have potential value in cholesterol-lowering diets.

As described above, various studies have demonstrated a hypocholesterolemic effect for milk whey proteins. Dairy products enriched with whey proteins and added with specific probiotic cultures could be designed to benefit individuals with high levels of plasma cholesterol.

Effects on the Nervous System

Some peptides have been described with regulatory functions on the central or peripheral nervous systems. These are called opioid peptides and may act as opioid agonists (casormorphins) or antagonists (casoxins).

Peptides with Opioid and Antiopioid Activity

Opioid peptides are defined as peptides such as enkephalins, which have an affinity for an opiate receptor as well as opiate-like effects, inhibited by naloxone (Pihlanto-Leppälä, 2001). They are short peptides (5-10 amino acids) termed casomorphins or exorphins. When opioid peptides are injected into the bloodstream they induce an analgesic and sedative effect due to their action on the nervous system (Shah, 2000).

Typical opioid peptides originate from three precursor proteins: proopiomelanocortin (endorphins), proenkephalin (enkephalin) and prodynorphin (dynorphins) (Höllt, 1983). All these typical opioid peptides have the same N-terminal sequences, Tyr-Gly-Gly-Phe (YGGF). Milk-protein derived opioid-like peptides have been designated as "atypical" opioid peptides because they do not derive from those three precursor proteins (Höllt, 1983).

Many opioid peptides are biologically very potent. Therefore, even nutritionally insignificant amounts of liberated peptides might be sufficient to exert physiological effects (Meisel and FitzGerald, 2000).

Opioid activity has been found especially in β-casein, but also α_{s1}-casein, κ-casein and whey proteins have been reported as potential precursors of opioid peptides. Opioid peptides from α_{s1}-casein (f90-96), β-lactoglobulin (f102-105) and α-lactoalbumin (f50-f53), are called α-casomorphin, β-lactorphin and α-lactorphin, respectively (Schanbacher *et al*., 1998).

Opioid peptides exert their activity by binding to specific receptors of the target cell. Both peptides with opioid activity (agonist), or anti-opioid (antagonist) bind to target receptors displaying a specifically opioid action or inhibition. The individual receptors are responsible for specific physiological effects – e. g. the μ-receptor for emotional behavior and suppression of intestinal motility, the σ-receptor for emotional behavior, and the κ-receptor for sedation and food intake.

According to Drewnowski (1992), the pleasure response to palatable foods, notably those rich in sugar and fat, may be mediated through the endogenous opioid peptide systems. Preferences for sweet taste appeared to be under opioid control. Studies on rats showed that opioids most effectively reduce intakes of preferred sweet liquids such as glucose, sucrose or saccharin solutions. Studies on animals have suggested that preferences for dietary fats may also be under opioid control.

The main effect of opioid peptides regarding food intake refers to foods rich in sugar and fats. Therefore, it might be expected that opioid peptides act interfering with pleasure associated with the eating of this type of palatable foods, perhaps selectively contributing to diminish their consumption (Sgarbieri e Dias, 2007).

The opioid peptide's activity is comparable to that of the endogenous opioids (like endorphins), acting both on the central nervous system and in peripheral organs. On the central nervous system they can cause analgesia, catalepsy, sedation, hypotension, respiratory depression, body temperature dysregulation, appetite and food intake dysregulation, suppression of gastric secretion, increase in the secretion of growth hormone (GH), prolactin (PRL), antidiuretic hormone (ADH), luteinizing hormones (LH), follicle-stimulating hormone (FSH), thyroid-stimulating hormone (TSH), adrenocorticotropic hormone (ACTH), sexual behavior regulation (Sgarbieri and Dias, 2007). The peripheral effects may include suppression of intestinal motility and improvement of melanocyte-stimulating hormone (MSH) activity (Sgarbieri, 1999).

All bioactive peptides originated from bovine α-casein behave as opioid antagonist (Shah, 2000). Milk also contains latent opioid antagonist peptides, termed casoxins (shown in Table 3) and lactoferroxins, derived from κ-casein and lactoferrin, respectively (Fiat *et al.*, 1993). These peptides would presumably be capable of antagonizing the inhibition of gut motility induced by casomorphins, although it is uncertain that theses peptides are actually produced during *in vivo* digestion of milk (Schanbacher *et al.*, 1998). Besides, α- and β-caseins are present in higher concentration in normal milk than κ-casein and lactoferrin.

Mammalian opioidergic systems consist of opioid receptors and their endogenous ligands-endogenous opioid peptides. Depending on their location, their physiological significance appears to be related to a considerable number of neuroendocrine regulatory functions (Meisel and FitzGerald, 2000).

Opioids have been used for centuries to modify gastrointestinal function and motility (Allescher *et al.*, 2000). Opioids are reported to influence gastrointestinal function in two ways: first, they affect the smooth muscles – which reduces transit time of intestinal mass – and second, they affect the intestinal transport of electrolytes, which explains their antisecretory properties (Pihlanto-Leppälä, 2001). These peptides may enter peripheral blood intact and exert systemic effects or they may produce local effects in the gastrointestinal tract (Meisel and FitzGerald, 2000). The presence of opioid activity may, for example, interfere with casein digestion and absorption, because during casein hydrolysis opioid peptides are released (β-casomorphins) that bind to the digestive system opioid receptors decreasing gastrointestinal motility, what may contribute to increasing the stomach emptying time and the intestinal content transit time (Daniel *et al.*, 1990).

Casomorphins have been found to prolong gastrointestinal content transit time and to exert antidiarrhoeal action (Shah, 2000). Examples of casomorophins are shown in Table 4. The therapeutic use of casomorphins as antidiarrhoeal substances does not only cause change of secretion, but also changes motility patterns due to an action on reflexes mediated by the enteric nervous system (Allescher *et al.*, 2000). Evidence has accumulated that the enhancement of net water and electrolyte absorption by β-casomorphins in the small and large intestine is a major component of their antidiarrhoeal action (Tomé *et al.*, 1987). Furthermore, the potential for casomorphins to retard the rate of passage of the digesta throught the GI tract has important implications for other bioactivities depending on the time available within the GI tract for interaction with target bacteria or intestinal epithelium (Schanbacher *et al.*, 1998).

The opioid peptides fragments of β-casein are called β-casomorphins due to their morphine-like behavior. A tetrapeptide amide, morphiceptin, is the most active opioid agonist in the bovine β-casomorphin group (Shah, 2000).

Table 4. Opioid and anti-opioid peptides derived from α and β-casein (casomorphins) e from κ-casein (casoxins)

Casein / peptides	Amino acid sequence*	Activity
α_{s1} (δ bind)		
90-95	RYLGYL	opioid
90-96	RYLGYLQ	opioid
91-96	YLGYLQ	opioid
β (μ bind)		
60-63	YPF	opioid
60-64	YPFP	opioid
60-65	YPFPG	opioid
60-66	YPFPGP	opioid
60-70	YPFPGPIPNS	opioid
κ (μ,δ bind)		
25-34	YIPIGYVLSR	antiopioid
34-38	RYPSY	antiopioid
33-38	SRYPSY	antiopioid
35-38	YPSY	antiopioid
35-41	YPSYGLN	antiopioid
58-61	YPYY	antiopioid

* Amino acids: R – arginine; Y – tyrosine; L – leucine; G – glycine; Q – glutamine; P – proline; F – phenylanaline; I – isoleucine; N – asparagine; S – serine.

Source: Sgarbieri and Dias (2007).

The neonatal intestine appears to be permeable to (pre-)casomorphins; in adult systems, the intestinal brush border membrane seems to be the main target site for the physiological effects of food-derived opioid peptides (Meisel and FitzGerald, 2000). β-casein-derived peptides may pass through the intestinal mucosa in neonates via passive transport, so babies may become calm and sleepy (Sturner and Chang, 1988). Besides, casomorphin may participate in the endocrine regulation of pregnancy, e.g. by stimulation of prolactin release (Yen *et al.*, 1985).

Rokka *et al.* (1997) characterized bioactive peptides formed in UHT milk fermented with the probiotic *Lactobacillus GG* strain, following hydrolysis by pepsin and trypsin *in vitro*. The peptide fractions obtained were analyzed by fast atom bombardment mass spectroscopy and the spectrum of the molecular weights of the peptides was compared to the molecular weights of the bioactive peptides presented in literature. Weak molecular ion signals corresponding to opioid, opioid antagonist, immunostimulating and antihypertensive peptides were also detected in the fractions. The following opioid peptides were observed and identified according to literature in different fractions: Tyr-Pro-Phe-Pro (YPFP) from bovine β-casein, Arg-Tyr-Leu-Gly-Tyr-Leu-Glu (RYLGYLE) from bovine α-lactalbumin and Tyr-Pro-Phe-Pro-Gly-Pro-Ile-Pro-Asn-Ser-Leu (YPFPGPIPNSL) from bovine β-casein.

β-casomorphins (β-CMs) produced during the Edam cheese ripening process, with and without added bifidobacteria, were analyzed by Sabikhi and Mathur (2001). Results indicated that β-CM 3 was present (~35mg%) in all Edam cheese samples during ripening. The presence of bifidobacteria did not have any effect on the formation of opioid peptides in that study.

Effects on Immune System

Milk-derived peptides exert immunomodulatory effects through different mechanisms, such as antioxidant, antimicrobial and antiviral activities and the stimulation of glutathione synthesis and the antibody response system.

Immunomodulation by Milk Proteins

There are many milk components with immunomodulatory properties which can act synergistically providing protection. According to Gill and Rutherfurd (1998) these milk component are: α-lactalbumin, β-lactoglobulin, immunoglobulins, bovine serum albumin, lactoferrin, lactoperoxidase, complement, cytokines, hormones and growth factors, casein and casein fractions, whey and casein derived peptides, nucleotides, milkfat, glycoproteins and glycopeptides, vitamins, minerals and leukocytes.

The immune system employs both non-specific and specific immune responses to confer protection against disease. Non-specific components of the host defense include physicochemical barriers (for example skin, mucus, lysozyme, complement, interferons), natural killer cells and phagocytic cells (for example, neutrophils, monocytes/macrophages) (Gill and Rutherfurd, 1998).

Perhaps the class of food proteins more directly related to immunity are the immoglobulins (IgG1, G2, A and M) which occur in the blood serum and other fluids such as colostrum and milk whey (Sgarbieri, 1999). Immunoglobulins are not transferred across the placenta to the mammalian fetus and hence babies and calves are born with very low concentrations of serum immunoglobulins (Gill and Rutherfurd, 1998, Shah, 2000). However, immunoglobulins occur in high concentration in colostrum, the first fluid secreted by the mammary gland of mammals immediately post-partum. In the colostrum the immunoglobulins have an important role in the passive immunity of the neonates. Besides immunomodulation, the proteins found in the colostrum (whey proteins) are responsible for the development of the gastrointestinal tract of the newborn.

Breast feeding facilitates physical transmission of passive immunity via a number of multifunctional factors, which have a direct effect on the neonate's resistance to bacterial and viral infections; caseins are included among these factors. In fact, during enzymatic digestion of human and bovine milk, caseins and peptides with immunomodulating capacities are released (Silva and Malcata, 2005).

IgA is the predominant class of immunoglobulins in human milk, whereas IgG_1 is predominant in cow's milk (Goedhart and Bindels, 1994). Even though it is structurally different, it appears to have the same function (Facon *et al.* 1993). Secretory IgA antibodies in human milk neutralize bacterial toxins and virulence factors and inhibit adherence and proliferation of bacteria on epithelial surfaces by binding to the bacterial adhesins (Davin *et al.*, 1991).

Bovine immunoglobulin concentrate (BIC) can be obtained as described by Kelly *et al.* (1996). The reported paper describes a BIC containing high levels of specific IgG directed against *Clostridium difficile* toxins used for the immuzination of the cow. According to Steijns (2001), BIC for human application is only in its infancy, because there are still uncertainties with respect to dosage safety and timing of administration. Natural antimicrobials like BIC may have a number of advantages over synthetic antibiotics: lower cost; their polyclonal nature (in principle) enhances their resistance against bacterial inactivation; their activity is most likely to be limited to the intestinal lumen; and they place less of a burden on the microbial gut ecology (Steijns, 2001).

Pacheco and Sgarbieri (2005), confirmed the ability of undenatured whey proteins to stimulate glutathione synthesis in mouse liver to a much greater extent than casein. Besides, they observed that hydrolysis of whey protein using the proteases pancreatin and Protamex™ (Novo Nordish) yields hydrolysates rich in peptides with biological activity to stimulate synthesis of gluthatione by liver cells. In another paper, Sgarbieri *et al.* (2000), found that the concentration of plasma protein and sulfur amino acids of rats intubated with WPC (whey protein concentrate) was substantially higher than that of rats fed with casein after 10 minutes, providing an additional reason to speculate that such a whey protein provides amino acids or sequences that might promote fast increase of hepatic gluthatione synthesis.

An increase in gluthathione levels has been observed in HIV-infected adults under oral whey protein supplementation (Bounous, *et al.*, 1988). Moreno *et al.* (2005), conducted a prospective double-blind clinical trial for 4 months with 18 vertically HIV-infected children, under therapy, who had received whey protein, maltodextrin (placebo), or none. Erythrocyte glutathione concentration, T lymphocyte counts (CD4+ and CD8+) and occurrence of associated co-infections were evaluated. Antiretroviral therapy associated with WPC supplementation, rich in glutamyl-cysteine, increased the erythrocyte GSH levels, and suggested an improvement in the TCD4/CD8 ratio and a decrease in the occurrence of acute co-infections.

The exact mechanism through which whey proteins stimulate glutathione synthesis is still not completely understood. There is growing evidence for the hypothesis that tissue glutathione synthesis is dependent on the supply of dietary amino acids, where the sulfur amino acid content would be specifically a major determinant of hepatic glutathione content (Cross and Gill, 1999). Another hypothesis, according to Pacheco and Sgarbieri (2005) is based on the observation that some whey proteins, such as bovine serum albumin, lactoferrin, immunoglobulins, and β-lactoglobulin, are rich sources of sulfur amino acids, mainly cysteine. Furthermore, these proteins contain in their primary sequences various segments of glutamic acid-cystine (Glu-Cys_2) forming a large number of disulfide bridges. These segments seem to be preserved in the digestion of undenatured whey protein by monogastric animals, but they can be broken down into severely heat-denatured protein.

Important variations can occur in the immunomodulating activity of digested milk products. In general, pepsin, chymotrypsin or neuraminidase treatment of casein or whey proteins reduce their immunomodulating effects while trypsin, pancreatin and pronase have no effect or enhanced the activity of some milk proteins on the immune system (Otani and Odashima, 1997).

Gill and Rutherfurd (1998), observed that WPC-based diets enhanced mucosal antibody responses to oral immunization; mice fed WPC and orally immunized with cholera toxin and ovalbumin showed significant enhancement of mucosal antibody responses.

Patients with chronic inflammatory diseases of the gastrointestinal tract may benefit from dietary products that can counteract the host cellular responses, which contribute to localized inflammation (Cross and Gill, 1999). The immunomodulating effect of whey protein products was evaluated *in vitro* by measuring their impact on lymphocyte proliferation (Mercier *et al.*, 2004). Microfiltered-whey protein isolates significantly increased lymphocytes proliferation at a concentration of $100\mu g.mL^{-1}$.

Lactoferrin administered orally to mice has been demonstrated to increase the number of CD4 cells in spleen and Peyer's Patches, and enhance antigen-specific proliferation of Peyer's patch and spleen cells (Debabbi *et al.*, 1998).

Rokka *et al.* (1997), identified the immunostimulating peptides Thr-Thr-Met-Pro-Leu-Trp (TTMPLW) liberated from bonive α_{s1}-casein and Gly-Leu-Phe (GLF) derived from α-lactalbumin. This hexapeptide and tripeptide were released by enzymatic proteolysis of UHT milk by *Lactobacillus* GG. Similarly, LeBlanc *et al.* (2002), identified peptides released during fermentation of milk by *Lactobacillus helveticus* which has immunoenhancing and antitumor properties.

Moreover, bovine milk components have been shown to possess anti-cancer and anti-infection properties (Parodi, 1996; Bounous *et al.*, 1991; Neurath *et al.*, 1996). The immune system stimulatory peptides activate cellular components and their chemical mediators against various types of antigens including pathogenic microorganisms and tumoral cells, preventing various pathologies (Sgarbieri e Dias, 2007). Therefore functional foods may act not only on immune system stimulation but also in the prevention of tumor development

(Park *et al.*, 1997) and induction of immune tolerance, diminishing the levels of allergenic reactions (Tomé and Debalbi, 1998).

Antimicrobial and Antiviral Activity

Minor whey proteins such as lactoferrin (Lf), lactoperoxidase (Lp), lysozyme and immunoglobulins, structurally modified α-lactalbumin (HAMLET - human alpha-lactalbumin made lethal to tumor cells and BAMLET – bovine α-lactalbumin made lethal to tumor cells), are considered antimicrobial proteins (Shah, 2000, Sgarbieri and Dias, 2007). The antibacterial activity of milk is attributed mainly to the immunoglobulins, but the non-immune proteins also exhibit distinct antibacterial activities. The antimicrobial actions of lactoperoxidase, lysozyme and lactoferrin may be asserted either in the mammary gland or in the GI tract of babies (Schanbacher *et al.*, 1998).

Apart from immunoglobulins, antibodies and plasma cells, milk also contains antibiotic/host protective peptides that are of value not only for maintenance of its integrity, but also for protection of the newborn and, possibly, protection of the lactating mother (Lahov and Regelson, 1996). Casecidin was among the first peptides identified as having antimicrobial and antiviral activity. A compound generated by chymosin digestion of casein at pH 6 or 7, casecidin inhibited *in vitro* staphylococci, sarcina, *Bacillus subtilis, Diplococcus pneumoniae* and *Streptococcus pyogenes* (Lahov and Regelson, 1996).

Lactoferrin is a glycoprotein present in human colostrum and human mature milk at concentrations of about 5-7 and 1-2g.L^{-1}, respectively (Goldman *et al.,* 1982). The concentration of lactoferrin in bovine colostrum and milk is about 1.5-5 mg.mL^{-1} and 0.1 mg.mL^{-1}, respectively (Tsuji *et al.*, 1990). This indicates that lactoferrin is even more important in humans than bovine species (Shah, 2000).

Two basic biochemical properties of lactoferrin contribute to its involvement in the host defense: the extremely powerful iron-binding capacity and the strong interaction with other molecules and surfaces (Hooijdonk *et al.*, 2000). Lactoferrin is both bacteriostatic and bactericidal, due respectively to its high affinity for Fe^{3+} and the ability of the intact molecule or an *N*-terminal peptide of bovine or human lactoferrincin, to bind bacterial lipopolisaccharide and disrupt bacterial membranes (Nuijens *et al.,* 1996). Lactoferrin also causes release of lipopolysaccharide molecules from the outer membrane of the Gram-negative bacteria and acts directly as an antibiotic (Shah, 2000).

Pepsin digestion of either bovine lactoferrin or human lactoferrin releases peptides from the *N*-terminus (17-41 and 1-47, respectively) with 100 to 1000 times the bactericidal activity of intact lactoferrin, suggesting that lactoferrin latent contains at least one antibacterial peptide region that is released when the molecule is hydrolyzed (Goedhart and Bindels, 1994, Nuijens *et al.,* 1996).

Lactoferrin plays a role in the cellular defense system, which comprises a close interaction between neutrophils, lymphocytes, macrophages and their secretory products upon microbial invasion (Hooijdonk *et al.*, 2000). Besides, lactoferrin can act synergistically with IgG and lysozyme for enhanced antimicrobial protection (Nuijens *et al.,* 1996, Iyer and Lönnerdal, 1993).

The bacteriostatic and bactericidal effects of lactoferrin *in vitro* have been studied in a wide range of microorganisms (Goedhart and Bindels, 1994). McCann *et al.* (2005), isolated and characterized five antibacterial peptides derived from the f(164-207) region of bovine α_{s2}-casein. The minimal inhibitory concentration (MIC) of these peptides against the Gram-positive bacteria, *Listeria innocua, Bacillus subtilis* and *Listeria monocytogenes* ranged from 4.8 to 21μg.mL^{-1} and against the Gram-negative bacteria, *Escherichia coli* and *Salmonella enteritidis* was >76μg.mL^{-1} . The MIC of the partially purified chymosin digest of sodium caseinate was very high against all bacteria tested and was most active against *L. innocua* (625μg.mL^{-1}), *B. subtilis* (312.5μg.mL^{-1}), *Salmonella typhimurium* (625μg.mL^{-1}), *Salmonella salford* (625μg.mL^{-1}), and two strains of *L. monocytogenes* (625μg.mL^{-1}).

Lactoferricin B, a peptide derived from bovine lactoferrin due to action of pepsin has been found to have antimicrobial activity against various bacteria and *Candida albicans* (Jones *et al.*, 1994). *Proteus* spp., *Pseudomonas cepacia* and *Serratia* spp. were resistant. Lactoferricin exerts a transient antimicrobial effect at mucosal surfaces. Lactoferricin is more bactericidal than undigested lactoferrin because the peptide's smaller size may facilitate access to susceptible sites on the microbial surface (Tomita *et al.* 1991).

Lysozyme is an antimicrobial enzyme isolated from colostrum, milk, tear, and saliva. Human milk is a better source of lysozyme (2.5 mg.100mL^{-1}) than bovine milk (0.025 mg.100mL^{-1}). As a conclusion, the lysozyme concentration in bovine milk is probably too low to significantly contribute to the overall bacteriostatic and bactericidal activity (Hooijodonk *et al*, 2000; Aimutis, 2004). Lysozyme is a 15 kDa single-chain protein. The amino acid content of bovine milk lysozyme is different from that of human milk and egg white lysozyme. The enzyme hydrolyzes β - 1 $\rightarrow$ 4 linkages between *N*-acetylmuramic acid and 2-acetyl-amino-2-deoxy-D-glucose residues in bacterial cell walls, resulting in cell lysis (Shah, 2000).

Lactoperoxidase is a major antibacterial agent in colostrum. Its primary function is to catalyze the oxidation of certain molecules, at the expense of hydrogen peroxide, in order to generate reactive products with a wide antimicrobial activity (Kussendrager and Hooijdonk, 2000). The enzyme, in the presence of H_2O_2 catalyses the oxidation of thiocyanate (SCN^-) and produces an intermediate product with antimicrobial properties (Shah, 2000, Aimutis, 2004). Different groups of bacteria show a varying degree of resistance to lactoperoxidase. Gram-negative, catalase positive organisms such as pseudomonas, coliforms, salmonellae and shigellae are not only inhibited by lactoperoxidase, but depending on medium conditions (pH, temperature, incubation time, cell density, etc.) may be killed, provided H_2O_2 is supplied exogenously (Kussendrager and Hooijdonk, 2000).

Besides promoting bifidobacterium growth and modulating immune system response, κ-casein glycomacropeptide (GMP) has the ability to bind cholera and *Escherichia coli* enterotoxins, inhibit bacterial and viral adhesion, and suppress gastric secretions (Brody, 2000).

Inhibition of adhesion may prevent the colonization of pathogens at an early stage of infection. Ouwehand *et al.* (1997), studied the effect of β-lactoglobulin on the adhesion of pathogens to human ileostomy glycoproteins. Of all tested strains, the adhesion of *E. coli* expressing SfaII fimbriae was influenced most by β-lg; inhibitory activity was also detected for SfaI-mediated adhesion and against a *Klebsiella oxytoca* strain isolated from faeces. They observed that β-lg inhibited SfaII fimbriae-mediated adhesion in a concentration-dependent manner, with optimal activity around 50μg.mL^{-1}. The adhesion of *S. typhimurium* and *S. enteritidis* was also significantly reduced ($0.05 < P < 0.01$), but much less than that of the *K. oxytoca* strain and SfaI and II-mediated adhesion. This research team has observed previously that the adhesion of the mentioned strains was not affected by a fraction from bovine colostrum containing components with a molecular weight of less than 30kDa (Ouwehand *et al.*, 1995). It may be that other components present in this fraction, mainly α-lactalbumin and lactose, influence the adhesion inhibitory activity of β-lg.

Casein-derived immunopeptides have been shown to stimulate the phagocytic activities of murine and human macrophages and to protect against *Klebsiella pneumoniae* infection in mice. The peptides may stimulate the proliferation and maturation of T cells and natural killer cells for the defense of the newborn against a large number of bacteria, particularly enteric bacteria (Shah, 2000).

It has been shown, according to Neurath *et al.* (1996), that bovine β-lactoglobulin modified by 3-hydroxyphthalic anhydride (3HP-β-lg) is a potent inhibitor of infection by human (HIV-1, HIV-2) and simian (SIV) immunodeficiency viruses. Because of its ease of preparation and being from a widely available and inexpensive source (whey), topically applied 3-HP-β-lg has the potential to diminish sexual transmission of HIV worldwide in an economically feasible way. Peptide fragments of β-lg have been shown to be bacteriostatic against *E. coli, B. subtiles* and *Staphylococcus aureus* (El-Zahar *et al.*, 2004).

A study conducted by Micke *et al.* (2001) evaluated the effect of oral supplementation with two different cysteine-rich whey protein formulas on plasma glutathione (GSH) levels and parameters of oxidative stress and immune status in HIV-infected patients. In a prospective blind clinical trial, 30 patients with stable HIV infection were randomized to a supplemental diet with a daily dose of 45g whey proteins of either Protectamin (Fresenius Kabi, Germany) or Immunocal (Immunotec, Canada) for two weeks. Plasma concentrations of total, reduced and oxidized glutathione, superoxide anion ($O2^-$) released by blood mononuclear cells, plasma levels of TNF-alpha and interleukins 2 and 12 were quantified with standard methods at baseline and after therapy. Plasma total GSH increased in the Protectamin group, while the difference in the Immunocal group did not reach significance. This study was extended for six months (Micke *et al.*, 2002). All patients continuing the trial were openly switched to Protectamin. The supplementation with whey proteins increased plasma glutathione levels in patients with advanced HIV-infection.

The kinetically trapped protein-lipid complex named HAMLET/BAMLET consists of the calcium depleted apo form of α-la in the molten globule state, which combines with a mono-unsaturated fatty co-factor (Chatterton *et al.*, 2006). The HAMLET has been shown to exhibit antimicrobial activity, in particular against *Streptococcus pneumoniae* (both antibiotic sensitive and resistant strains) and *Haemophilus influenza* (Chatterton *et al.*, 2006).

Milk proteins and peptides presenting antimicrobial and antiviral activity may find interesting applications in the field of food safety and as pharmaceuticals (Korhonen and Pihlanto, 2006).

Antioxidant Properties

In addition to binding of bivalent ions and enhance the solubility of many important minerals, CPPs can also prevent oxidation. Besides, the glutathione antioxidant system is indispensable for the cellular protective mechanisms (Bounous, 2000).

According to Kitts (2005), CPP is effective in suppressing .OH-induced damage of deoxyribose in both site-specific and non-specifc assays. The potential for CPP to have primary antioxidant activity is evidenced by direct scavenging of .OH free radicals in the non site-specific assay, as well as an affinity to exhibit secondary characteristics, as shown by the prooxidant sequestering activity in the site-specific assay. No difference in the extent of protection of deoxyribose from .OH radical generated degradation occurred at the highest CPP concentration (e. g. 1mg/mL), indicating that at this concentration a sufficient amount of CPP was present to provide both direct scavenging of free radicals as well as sequestering of the prooxidant iron catalyst.

The antioxidant activity of caseins and casein-derived peptides was evaluated by Rival *et al.* (2001). They used three free radical producing reactions: (1) lipoxygenase, (2) AAPH-catalyzed oxidation of linoleic acid, and (3) hemoglobin-catalyzed oxidation of linoleic acid hydroperoxide. The conclusion was that caseins and casein-derived peptides were able to inhibit enzymatic and non-enzymatic lipid peroxidation, suggesting they were preferred targets for the free radical intermediates.

According to Bounous (2000), depletion of glutathione is a common consequence of increased formation of reactive oxygen species during increased cellular activities. This phenomenon can occur in the lymphocytes during the development of the immune response and in the muscular cells during strenuous exercises.

Whey proteins seem to enhance immune responses and slow down the aging process, probably due to the increase in glutathione concentration (Pacheco and Sgarbieri, 2005). Liver glutathione plays a critical role in detoxification of reactive intermediates of oxidative metabolism, including those generated from endogenous molecules, metabolic stress, drugs, or carcinogens. When hepatic glutathione is depleted, animals and humans are more susceptible to tissue damage caused by free radical-generating compounds and toxic metabolites (Bounous and Gold, 1991).

MUSCULAR SYSTEM

Scientific studies have shown that proteins are not identical as far as health benefits are concerned. Depending on the prevalence of certain amino acids and amino acid sequences, molecular size and structural features and configuration, the proteins may present differentiated physiological functional properties (Burke, 2000).

Milk whey proteins are ingredients frequently used in sports nutrition due to the fact that important physiological benefits have been demonstrated in this group related to improved performance and quick recovery after the exercise.

Cell Restoration

A series of characteristics intrinsic to whey proteins make them especially useful for optimizing recovery from strenuous exercise, such as rapid and high digestibility, excellent metabolic efficiency and high biological value (Borges *et al.,* 2001). Whey proteins are known to have a perfectly balanced amino acid composition, in addition to having especially high concentrations of branched chain amino acids (BCAA) and sulfur-containing amino acids. Furthermore, whey proteins provide a steady release of amino acids into the bloodstream, stimulating anabolic pathways, particularly protein synthesis. These characteristics make whey proteins especially suited for applications in the field of nutrition for athletes and people that engage in regular physical exercise, particularly to meet the physiological needs of the body during and after prolonged and/or severe exercise (Boirie *et al.,* 1997).

There is evidence that during exercise changes occur in the concentrations of some free amino acids since they are readily metabolized to maintain a high flux of the Kreb's or citric acid cycle. Maintenance of the synthesis of citric acid cycle intermediates retards muscle fatigue, with a correspondingly high depletion of muscle glycogen. Muscle is the main site of conversion of the carbon chains of amino acids into glutamine, an amino acid that performs important functions in several body tissues. Differently from the liver, where most amino acids are normally oxidized, skeletal muscle cells are able to metabolize only six amino acids, i.e. the branched-chain amino acids (leucine, isoleucine and valine) and glutamate, aspartate and asparagine (Wagenmarkers, 1998).

During exercise, protein synthesis is depressed and this change leaves the branched-chain amino acids in the blood (mostly derived from muscle catabolism) available to be taken up by skeletal muscle cells that require large amounts of energy-providing compounds to fuel muscle contraction. Thus, BCAA's are metabolized into glucose to meet muscle energy requirements (Biolo *et al.,* 1997).

Bovine whey contains about 6,05g isoleucine and 5,95g valine per 100g protein, though separation, concentration and isolation processes may influence these levels. Processed whey products high in β-lactoglobuline will also contain high levels of BCAA's as a result of the higher concentration of BCAA's in this protein fraction (Pacheco *et al.,* 2002).

During exercise there is a marked increase in the oxidation of BCAA's such as leucine, isoleucine and valine, which are degraded by the active skeletal muscles with release of nitrogen, which, along with pyruvate, will be converted to alanine in the muscle and transferred to the liver. Next, the liver extracts nitrogen from alanine and converts some of it back into pyruvate, which can then be used to produce more glucose that will be launched into the bloodstream. For that reason, the capacity of BCAA's to act as energy providers during resistance training or prolonged physical exercise, as well as the protein/BCAA ratio in whey, make these proteins an option of choice among conventional protein sources (Rasmussen *et al.,* 2000).

A reduction in the availability of amino acids during exercise may limit the stimulating effect of insulin on tissue protein synthesis. A diet supplemented with a blend of partially hydrolyzed whey proteins was found to stimulate insulin secretion and increase plasma amino acid concentrations more efficiently than diets supplemented with intact (i.e. non-hydrolyzed) or with carbohydrates alone (Van Loon *et al.* 2000). In another study, a diet added with hydrolyzed whey proteins did not increase glycogen synthesis, although it did result in increased plasma insulin levels and higher glucose uptake and enhanced protein synthesis in muscle and liver cells (Jentjens *et al.*, 2001).

Increased levels of the amino acids arginine and lysine may have a modulating effect on the release of growth hormone – an effect considered potentially beneficial to bodybuilders and athletes and people who wish to increase muscle mass. Whey protein isolates are produced by ultrafiltration of fluid whey and are considered an ideal source of these amino acids – ingredients used to supplement high-carbohydrate cereal-based products.

Immunomodulator

In recent years, scientists have begun to investigate the ability of certain proteins to improve immunity, prevent diseases, and indirectly extend longevity. Whey in the form of whey protein concentrates and isolates contains high concentrations of sulfur amino acids, particularly cystine. This semi-essential amino acid is an indispensable precursor for glutathione synthesis, a multifunctional molecule that helps to metabolize and eliminate xenobiotic compounds, some of which carcinogenic, in addition to being a natural cell antioxidant. Glutathione acts as a free radical-destroying antioxidant, such as reactive oxygen species (ROS) and reactive nitrogen species (RNS0, and protects the cell against harmful effects (McIntosch, 1998).

It is well documented that exhaustive exercise may cause immune depression (Smith, 2003, Elphick *et al.*, 2003), increased free radical production and higher than normal protein catabolism. Whey proteins and their hydrolysates play an important part in the body's defense, acting on the immune system (humoral and cellular) by stimulating glutathione synthesis and the production of lymphocytes and antibodies. In addition to being rapidly digested and absorbed, several of these proteins and their metabolites are antioxidant and free radical complexing agents.

There is currently a consensus that most people prefer to get their nutrients and other healthy substances from foods. As for proteins, a series of benefits directly related to important nutritional aspects has been explored to guide consumers to make healthy choices and obtain the optimum benefits from a healthy diet. Within this context, whey proteins hold a privileged position since research aimed at identifying and associating specific food components to functional-physiological properties have confirmed the ability of whey proteins to modulate a series of metabolic responses. This fact has motivated better-informed consumers to turn to whey proteins to help them improve their physical performance faster and more efficiently.

Although many people are looking for immediate benefits or fast results when they decide to adopt a healthy diet, it should be emphasized that there is no such magic nutritional solution. The benefits of a healthy diet are associated with both the amounts of proteins and the frequency with which they are consumed since excessive consumption of proteins can have catastrophic health consequences if they are consumed in amounts higher than recommended for prolonged periods of time. Therefore, in view of their extreme importance, health claims should be solidly based on reliable scientific evidence and be clear, simple and easy to understand for the consumer.

OTHER RELATED EFFECTS

Other functional properties of milk protein described and commented in literature and that do not exactly fit in the categories already described above.

Growth Factor Activities

Growth factors are composed of a heterogeneous group of proteins and peptides that initiate cellular growth and expression of differentiated function. Circulating growth factors may act in an analogous manner to classic endocrine hormones (Donovan and Odle, 1994). In addition, most factors are synthesized in multiple cell types throughout the body and may act locally within a particular cell or upon adjacent via autocrine or paracrine mechanisms, respectively (Donovan and Odle, 1994).

Mammary secretions and milk contain many growth factors which modulate mammary development, growth and differentiation of a variety of cell types, and probably also modulate neonatal development of the gastrointestinal tract and other organ systems (Dovovan and Odle, 1994; Koldovsky, 1989).

Bovine growth factors include insulin-like growth factor (IGF-1), epidermal growth factor (EGF), transforming growth factor beta (TGF-β) and fibroblast growth factor (acid or basic). The envisaged (medical) food applications of growth factors include gut repair after/during radiation or chemotherapy, according to Steijns (2001).

An interesting research study measured growth factor concentrations in human milk from mothers of term and premature infants to identify any adaptive responses to premature delivery (Read *et al.*, 1985). Mothers of premature infants produced milk containing higher concentrations of epidermal growth factor and increased growth-promoting activity *in vitro*. It was suggested that changes in milk growth factor concentrations may occur as a compensatory mechanism to accelerate growth and development in pre-term infants, and if so, it follows that premature infants could benefit more from their own mother's milk than from pooled human milk or from cow's milk-based formulae.

Management of Stress

A series of studies have been conducted to investigate whether α-lactoalbumin might alleviate symptoms of stress in adult subjects. Tryptophan is a precursor of brain serotonin (a neurotransmitter believed to play an important role in the regulation of body temperature, mood, sleep, vomiting, appetite and sexuality). α-lactalbumin has a high tryptophan content.

Markus *et al.* (2000) tested whether α-la may increase the plasma Trp-LNAA ratio and reduce depressive mood and cortisol concentrations in stress-vulnerable subjects under acute stress. The plasma Trp-LNAA ratio was 48% higher after the α-la diet than after the casein diet, indicating that during the α-la diet, more tryptophan was available for uptake into the brain. With the experimental diet, higher prolactin concentrations and improvement in mood were observed, in addition to a reduced cortisol stress response in subjects who were highly vulnerable to stress.

Minet-Ringuet *et* al. (2004), showed that sleep recovery, in particular slow-wave sleep, was improved in rats re-fed with α-lactalbumin. This result confirms the close relationship between feeding and sleep and suggest that α-la could be used to improve sleep in adults submitted to nutritional disturbances.

Anti-Cancer Properties

Animal and cell culture studies suggest that milk proteins, especially whey proteins, are involved in mechanisms that provide protection against cancer development.

Whey proteins have been shown to have protective properties against a number of neoplasias. Tests have been performed to study a wide variety of cancer cell types, such as mammary cancer cells (Laursen *et al.*, 1990), apoptosis in human leukemia cells (Roy *et al.*, 2002), apoptotic action in various types of cancer cells (Svensson, 2000, Häkanson *et al.*, 1995), protective action on human prostate cells (Kent *et al.*, 2003), inhibitory action in mammary cells and rat mammary carcinoma model (Baruchel and Viaux, 1996), inhibition of growth and metastasis in murine melanoma cells (Bezault, 1994), inhibition of hepatocellular carcinoma caused by the hepatitis C virus (HCV) (Ikeda *et al.,* 1998), tumor inhibition in animal model systems (McIntosh *et al.* 1995, Bounous *et al.*, 1988, Dias *et al*, 2006) and in humans submitted to head and neck cancer surgery (Chmiel, 1997), treatment of metastatic mammal cancer and metastatic liver carcinoma (Kennedy *et al*, 1995), and skin papillomas (Gustafsson *et al*, 2004).

Apoptosis is a form of cell death in which a programmed sequence of events leads to the elimination of cells without releasing harmful substances into the surrounding area. This mechanism is important in developing and maintaining health by eliminating old cells, unnecessary cells, and unhealthy cells. In sepsis, several cell types (e.g., lymphocytes) undergo apoptosis. For full engulfment of these cells, milk fat globule epidermal growth factor-factor VIII (MFG-E8) is necessary (Miksa *et al.*, 2006).

Abdelali *et al.* (1995) studied the modulating effect of some dairy products on initiation of putative preneoplastic lesions in rat colon (aberrant crypts) by 1,2-dimethylhydrazine dihydrochloride. They evaluated uninoculated skim milk, skim milk fermented with *Bifidobacterium* sp Bio (Danone strain 173010), and a suspension of the same lactic acid bacteria were incorporated in the animals' diet. The tested diets significantly reduced the incidence of aberrant crypts compared with the control diet by 51%, 49%, and 61%, respectively.

The human α-lactalbumin made lethal to tumor cells (HAMLET) is a protein lipid complex of apo-α-la and oleic acid (C18:1:9 cis) that induces cell death selectively in tumor cells and immature cells, but spares healthy cells (Häkansson *et al.,* 1995, Svensson *et al.*, 2003, Baltzer *et al.*, 2004, Häkanson *et al.*, 2000, Fischer *et al.*, 2004, Gustafsson *et al.*,

2004). On binding to the surface of tumor cells, HAMLET was internalized and then transported to the cells nucleus causing chromatin rupture and DNA fragmentation (Häkansson *et al.,* 1995, Svensson *et al.*, 2003). Transformed bovine α-la (BAMLET) exhibited similar properties in terms of tumor cells lethality. Both bovine α-la and human α-la (BAMLET and HAMLET) caused death to some types of bacteria, in addition to inducing apoptosis in cancer cells but did not have any effect on normal adult cells (Häkanson *et al.*, 2000).

HAMLET shows broad activity against many different tumor cell lines in vitro, embryonal cells are moderately sensitive, while healthy, differentiated cells are resistant to the apoptosis-inducing effect (Baltzer *et al.*, 2004). Fischer *et al.* (2004) and Gustafsson *et al.* (2004), demonstrated that HAMLET was capable of causing the dealth of human glioblastoma cells and regression of tumors transplanted to the brain of immunodepressed rats. The efficacy of the treatment was compared with infusion of natural (i.e. non-transformed) α-la that did not exert any effect on the transplanted cells. Magnetic resonance images showed considerable differences in size between the brain tumors of the rats (456 versus 63 mm^3) treated with native α-la and HAMLET, respectively.

The protective effect of dairy, red meat and soybean proteins against the development of intestinal cancer in rats was compared by Mclntosh *et al.* (1995). The findings show that dietary whey protein and casein appear to play a much more significant role in cancer prevention than meat and soy proteins.

A study by Mclntosh *et al.* (1998), compared the efficacy of a 15% whey protein (WPC) diet with a soy protein isolate (SPI) diet at the same protein concentration. They found a highly statistical advantage for WPC relative to inhibition of aberrant crypt foci (ACF) formation. Addition of 5% of either lactoferrin or β-lactoglobulin to the SPI diet emphasized the difference between the two dietary proteins and highlighted the antitumoral properties of the two added whey proteins. The precancer lesions (ACF) were induced by administration of the colon cancer drug 1,2-dimethyhydrazine (DMH). The mechanism behind the apparent anti-cancer activity of dietary whey proteins may be related to their sulfur amino acid content, which promote gluthatione synthesis and may stabilize DNA by methylation thereby decreasing susceptibility to tumorigenesis.

Antitumor effects often involve a number of actions that directly involve stimulation of glutathione (GSH), lymphocyte and immunoglobulin synthesis. Components rich in the sulphur amino acids cysteine and methionine and containing the γ-glutamylcysteine residue which makes cysteine readily bioavailable for synthesis of glutathione, a strong xenobiotic deactivating and anti-neoplastic agent. Methionine, which can also be utilised for glutathione synthesis in times of cysteine deficiency, is also important as a methyl donor (Parodi, 1998). It is considered that oxygen radical generation is frequently a critical step in carcinogenesis, hence the effect of glutathione on free radicals as well as carcinogen detoxification, could be important in inhibiting carcinogenesis induced by a number of different mechanisms (Bounous, 2000).

Whey proteins have been shown to have protective properties against a number of pathological conditions and they have been shown to represent an effective and safe cysteine donor for GSH replenishment during GSH depletion in immune deficiency states (Bounous,

2000). Glutathione concentration is high in most tumor cells and this may be an important factor in resistance to chemotherapy.

A variant of human α-la was shown to selectively enter tumor cells and induces an apoptosis-like mechanism, probably binding to histones and thereby disrupting the chromatin in tumor cell nuclei (Duringer *et al.,* 2003).

Bovine serum albumin has been shown *in vitro* to inhibit growth of the human breast cancer cell line MCF-7 (Laursen *et al.,* 1990).

Tsuda *et al.* (2000) indicate bovine lactoferrin (bLF) as natural ingredient of milk with potential for chemoprevention of colon and other cancers. They observed that bovine lactoferrin inhibits colon carcinogenesis in the post-initiation stage in male rats treated with azoxymethane without any overt toxicity. The incidence of adenocarcinomas in the groups receiving 2% and 0.2% lactoferrin were thus 15% and 25%, respectively, in contrast to the 57.5% control value. Accordingly, Sternhagen and Allen (2001), showed that bovine alpha-lactalbumin also has antiproliferative effects in human colon adenocarcinoma cell lines.

The anticarcinogenic as well as the hypocholesterolemic effects of whey proteins can be enhanced with the use of specific probiotic cultures in the manufacture of dairy functional products.

Tavan *et al.* (2002), investigated the protective effects of fermented milk against heterocyclic aromatic amines (HAA) – initiating agents of colon carcinogenesis formed during food preparation and that present a very high mutagenic potential. Four different diets were given to four rat groups: supplemented with 20% water, 30% non-fermented milk, 30% *Bifidobacterium animalis* DN-173 010 fermented milk and 30% *Streptococcus thermophilus* DN-001 158 fermented milk. The results showed that dairy products decreased the incidence of aberrant crypts in rats: 66%, 96% and 93% inhibition with the milk-supplemented diet, the *B. animalis* fermented milk-supplemented diet and the *S. thermophilus* fermented supplemented diet, respectively.

Proteins and Peptides with Several Functionalities

Several milk proteins and peptide fraction were shown to present more than one functional property (see Figure 2).

According to Kampa *et al.* (1996), human α_{s1}-casein contains a potent opioid peptide - α_{s1}-casomorphin (f158-162) with the sequence Tyr-Val-Pro-Phe-Pro (YVPFP). Both α_{s1}-casomorphin and its amide inhibit in a dose-dependent and reversible manner the proliferation of T47D human breast cancer cells. Casomorphin peptides could, if they are well tolerated, represent a possible new, and perhaps more physiological, approach to cancer chemotherapy.

The tetrapeptide α-lactorphin (YGLF), released by gastric or pancreatic hydrolysis from α-lactoalbumin, possess ACE-inhibitory activity, as well as opioid activity (Nurminen *et al.,* 2000).

Besides mineral-binding activity, CPP is beneficial to the enhancement of mucosal immunity. Otani *et al.* (2000), evaluated the effect on serum and intestinal IgG or IgA production of a dietary casein phosphopeptide preparation (CPP-III, mainly consisting of residues 1–32 of bovine α_S2-casein and residues 1–28 of bovine β-casein) in mice.

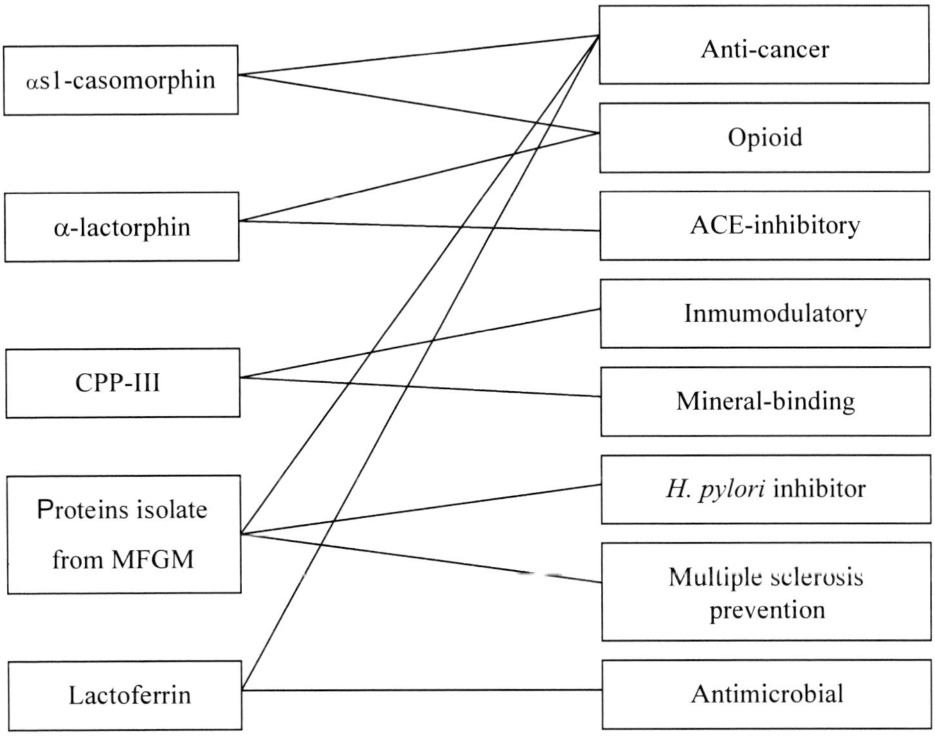

Figure 2. Examples of bioactive proteins and peptides presenting more than one functional activity.

The CPP-III-added diets resulted in high levels of serum IgA compared with those of rats fed the control diet. Mice fed CCP-III added diets exhibited a similar serum IgG response towards β-lactoglobulin injected peritoneally and ovalbumin ingested orally to those fed the control diet.

The whey protein fraction contains a number of high-affinity binding proteins that bind iron, folic acid, vitamin B12, riboflavin, retinol and vitamin D (Parodi, 1998). According to Bezault *et al.* (1994), lactoferrin is a milk protein that inhibits the growth of solid tumors and, in addition, reduces melanoma colonization of the lung. The iron-binding property of lactoferrin may make this potential pro-carcinogen unavailable for intestinal damage, whereas the vitamin B-binding proteins make their vitamins (potential anti-cancer agents) more bioavailable and protect them from utilization by intestinal microorganisms (Parodi, 1998). Free iron may act as a mutagenic promoter by inducing oxidative damage to nucleic acid structure (Weinberg, 1996).

The components of the MFGM (Milk Fat Globule Membrane) are multifunctional. They are cholesterolemia-lowering factor, inhibitors of cancer cell growth, vitamin binders, inhibitor of *Helicobacter pylori*, inhibitor of beta-glucuronidase of the intestinal *Escherichia coli*, bactericidal agent, possible suppressor of multiple sclerosis (MS), and phospholipids as agents against colon cancer, gastrointestinal pathogens, Alzheimer's disease, stress and depression (Spitsberg, 2005). The following effects are atributed to the MFGM protein fraction: prevention of the development of breast cancer in humans (Spitsberg and Gorewit, 1997), potential colon cancer prevention (Spitsberg, 2005), inhibition of *H. pylori* infection (Wang *et al.* 2001) and suppression of human multiple sclerosis. However, there are controversies about the properties of proteins and nonprotein components of the bovine milk fat globule membrane (MFGM).

Riccio (2004), listed three human pathologies that are apparently linked to cow milk consumption and to the presence of immune response against the MFGM proteins: multiple sclerosis, autism and coronary hearth disease. New investigations are necessary to elucidate if the MFGM can be considered a nutraceutical or if it should be eliminated from the products. Let's take the MS (Multiple Sclerosis) as an example. MS is a multifactorial autoimmune disease in which environmental factors, such as diet, may disrupt immunological self-tolerance to central nervous system (CNS) myelin antigens in genetically susceptible individuals. The structure being destroyed in the course of MS is myelin, which is a lipid-rich, multilamellar membrane discontinuously wrapped around nerve axons to facilitate nerve impulse conduction (Riccio, 2004). It should be clarified if the butyrophilin, present in the MFGM, can act as a possible suppressor of multiple sclerosis (Spitsberg, 2005), or if it acts as a putative auto-antigen in MS that can induce an MS-like disease (Stefferl *et al.,* 2000).

CONCLUSION

At present, a number of functions is attributed to dietary proteins in addition to contributing the amino acids for construction and restoration of body tissues. By the traditional nutritional concept, food proteins after having been ingested had to be digested and broken down into their amino acids, which are then absorbed by an energy-dependent active process before reaching the general circulation through the portal system.

After reaching the systemic circulation, two main metabolic pathways would follow: 1. the anabolic route with "de novo" synthesis of body protein; 2. the anabolic route where excess amino acids are converted to glucose (gluconeogenesis) or TCA cycle intermediates (tricarboxylic acid cycle), participating in the body's energy production process. Excess nitrogen from amino acids is transformed into ammonia and urea (urea cycle) and excreted via urine.

Advances in biochemistry and biological sciences opened up the way to demonstrate an array of newly discovered structure-function properties of numerous food proteins.

In the last decades, it has been shown that amino acid profile, molecular mass and molecular structure determines how food proteins can be digested, absorbed into the body and modulate protein metabolism and other aspects of metabolic processes.

The speed and specificity of digestion and absorption of the protein components will modulate the general metabolism in different ways.

It has been known for some time that food proteins do not have to be hydrolyzed to free amino acids before they can be absorbed by the body. Peptides of different molecular mass and even intact proteins can be absorbed into the body's general circulation and perform a variety of different functions in the body.

Many studies have demonstrated both *in vitro* and *in vivo* that proteolysis may generate peptides of various sizes and amino acid sequences with various physiological properties and capable of modulating animal and human body metabolism in very different ways, such as proteins and peptides capable of modulating immune reactions and functions; antihypertensive peptides, capable of lowering blood pressure; opioid-like peptides (both agonist and antagonists) potentially capable of regulating central nervous system functions and peripherally influencing gastrointestinal functions; caseinophosphopeptides, which bind and transport essential minerals, in addition to their reducing and antimicrobial properties; protein and peptides with antibiotic properties, acting mainly as regulatory factors of gastrointestinal functions. Many other proteins and peptides have been described with special properties having a strong potential impact on human health.

The future points to the development of nutrigenomic sciences. Increased knowledge and understanding of the role of human genomics make it now possible to study the interaction of food components, including proteins and food derived peptides, with an array of genes which might be over expressed or repressed with potential consequences to human health and diseases.

Milk protein functionality can be considered a new field of research. Most of the mechanisms involved in the activity of these proteins remain to be explored.

References

Abdelali, H., Cassand, P., Soussotte, V., Daubeze, M., Bouley, C., Narbonne, J. F. (1995). Effect of dairy products on initiation of precursor lesions of colon cancer in rats. *Nutrition and Cancer, 24*: 121-132.

Abdel-Salam, D. M. E., Czimmer, J., Debrenceni, A., Szolesanyi, J., Mózsik, G. (2001). Gastric mucosal blood flow and microcirculation. An overview. *Journal of Physiology, 95*: 105-127.

Addeo, F. Chianese, L., Salzano, A., Sacchi, R., Cappuccio, U., Ferranti, P., Malorni, A. (1992). Characterization of the 12% trichloroacetic acid-insoluble oligopeptides of Parmigiano-Reggiano cheese. *Journal of Dairy Research*, *59*: 401-411.

Ahmed, S., Mozaffar, Z., Saxena, V., Miranda, Q. R. (1998). Sepralac TM process for the separation of whey proteins. *Proceedings of the second International Whey Conference*, Chicago, USA, 27 29, IDF Special Issue 9804: 100.

Aimutis, W. R. (2004). Bioactive propertied of milk proteins with particular focus on anticariogenesis. *The Journal of Nutrition, 134:* 989S-995S.

Allen, A., Flemström, G., Garner, A., Kivilaasko, E. (1993). Gastroduodenal mucosal protection. *Physiological Reviews, 73*: 823-857.

Allescher, H. D., Storr, M., Piller, C., Brantl, V., Schusdziarra, V. (2000). Effect of opioid active therapeutics on the ascending reflex pathway in the rat ileum. *Neuropeptides*, *34*: 181-186.

Antunes, A. E. C., Silva, E. R. A., Marasca, E. T. G., Moreno, I., Lerayer, A. L.S. (2007). Probióticos: agentes promotores de saúde. *Nutrire, 32*: 103-122.

Ariyoshi, Y. (1993). Angiotensin-converting enzyme inhibitors derived from food proteins. *Trends in Food Science and Technology*, *4*: 139-144.

Ashar, M. N. and Chand, R. (2004). Fermented milk containing ACE-inhibitory peptides reduces blood pressure in middle aged hypertensive subjects. *Milchwissenschaft, 59:* 363-366.

Baltzer, A., Svanborg, C., Jaggi, R. (2004). Apoptotic cell death in the lactating mammary gland is enhanced by a folding variant of α-lactalbumin. *Cellular and Molecular Life Sciences, 61*: 1221-1228.

Baró, L., Jiménez, J., Martínez-Férez, A., Bouza, J. J. (2001). Péptidos y proteínas de la leche con propriedades functionales. *Ars Pharmaceutica, 42*: 135-145.

Baruchel, S. and Viaux, G. (1996). *In vitro* modulation of cellular gluthathione by a humanized native milk protein isolate in mammal cells and rat mammary carcinoma model. *Anticancer Research, 15:* 1095-1100.

Beena, A. and Prasad, V. (1997). Effect of yogurt and bifidus yogurt fortified with skim milk powder, condensed whey and lactose-hydrolysed condensed whey on serum cholesterol and triacylglycerol levels in rats. *Journal of Dairy Research, 64:* 453-457.

Bennett, T., Desmond, A., Harrington, M., McDonagh, D., FitzGerald, R., Flynn, A., Cashman, K. D. (2000). The effect of high intakes of casein and casein phosphopeptide on calcium absortion in the rat. *British Journal of Nutrition, 83*: 673-680.

Bezault, J., Bhimani, R., Wiprovnich, J., Furmanski, P. (1994). Human lactoferrin inhibits the growth of solid tumours and development of experimental metastesis in mice. *Cancer Research, 54*: 2310-1312.

Biolo, G., Tipton, K.D.; Klein, S.: Wolfe R.R. (1997). Anabundance supply of amino acids enhances the metabolic effect of exercise on muscle protein. *American Journal of Phsiology-Endocrinology and Metabolism,* 36:E122-E129.

Boirie, Y., Dangin, M., Gachon, P., Vasson, M. P., Maubois, J. L., Beaufrere, B. (1997). Slow and fast dietary proteins differently modulate postprandial protein accretion. *Proceedings of National Academy of Sciences of the USA, 94:* 65-71.

Borges, P. F. Z., Sgarbieri, V. C., Dias, N. F. G. P., Jacobucci, H. B., Pacheco, M. T. B., Baldini, V. L. S. (2001). Produção piloto de proteínas de leite bovino: Composição e valor nutritivo. *Brazilian Journal of Food Technology, 52:* 1-8.

Bouhallab, S., Mollé, D., Léonil, J. (1992). Tryptic hydrolysis of caseinomacropeptide in membrane reactor: preparation of active peptides. *Biotechnology Letters*, 14: 805-810.

Bounous, G. (2000). Whey protein concentrate (WPC) and glutathione modulation in cancer treatment. *Anticancer Research, 20*: 4785-4792.

Bounous, G., Batist, G., Gold, P. (1991). Whey proteins in cancer prevention. *Cancer Letters, 57*: 91-94.

Bounous, G., Papenburg, R., Kongshanv, P. A., Gold, P., Fleizer, D. (1988). Dietary whey protein inhibits the development of dimethylhydrazine induced malignancy. *Clinical and Investigative Medicine, 11:* 213-217.

Bounous, G. and Gold, P. (1991). The biological activity of undenaturated dietary whey proteins: role of glutathione. *Clinical and Investigative Medicine, 14*: 296-309.

Brody, E. P. (2000). Biological activities of bovine glycomacropeptide. *British Journal of Nutrition, 84*: 39S -46S.

Bronner, E. Calcium in exercise and sports In: Driskell, A. and Wolinsky, I. *Macroelements, water and electrolytes in sports nutrition.* Boca Raton: CRC Press; 1999; 17-27.

Burke, L. (2000). Sports foods – a new market for the food industry. *Food Australia, 52*: 405.

Carrol, K. K. (1991). Review of clinical studies on cholesterol lowering response of soy bean protein. *Journal of the American Dietetic Association, 91*: 820-827.

Carrol, K. K. and Kurowska, E. M. (1995). Soy consumption and cholesterol reduction: review of animal and human studies. *Journal of Nutrition, 125*: 594S-597S.

Cayot, P. and Lorient, D. Structure-function relationships of whey proteins. In: Damodaran, S., Paraf, A. (Eds) *Food Proteins and their Application*, New York: Marcell Dekker; 1997; 225-255.

Chabance, B., Jollés, P., Izquierdo, C., Mazoyer, E., Francoual, C., Drouet, L, Fiat, A. –M. (1995). Characterization of an antithrombotic peptide from κ-casein in newborn plasma after milk ingestion. *British Journal of Nutrition, 73*: 583-590.

Chatterton, D. E. W., Smithers, G., Roupas, P., Brodkorb, A. (2006). Bioactivity of β-lactoalbmuni and α-lactoalbumin – technological implications for processing. *International Dairy Journal, 16*: 1229-1240.

Chiu, C. –H., Lu, T. –Y., Tseng, Y –Y, Pan, T. –M. (2005). The effect of *Lactobacillus*-fermented milk on lipid metabolism in hamsters fed on high-colesterol diet. *Applied Microbiology and Biotechnology, 71*: 238-245.

Chmiel, J. F. (1997). Anti-tumor effect of dietary whey protein and its value for head and neck cancer patients. *International Dairy Federation, 4*: 310-314.

Clemente, A. (2000). Enzymatic protein hydrolysates in human nutrition. *Trends in Food Science and Technology, 11*: 254-262.

Collingnon, A. and Butel, M. J. Établissement et composition de la flore microbienne intestinale. In: Rambaud, J. C., Buts, J. P, Corthier, G., Flourié, B. Ed. Flore microbienne intestinale. Montroug: John Libbey; 2004; 19-35.

Costa, E. L., Almeida, A. R., Netto, F. M., Gontijo, J. A. R. (2003). Effect of intraperitoneally administered hydrolysed whey protein on blood pressure and renal sodium handling in awake spontaneously hypertensive rats. *Brazilian Journal of Medical and Biological Research, 38*: 1817-1824.

Cross, M. and Gill, H. (1999). Modulation of immune function by a modified bovine whey protein concentrate. *Immunol Cell Biol, 77*: 345-350.

Daniel, H., Vohwinkel, M., Rehner, G. (1990). Effect of casein and β-casomorphins on gastrointestinal motility in rats. *Journal of Nutrition, 120*: 252-257.

Davin, J. –C., Seterre, J., Mahieu, P. R. (1991). The high lectin-binding capacity of human secretory IgA protects nonspecifically mucoseae against environmental antigens. *Biology of the Neonate, 59*: 121-125.

Debabbi, H., Dubarry, M., Rautureau, M., Tome, D. (1998). Bovine lactoferrin induces both mucosal and systemic immune response in mice. *Journal of Dairy Research, 65*: 283-293.

Dias, N. F. G. P., Sgarbieri, V. C., Jacobucci, H. B., Rangel, H. A., Tanikawa, C. (2006). Dietary protein, immune function and colon carcinogenisis in mouse. *Lait, 86*: 11-14.

Donovan, S. M. and Odle, J. (1994). Growth factors in milk as mediators of infant development. *Annual Review of Nutrition, 14*: 147-167.

Drewnowski, A., Krahn, D. D., Demitrack, M. A., Nairn, K., Gosnell, B. (1992). Taste responses and preferences for sweet high-fat foods: evidence for opioid involvement. *Physiology and Behaviour, 51:* 371-379.

Düringer, C., Hamiche, A., Gustafsson, L., Kimura, H. S. C. (2003). HAMLET interacts with histones and chromatin in tumor cell nuclei. *The Journal of Biological Chemistry, 278*: 42131-42135.

Ellegard, K. H., Gammelgard-Larsen, C., Sorensen, E. S., Fedosov, S. (1999). Process scale chromatographic isolation, characterization and identification of tryptic bioactive casein phosphopeptides. *International Dairy Journal, 9*: 639-652.

Elphick, G. F., Greenwood, B. N., Campsi, J., Pleshner, M. (2003). Increased serum nigM in voluntary physically active rats: a potential role for β_1 cells. *Journal of Applied Physiology, 94:* 660-667.

El-Zahar, K., Sitohy, M., Choiset, Y., metro, F., Haertlé, T., Chobert, J. M. (2004). Antimicrobial activity of ovine whey protein and their peptic hydrolysates. *Milchwissenschaft, 59*: 653-656.

Etzel, M. R. (2004). The emerging role of dairy proteins and bioactive peptides in nutrition and health. *The Journal of Nutrition, 134*: 996S-1002S.

Facon, M., Skura, B. J., Nakai, S. (1993). Potential for immunological supplementation of foods. *Food and Agricultural Immunology*, *5*: 85-91.

Fiat, A. M., Migliore-Samour, D., Jolles, P., Drouet, I., Sollier, C. B. D., Caen, J. (1993). Biologically active peptides from milk proteins with emphasis on two examples concerning antithrombotic and immunomodulating activities. *Journal of Dairy Science, 76*: 301-310.

Finot, P. A. Effects of processing and storage on the nutritional value of food proteins. *In:* Damodaran, S., Paraf, A. (Eds). *Food Proteins and their Applications.* New York; Marcel Dekker; 1997; 551-577.

Fisher, W., Gustafsson, L., Mark, S., Bjerkvig, R., Svanborg, C. (2004). Human α-lactalbumin made lethal to tumor cells (HAMLET) kills human glioblastoma cells in brain xenographs by an apoptosis-like mechanim and prolongs survival. *Cancer Research, 64*: 2105-2112.

FitzGerald, R. J. (1998). Potential uses of caseinophophopeptides. *International Dairy Journal, 8:*451-457.

FitzGerald, R. J. Caseinophosphopeptides (CPPs) as functional ingredients. In: Mattila-Sandholm, T., Saarela, M. (Eds). *Functional Dairy Products.* Boca Raton; Woodhead Publishing; 2003; 187-202.

FitzGerald, R. J., Murray, B. A., Walsh, D. J. (2004). The emerging role of dairy proteins and bioactive peptides in nutrition and health. *American Society of Nutritional Science,* 980S-988S.

Fox, P.F. and McSweeney, P. L. H. *Dairy Chemistry and Biochemistry*. London; Blackie Academic and Professional; 1998.

Fuglsand, A., Nilsson, D., Nyborg, N. C. B. (2002). Cardiovascular effects of fermented milk containig angiotensin-converting enzyme inhibitors evaluated in permanently catheterized, spontaneously hypertensive rats. *Applied and Environmental Microbiology, 68*: 3566-3569.

Gill, H. S., Rutherfurd, K. J. (1998). Immunomodulatory properties of bovine milk. *Bulletin of the IDF, 336*: 31-35.

Gilliland, S. E., Nelson, C. R., Maxwell, C. (1985). Assimilation of cholesterol by *Lactobacillus acidophilus. Applied and Environmental Microbiology, 49*: 377-381.

Gobbetti M., Ferranti P, Smacchi E, Goffredi F, Addeo F (2000). Production of angiotensin-I-converting-enzyme-inhibitory peptides in fermented milk started by *lactobacillus delbrueckii* ssp *bulgaricus* SS1 and *Lactococcus lactis* ssp *cremoris* FT4. *Applied Enviromental Microbilogy, 66*: 3898-3904.

Goedhart A. C. and Bindels, J. G. (1994). The composition of human milk as a model for the desing of infant formulas: recent findins and possible applications. *Nutrition Research Reviews, 7*: 1-23.

Goldman, A.S., Garza, C., Nichols, B. L., Golblum, R. M. (1982). Immunologic factors in human milk during the first year of lactation. *Journal of Pediatrics, 100*: 563-567.

Gurgel, P.V., Carbonell, R.G.; Swaisgood, H.E. (2000). Fractionation of whey proteins with a hexapeptide ligand affinity resin. *Bioseparation*, 9: 385-392.

Gustafsson L., Leijonhufond, I., Aronsson, A., Mossberg, A. –K., Svanberg, C. (2004). Treatment of skin papillomas with topical α-lactalbumin-oleic acid. The *New England Journal of Medicine, 350:* 2663-2672.

Häkanson, A., Sevensson, M., Mossberg, A. –K., Sablhaewad, H., Linse, S., Lazou, I., Lönnerdal B., Svanberg, C. (1995). Apoptosis induced by a human milk protein. *Proceedings of the National Academy Sciences 92:* 8064-8068.

Häkanson, A., Sevensson, M., Mossberg, A. –K., Sablhaewad, H., Linse, S., Lazou, I., Lönnerdal B., Svanberg, C. (2000). A folding variant of α-lactalbumin with bactericidal activity against *Streptococcus pneumoniae. Molecular Microbiology, 35*: 589-600.

Hansen, M. (1995) The effect of casein phosphopetides on zinc and calcium absorption evaluated in experimental models and humans. PhD thesis, The Royal and Veterinary and Agricultural University, Denmark.

Hata, Y., Yamamoto, M., Ohni, M., Nakamura, K., Takano, T. (1996). A placebo-controlled study of the effect of sour milk on blood pressure in hypertensive subjects. *American Journal of Clinical Nutrition, 64*: 767-771.

Heaney, R. P., Saito, Y., Orimo, H. (2006). Effect of caseinphosphopeptide on absorbability of co-ingested calcium in normal postmenopausal women. *Journal of Bone and Mineral Metabolim, 12:* 77-81.

Hernández-Ledema, B., Amigo, L., Ramos, M., Recio, I. (2004). Angiotensin converting enzyme inhibitory activity in commercial fermented products. Formation of peptides under simulated gastrointestinal digestion. *Journal of Agriculture and Food Chemistry, 52*: 1504-1510.

Hernández-Ledema, B., Miralles, B., Amigo, L., Ramos, M., Recio, I. (2005). Identification of antioxidant and ACE inhibitory peptides in fermented milk. *Journal of the Science of Food and Agriculture, 85*: 1041-1048.

Höllt, V. (1983). Multiple endogenous opioid peptides. *Trends in Neuroscience, 6*, 24-26.

Hooijdonk, A. C. M., Kussendrager, K. D., Steijns, J. M. (2000) *In vivo* antimicrobial and antiviral activity of components in bovine milk and colostrums involved in non-specific defence. *British Journal of Nutrition, 84*: 127S-134S.

Ijäs, H., Collin, M., Finckenberg, P., Pihlanto-Leppälä, A., Korhonen, H., Korpela, R., Vapaatalo, H., Nurminen, M. –L. (2004). Antihypertensive opioid-like milk peptide α-lactorphin: lack of effect on behavioural test in mice. *International Dairy Journal, 14*: 201-205.

Ikeda, M., Sugiyama, K., Tanaka, T, Tanaka, K., Sekihara, H, Shimotohno, K, Kato, N (1998). Lactoferrin markedly inhibits hepatitis C virus in cultures human hepatocytes. *Biochemical and Biophysical Research Communications, 245:* 549-553.

Isolari, E. (2001). Probiotics in human disease. *American Journal of Clinical Nutrition, 73*: 1142S-1146S.

Iyer, S. and Lönnerdal, B. (1993). Lactoferrin, lactoferrin receptors and iron metabolism. *European Journal of Clinical Nutrition, 47*: 232-241.

Jacobucci, H. B., Sgarbieri, V. C., Dias, N. F. G. P. (2001). Impact of different dietary protein on rat growth, blood serum lipids and protein and liver cholesterol. *Nutrition Research, 21*: 905-915.

Jentjens, R. L. P. G., van Loon, L. J. C., Mann, C. H., Wagenmakers, A. J. M., Jeukendrup, A. E. (2001). Addition of protein and amino acids to carbohydrates does not enhance postexercise muscle glycogen synthesis. *Journal of Applied Physiology, 91*: 839-846.

Johansson, I. (2002). Milk and dairy products: possible effects on dental health. *Scandinavian Journal of Nutrition, 46*: 119-122.

Jones, E. M., Smart, A., Bloomberg, G., Burgess, L., Millar, M. R. (1994). Lactoferricin, a new antimicrobial peptide. *Journal of Applied Microbiology, 77*: 208-214.

Kampa, M., Loukas, S., Hatzoglou, A., Martin, P., Martin, P. –M, Castanas, E. (1996). Identification of a novel opioid peptide (Tyr-Val-Pro-Phe-Pro) derived from human α_{s1}-casein (α_{s1}-casomorphin, and α_{s1}-casomorphin amide. *Biochemical Journal, 319*: 903-908.

Kelly, C. P., Pothoulakis, C., Vavva, F., Castagliuolo, I., Bostwick, E. F., O'Keane, J. C., Keates, S., Lamont, J. T. (1996). Anti-*Clostridium difficile* bovine immunoglobulin concentrate inhibit cytotoxicity and enterotoxicity of *C. difficile* toxins. *Antimicrobial Agents and Chemotherapy, 40*: 373-379.

Kennedy, R. S., Konok, G. P., Bounou, G., Baruchel, S., Lee, T. D. (1995). The use of a whey protein concentrate in the treatment of patients with metastatic carcinoma: phase I-II clinical study. *Anticancer Research, 15*: 2643-2649.

Kent, K. D., Harper, W. J., Bomser, J. A. (2003). Effect of whey protein isolate in intracellular glutathione and oxidant-induced cell death in human prostate epithelial cell. *Toxicology in Vitro, 17:* 27-33.

Kitts, D. D. (2005). Antioxidant properties of casein-phosphopeptides. *Trends in Food Science and Technology, 16*: 549-554.

Kitts, D. D. and Weiler, K. (2003). Bioactive proteins and peptides from food sources. Applications of bioprocess used in isolation and recovery. *Current Pharmaceutical Design, 9*: 1309-1323.

Koldovsky, O. (1989). Critical review: search for role of milk-borne biologically active peptides for the suckling. *Journal of Nutrition, 119:* 1543-1551.

Korhnen, H., Pihlanto-Leppälä, A., Rantamäki, P., Tupasela, T. (1998). Impact of processing on bioactive proteins and peptides. *Trends in Food Science and Technology, 9:* 307-319.

Korhonen, H. (2002). Review: Technology options for new nutritional concepts. *International Journal of Dairy Technology, 55*: 79-88.

Korhonen, H. and Pihlanto, A. (2003). Food-derived bioactive peptides-opportunities for designing future foods, *Current Pharmaceutical Desing, 9*: 1297-1308.

Korhonen, H. and Pihlanto, A. (2006). Bioactive peptides: production and functionality. *International Dairy Journal, 16*: 945-960.

Kussendrager, K. D. and Hooijdonk, A. C. M. (2000). Lactoperoxidase: physico-chemical properties, occurrence, mechanism of action and applications. *British Journal of Nutrition, 84*: 19S-25S.

Lahov, E. and Regelson, W. (1996). Antibacterial and immunostimulanting casein-derived substances from milk: caseicidin, isracidin peptides. *Food and chemical Toxicology, 34*: 131-145.

Laursen, I., Briad, P., Lykkesfeldt, A. E. (1990). Serum albumin as a modulator on growth of the human breast cancer cell line, MCF-7. *Anticancer Research, 10*: 343-351.

LeBlanc, J. G., Matar, C, Valdez, J. C., LeBlanc, J., Perdigon, G. (2002). Immunomodulating effect of peptidic fractions issued from milk fermented with *Lactobacillus helveticus. Journal of Dairy Science, 85*: 2733-2742.

Leclcrc, P. –L., Gauthier, S. F., Bachelard, H., Santure, M., Roy, D. (2002). Antihypertensive activity of casein-enriched milk fermented by *Lactobacillus helveticus. International Dairy Journal, 12:* 995-1004.

Li, G. –H., Le, G. –W., Shi, Y. –H., Shrestha, S. (2004). Angitensin I-converting enzyme inhibitory peptides derived from food proteins and their physiological and pharmacological effects. *Nutrition Research, 24*: 469-486.

Lovegrove, J.and Jackson, K., Coronary heart disease. In: Mattila-Sandholm, T., Saarela, M. (Eds). *Functional Dairy Products,* Boca Raton: Woodhead Publishing; 2003.

Maeno, M., Yamamoto, N., Takano, T. (1996). Identification of antihypertensive peptides from casein hydrolysate produced by a proteinase from *Lactobacillus helveticus* CP790. *Journal of Dairy Science, 73*: 1316-1321.

Marshall, W. E. Amino Acids, Peptides and Proteins in Functional Foods. In: Goldberg, I. (Ed.) Functional Foods. New York: Chapman and Hall; 1994; 424-260.

Markus, C. R., Olivier, B. Panhuysen, G. E. M., Gusten, J. V., Alles, M. S., Tuiten, A, Westenberg, H. G. M., Fekkes, D., Koppeschaar, H. F., Haan, E. E. H. F. (2000). The bovine protein α-lactoalbumin increases the plasma ratio of tryptophan to the other large neutral amino acids, and in vulnerable subjects raises brain serotonin activity, reduces cortisol concentration, and improves mood under stress. *American Journal of Clinical Nutrition, 71*: 1536-1544.

Marteau, P., Cuillerier, E., Meance, S., Gerhardt, M. F., Myara, A., Bouvier, M., Bouley, C., Tondu, F., Bommelaer, G., Grimaud, J. C. (2002). Bifidobacterium animalis strain DN-173 010 shortens the colonic transit time in healthy women: a double-blind, randomized, controlled study. *Alimentary Pharmacology and Therapeutics, 16*: 587-593.

Maruyama, S., Mitachi, H., Tanaka, H., Tomizuka, N., Suzuki, H. (1987). Studies on the active site and antihypertensive activity of angiotensin I-converting enzyme inhibitors derived from casein. *Agricultural and Biological Chemistry, 51*: 1581-1586.

Masuda, O., Nakamura, Y., Takano, T. (1996). Antihypertensive peptides are present in aorta after oral administration of sour milk containing these peptides to spontaneously hypertensive rats. *Journal of Nutrition, 126*, 3063-3068.

Matsumoto, H., Shimokawa, Y., Ushida, Y., Toida, T., Hayasawa, H. (2001). New biological function of bovine α-lactoalbumin: protective effect against ethanol- and stress-induced gastric mucosal injury in rats. *Bioscience Biotechnology Biochemistry, 65:* 1104-1111.

McCann, K. B., Shiell, B. J., Michalski, W. P., Lee, A., Wan, J., Roginski, H., Coventry, M. J. (2005). Isolation and characterization of antibacterial peptides derived from the f(164-207) region of bovine α_{s2}-casein. *International Dairy Journal, 15*: 133-143.

McIntosh, G. H., Regester, G. O., Le Leu, R. K., Royle, P. J., Smither, G. W. (1995). Dairy proteins protect against dimethylhydarazine-induced intestinal cancers in rats. *Journal of Nutrition, 125:* 809-816.

McIntosh, G. H., Royle, P. J., Leu, R. K., Regester, G. O., Johnoson, M. A., Grinsted, R. L., Kenward, R. S., Smithers, G. W. (1998). Whey proteins as functional food ingredients? *International Dairy Journal, 8:* 425-434.

Meisel, H. (1998). Overview on milk protein-derived peptides. *International Dairy Journal, 8:* 363-373.

Meisel, H. and Bockelmann, W. (1999). Bioactive peptides encrypted in milk proteins: proteolytic activation and thropho-functional properties. *Antonie van Leeuwenhoek, 76*: 207-215.

Meisel, H. and FitzGerald, R. J. (2000). Opioid peptides encrypted in intact milk protein sequences. *British Journal of Nutrition, 84*: 27S -31S.

Meisel, H., Goepfert, A., Günther, S. (1997). ACE-inhibitory activity in milk products. *Milchwinssenschaft, 52*: 307-311.

Meisel, H. (2004). Multifunctinal peptides encrypted in milk proteins. *Biofactors, 21*: 55-61.

Meisel, H. and Bockelmann. (2004). Bioactive peptides encrypted in milk proteins: proteolytic activation and thropho-functional properties. *Antonie van Leeuwenhoek, 76*: 207-215.

Mercier, A., Gauthier, S. F., Fliss, I. (2004). Immunomodulation effects of they proteins and their enzymatic digests. *International Dairy Journal, 14*: 175-183.

Mezzaroba, L. F. H., Carvalho, J. E., Ponezi, A. N., Antônio, M. A., Monteiro, K. M., possenti, A., Sgarbieri, V. C. (2006). Antiulcerative properties of bovine α-lactoalbumin. *International Dairy Journal, 16*: 1005-1012.

Micke, P., Beeth, K. M., Buhl, R. (2002). Effects of long-term supplementation with whey proteins on plasma glutathione levels of HIV-infected patients. *European Journal of Nutrition, 41*: 12-18.

Micke, P., Beeth, K. M., Schlaak, J. F., Buhl, R. (2001). Oral supplementation with whey proteins increases plasma glutathione levels of HIV-infected patients. *European Journal of Clinical Investigation, 31*: 171-178.

Miksa, M., Wu, R., Dong, W., Das, P., Yang, D., Wang, P. (2006). Dendritic cell-derived exosomes containing milk fat globule epidermal growth factor-factor VIII attenuate proinflammatory responses in sepsis. *Shock, 25*:586-593.

Minet-Ringet, J., Ruyer, P. M., Tome, D., Even, P. C. (2004). A tryptophan-rich protein diet efficiently restores sleep after food deprivation in the rat. *Behavioural Brain Research, 152:*335-340.

Moreno, Y. F., Sgarbieri, V. C., Silva, M. N., Toro, A. A. D. C., Vilela, M. M. S. (2005). Features of whey protein concentrate supplementation in children with rapidly progressive HIV infection. *Journal of Tropical Pediatrics, 52*: 34-52.

Mullally, M. M., Meisel, H., FitzGerald, R. J. (1997a). Identification of a novel angiotensin-I-converting enzyme inhibitory peptide corresponding to a tryptic fragment of bovine β-lactoglobulin. *FEBS Letters, 402*: 99-101.

Mullally, M. M., Meisel, H., FitzGerald, R. J. (1997b). Angiotensin-I-converting enzyme inhibitory activities of gastric and pancreatic proteinase digests of whey proteins. *International Dairy Journal, 7:* 299-303.

Nagaoka, S., Futamura, Y., Miwa, K., Awano, T., Yamauchi, K., Kanamaru, Y., Tadashi, K., Kuwata, T. (2001). Identification of novel hypocholesterolemic peptides derived from bovine milk β-lactoglobulin. *Biochemical and Biophysical Research Communications, 281*: 11-17.

Nakamura, Y., Yamamoto, N., Sakai, K., Okubo, A., Yamazaki, S., Takano, T. (1995). Purification of angiotensin-I-converting enzyme inhibitors from sour milk. *Journal of Dairy Science, 78*: 777-783.

Narva, M., Kärkkäinen, M., Poussa, T., Lamberg-Allardt, C., Korpela, R. (2003). Caseinphosphopeptides in milk and fermented milk do not affect calcium metabolism acutely in portmenopausal women. *Journal of the American College of Nutrition, 22:*88-93.

Neurath, A. R., Li, Y. Y., Strick, N., Jiang, S. A. (1996). A herpesvirus inhibitor from bovine whey. *Lancet, 347*: 1703-1704.

Newburg, D. S. (1996). Oligosaccharides and glycoconjugates in human milk: their role in host defense. *Journal of Mammary Gland Biology Neoplasia, 1:* 271-282.

Noppe, W., Haezebrouck, P., Hanssens, I and De Cuyper, M. (1998). A simplified purification procedure of alfa-lactalbumin from milk using Ca^2 – dependent adsorption in hydrophobic expanded bed chromatography. *Bioseparation, 8:* 153-158.

Nuijens, J. H., van Berkel, P. H. C., Schanbacher, F. L. (1996). Structure and biological actions of lactoferrin. *Journal of Mammary Gland Biology Neoplasia, 1:* 285-294.

Nurminen, M. N, Sipola, M., Pihlanto-Leppälä, A., Piilola, K., Korpela, R., Tossavainen, O., Korhonen, H., Vappatalo, H. (2000). α-lactorphin lowers blood pressure measured by Radiotelemetry in normotensive and spontaneously hypertensive rats. *Life Sciences, 66:* 1535-1543.

Olander, M. A., Jakobsen, U. L., Hansen, M. B., Lihme, A. (2001). Fractionation of high-value whey proteins. *Scandinavian Dairy Information, 2*: 18-21.

Otani, H., Kihara, Y., park, M. (2000). The immunoenhancing property of dietary casein phosphopeptide preparation in mice. *Food and Agricultural Immunology, 12*: 165-173.

Otani, H., Odashima, M. (1997). Inhibition of proliferative responses of mouse spleen lymphocytes by lacto- and ovotranferrins. *Food and Agricultural Immunology, 9*: 193-201.

Ouwehand, A. C., Conway, P. L., Salminen, S. J. (1995). Inhibition of S-fimbria-mediated adhesion to human ileostomy glycoproteins by a protein isolated from bovine colostrums. *Infection and Immunity, 63*: 4917-4920.

Ouwehand, A. C., Salminen, S. J., Skurnik, M., Conway, P. L. (1997). Inhibition of pathogen adhesion by β-lactoglobulin. *International Dairy Journal, 7*: 685-692.

Pacheco, M. T. B., Bighetti, E., Antonio, M., Carvalho, J.E., Possenti, A., Sgarbieri, V. C. (2006). Antiulcerogenic activity of fraction and hydrolysate obtained from whey protein concentrate. *Brazilian Journal of Food Technology, 1:* 15-22.

Pacheco, M. T. B. and Sgarbieri, V. C. (2005). Effect of different hydrolysates of whey protein on hepatic glutathione content in mice. *Journal of Medical Foods, 8*: 337-342.

Pacheco, M. T. B., Amaya-Farfan, J., Sgarbieri, V. C. (2002). Partial characterization of a whey protein concentrate and its enzyme hydrolysates. *Journal of Food Biochemistry, 26:* 327-338.

Pan, Daodong, Luo, Y., Tanokura, M. (2005). Antihypertensive peptide from skimmed milk hydrolysate digested by cell-free extract of *Lactobacillus helveticus* JCM1004. *Food Chemistry, 91*: 123-129.

Park, Y. K., Kov, M. H., Carvalho, P. O. (1997). Recentes progressos dos alimentos funcionais. *Boletim da Sociedade Brasileira de Tecnologia de Alimentos: 31*: 2000-2006.

Parodi, P. W. (1996). Milk fat components: possible chemopreventive agents for cancer and other diseases. *Australian Journal of Dairy Technology, 51*: 24-32.

Parodi, P. W. (1998). A role for milk proteins in cancer prevention. *Australian Journal of Dairy Technology, 53*: p. 37-47.

Pelto, L., Laitinen, I., Lilius, E. –M. (1999). Current perspective on milk hypersensitivity. *Trends in food Science and Technology, 10*: 229-233.

Pihlanto-Leppälä, A., Koskinen, P., Piilola, K., Tupasela, T., Korhonen, H. (2000). Angiotensin I-converting enzyme inhibitory properties of whey protein digests: concentration and characterization of active peptides. *Journal of Dairy Research, 67*: 53-64.

Pihlanto-Leppälä, A. (2001). Bioactive peptides derived form bovine whey proteins: opioid and ace-inhibitory peptides. *Trends in Food Science and Technology, 11*: 347-356.

Pillanto-Leppälä, A, Rokka, T., Korhonen, H. (1998). Angiotensin I converting enzyme inhibitory peptides derived from bovine milk proteins. *Internationsl Dairy Journal, 8*: 325-331.

Pins, J. J. and Keenan, J. M. (2002). The antihypertensive effects of a hydrolysed whey protein isolate supplement (BioZate 1). *Cardiovascular Drugs Theraphy, 16*: 68.

Pouliot, Y., Wijers, M. C., Gauthier, S. F., Nadeau, L. (1999). Fractionation of whey protein hydrolysates using charged UF/NF membranes. *Journal of Membrane Science 158:* 105-114.

Qian, Z. Y., Jollés, P., Migliore-Samour, D., Schoentgen, F., Fiat, A. M. (1995). Sheep κ-casein peptides inhibit platelet aggregation. *Biochimia et Biophysica Acta, 1244*: 411-417.

Quirós, A, Hernández-Ledesma, B., Ramos, M., Amigo, L., Recio, I. (2005) Angiotensin-converting enzyme inhibitory activity of peptides derived form caprine kefir. *Journal of Dairy Science, 88*: 3480-3487.

Rasmussen, B. B., Tipton, K. D. Miller, S. L., Wolf, S. E., Wolfe, R. R. (2000). An oral assential amino acid carbohydrate supplement enhances muscle protein anabolism after resistance exercise. *Journal of Applied Physiology, 88:* 386-392.

Read, L. C., Francis, G. L., Wallace, J. C., Ballard, F. J. (1985). Growth factor concentration and growth-promoting activity in human milk following premature birth. *Journal of Developmental Physiology, 7*: 135-145.

Reynolds, E. C., Cain, C. J., Webber, F. L., Black, C. L., Riley, P. F., Johnson, I. H., Perich, J. W. (1995). Anticariogenicity of calcium phosphate complexes of tryptic casein phosphopeptides in the rat. *Journal of Dental Research, 74:* 1272-1279.

Riccio, P. (2004). The proteins of the milk fat globule membrane in the balance. *Trends in Food Science and Technology, 15*: 458-461.

Rival, S. G., Boeriu, C. G., Wichers, H. J. (2001). Casein and casein hydrolysates. 2. Antioxidative properties and relevance to lipoxigenase inhibition. *Journal of Agriculture and Food Chemistry, 49:* 295-302.

Rokka, T., Syvãoja, E. –L., Tuominen, J., Korhonen, H. (1997), Release of bioactive peptides by enzymatic proteolysis of *Lactobacillus* GG fermented milk. *Milchwissenschaft, 52:* 675-678.

Roy, M. K., Kuwabara, Y., Hara, K., Watanabe, Y., Tamai, Y. (2002). Peptides from N-terminal end of lactoferrin induce apoptose in human leukemic (HL-60) cells. *Journal of Dairy Science, 85:* 2065-2074.

Sabiki, L., Mathur, B. N. (2001). Qualitative and quantitative analysis of β-casomorphins in Edam cheese. *Milchwissenschaft, 56*: 198-200.

Sannier, F. S., Bordenave, Piot, J. M. (2000). Purification of goat beta-lactoglobulin from whey by an ultrafiltration membrane enzyme reactor. *Journal Dairy Research, 67*: 43-51.

Satake, M., Enjoh, M., Nakamura, Y., Takano, T., Kawamura, Y., Arai, S., Shimizu, M. (2002). Transepithelial transport of the bioactive tripeptide, Val-Pro-Pro, in human intestinal Caco-2 cell monolayers. *Bioscience Biotechnology and Biochemistry, 66*: 378-384.

Sautier, K., Dieng, K., Flament, C., Doucet, C., Suquet, J. P., Lemonier, D. (1983). Effects of whey protein, casein, soya-bean and sunflower proteins on the serum, tissue and faecal steroids in rats. *British Journal of Nutrition, 49*: 313-319.

Schanbacher, F. L., Talhouk, R. S., Murray, F. A., Gherman, L. I., Willett, L. B. (1998). Milk-borne bioactive peptides. *International Dairy Journal, 8*, 393-403.

Scholz-Ahrens, K. E. and Schrezeinmeir, J. (2000). Effect of bioactive substances in milk on mineral and trace element metabolism with special reference to casein phosphopeptides. *British Journal of Nutrition, 84*: S147-153.

Seppo, L., Jauhiainen T., Poussa, T., Korpela, R. (2003).A fermented milk high in bioactive peptides has a blood pressure-lowering effect in hypertensive subjects. *American Journal of Clinical Nutrition, 77:* 326-330.

Seppo, L., Kerojoki, O., Suomalainen, Korpela, R. (2002). The effect of a *Lactobacillus helveticus* LBK-16 H fermented milk on hypertension – a pilot study on humans. *Milchwissenschaft, 57*: 124-127.

Sgarbieri V. C. (1996). Proteínas em alimentos protéicos: propriedades, degradações, modificações. São Paulo: Livraria Varela; 1996.

Sgarbieri V. C. (1999). Food proteins and peptides presentig specific protection to human health. A review. *In*: Food and Nutrition Press Inc. (Org). *Food for Health in the Pacific Rim*. 1 ed. Trumbull. 1: 335-352.

Sgarbieri V. C., Rangel, H. A., Zinsly, P. F., Pacheco, M. T. B., Dias, N. F. G. P. (2000). Novel nutritional and physiological function of milk proteins. *In*: *Proceedings of 4th International Conference of Food Science and Technology*, China: Publisher Wuxi; 196-210.

Sgarbieri, V. C. (2004). Propriedades fisiológicas-funcionais das proteínas do soro de leite. *Revista de Nutrição, 17:* 397-409.

Sgarbieri, V. C. and Dias, N. F. G. P. Propriedades nutritivas e funcionais fisiológicas do leite e de suas proteínas. In: Angelis, R. and Tirapegui, J. *Fisiologia da Nutrição Humana: Aspectos Básicos, Aplicados e Funcionais*. São Paulo: Editora Atheneu; 2007.

Shah, N. (2000). Effects of milk-derived bioactives: an overview. *British Journal of Nutrition, 84*: 3S-10S.

Shuryo, N. and Modler, H. W. *Food Proteins: Processing Applications.* Canada: Wiley-VCH, 2000.

Silva, S. V. and Malcata, F. X. (2005). Caseins as source of bioactive peptides. *International Dairy Journal, 15*: 1-15.

Smith, L. L. (2003). Overtraining, excessive exercise and altered immunity. *Sports Medicine, 33*: 347-364.

Sipola, M., Finckenberg, P., Korpela, R., Vapaatalo, H., Nurminen, M, L. (2002). Effect of long-term intake of milk products on blood pressure in hypertensive rats. *Journal of Dairy Research, 69*: 103-111.

Spitsberg, V. L. (2005). Bovine milk fat globule membrane as a potential nutraceutical. *Journal of Dairy Science, 88*: 2289-2294.

Spitsberg, V. L. and Gorewit, R.C. (1997). Anti-cancer proteins found in milk. Anti-cancer proteins found in milk. Cals News, *3*. Coronell University, Ithaca, NY.

Stefferl, A., Schubart, A., Storch, M., Amini, a., Mather, I. H., Lassmann, H., Linington, C. (2000). Butyrophilin, a milk protein, modulates the encephalitogenic T cell response to myelin oligodendrocyte glycoprotein in experimental autoimmune encephalomyelitis. *Journal of Immunology, 165*: 2859-2865.

Steijns, J. M. (2001). Milk ingredients as nutraceuticals. *International Journal of Dairy Technology, 54*: 81-88.

Sternhagen, L. G. and Allen, J. C. (2001). Growth rates of human colon adenocarcinoma cell line are regulated by the milk protein alpha-lactalbumin. *Advances in Experimental Medicine and Biology, 501*: 115-120.

Sturner, R. A. and Chang, K. J. (1988). Opioid peptide content in infant formulas. *Pediatric Research, 23*: 4-10.

Svensson, M. (2000). Conversion of aplha lactalbumin to a protein inducing apoptosis. *Proceedings of the National Academy Sciences 97*: 4221-4226.

Svensson, M., Fast, J., Mossberg, A. –K., Düringer, C., Gustafsson, L., Hallgren, O., Brooks, C. L., Berliner, L., Linse, S., Svanberg, C. (2003). Alpha-lactalbumin unfolding is not sufficient to cause apoptosis but is required for the conversion to HAMLET (human α-lactalbumin made lethal to tumor cells). *Protein Science, 12:* 2794-2804.

Szabo, S., Nagy, L., Plebani, M. (1992). Glutathione, protein sulphydryls and cysteine proteases in gastric mucosal injury and protection. *Clinical Chimica Acta, 206*: 95-105.

Tauzin, J., Miclo, L., Gaillard, J. –L. (2002). Angiotensin-I-converting enzyme inhibitory peptides from tryptic hydrolysate of bovine α_{S2}-casein. *FEBS Letters: 531,* 369-374.

Tavan, E., Cayuela, C., Antoine, J., -M., Trugnan, G., Chaugier, C., Cassand, P. (2002). Effect of diary products on heterocyclic aromatic amine-induced rat colon carcinogenesis. *Carcinogenesis, 23*: 477-483.

Tomé, D. and Debalbi, H. (1998). Physiological effect of milk protein components. *International Dairy Journal, 8*: 383-392.

Tomé, D., Dumontier, A. M., Hautefeuille, M., Desjeux, J. F. (1987). Opiate activity and transepithelial passage of intact β-casomorphins in rabbit ileum. *American Journal of Physilogy - Gastrointestinal and Liver Physiology, 253*: G737-G744.

Tomita, M., Bellamy, W. R., Takase, M., Yamauchi, K., Wakabayashi, H., Kawase, K. (1991). Potent antibactericidal peptides generated by pepsin digestion of bovine lactoferrin. *Journal of Dairy Science, 74:* 4137-4142.

Tsuchita H., Sekiguchi, I., Kuwata, T., Igarashi, C., Ezawa, I. (1993). The effect of casein phosphopeptides on calcium utilization in young ovariectomized rats. *Zeitschrift fürErnährungswissenchaft, 32*: 121-130.

Tsuda, H., Sekine, K., Ushida, Y., Kuhara, T., Takasuka, N., Iigo, M., Han, B. S., Moore, M. A. (2000). Milk and dairy products in cancer prevention: focus on bovine lactoferrin. *Mutation Research, 462*: 227-233.

Tsuji, S., Hirata, Y., Mukai, F, Ohtagaki, S. (1990). Comparison of lactoferrin content in colostrums between different cattle breeds. *Journal of Dairy Science, 73*: 125-128.

Usman, Hosono, A. (1999). Viability of *Lactoacillus gasseri* and its cholesterol-binding and antimutagenic activities during subsequent refrigerated storage in nonfermented milk. *Journal of Dairy Science, 82*: 2536-2542.

Usman, Hosono, A. (2000). Effect of administration of *lactobacillus gasseri* on serum lipids and fecal steroids in hypercholesterolemic rats. *Journal of Dairy Science, 83*: 1705-1711.

Van Loon, L. J. C., Saris, W. H. M., Verhagen, H., Wagenmakers, A. J. M. (2000). Plasma insulin responses after ingestion of different amino acid or protein mixtures with carbohydrate. *American Journal of Clinical Nutrition, 72*: 96-105.

Wagenmakers, A. J. M. (1998). Muscle amino acid metabolism at rest and during exercise: role in human psysiology and metabolism. *Exercise and Sport Sciences Reviews:* 28: 287-314.

Wang, X. S., Hirmo, R., Millén, R., Wadstrom, T. (2001). Inhibition of *Helicobacter pylori* infection by bovine glycoconjugates in a BALB/cA mouse model. *Journal of Medical Microbiology, 50:* 430-435.

Weinberg, E. D. (1996). The role of iron in cancer. *European Journal of Cancer Prevention,* 5: 19-36

Wijers MC, Pouliot Y, Gauthier SF, Pouliot, M , Nadeau L. (1998). Use of nanofiltration for the desalting of peptide fractions from whey protein enzymatic hydrolysates. *Le Lait* 78: 621-632.

Yamamoto, N., Akino, A., Takano, T. (1994). Antihypertensive effects of different kinds of fermented milk in spontaneously hypertensive rats. *Bioscience Biotechnology and Biochemistry*, *58*: 776-778.

Yen, S. S. C., Quigley, M. E., Reid, R. L. Ropert, J. F., Cetil, N. S. (1985). Neuroendocrinology of opioid peptides and their role in the control of gonadotropin and prolactin secretion. *American Journal of Obstetrics and Gynecology, 152*: 485-493.

In: Protein Research Progress
Editor: A.B. Boscoe and C. R. Listow, pp. 169-187
ISBN 978-1-60456-663-3

Chapter 5

NUTRITIONAL CHARACTERISTICS OF DONKEY'S MILK PROTEIN FRACTION

Silvia Vincenzetti, Paolo Polidori and Alberto Vita
Dipartimento di Scienze Morfologiche e Biochimiche Comparate, via Gentile III da Varano, 62032 Camerino (MC), Italy
Dipartimento di Scienze Ambientali, Università di Camerino, Camerino (MC), Italy, via Circonvallazione 93/95, 62024 Matelica (MC), Italy.

ABSTRACT

Milk is one of the most common causes of food allergies among children under one year of age. No specific therapy exists for this allergy, and thus the only feasible response is to avoid assumption of milk and derived products. Studies conducted on the serum of children with hypersensitivity to milk have shown that caseins are the proteins with the greater allergenic potential. However, in some cases, children have also shown hypersensitivity to the β-lactoglobulines and to the α-lactoalbumines.

When food intolerance is diagnosed in an infant, it is often necessary to impose a period of total parenteral feeding, followed by breast feeding, considered the most correct method of re-feeding; when human milk cannot be given, alternative food sources must be sought.

Clinical studies have demonstrated that donkey milk could substitute breast feeding in infants affected by severe Ig-E mediated milk allergies. In these subjects, donkey milk is not only useful, but also safer than other types of milk. In fact donkey milk composition in lipids (high levels of linoleic and linolenic acid) and proteins (low caseins content) is very close to human milk.

In our studies on donkey milk we tried to deepen the knowledge about this product especially regarding the protein fractions. Among caseins, we identified mainly α_{S1}- and β-caseins, other types of caseins in such as κ- and γ-caseins were not found probably because of their low amount. Lysozyme content in donkey milk resulted to be very high (mean value 1.0 mg/ml) if compared to bovine (traces), caprine (traces) and human milk and varied at different stages of lactation. The high lysozyme content of donkey milk may be responsible of the low bacterial count reported in literature and also makes this milk suitable to prevent intestine infection to infants.

Among seroproteins, β-lactoglobulin and α-lactoalbumin content in donkey milk was 3.7 and 1.8 mg/ml respectively and remained invariable during the different stages of lactation.

The low caseins fraction amount as well as the high lysozyme content of donkey milk associated with its good palatability make this milk a valid alternative to powdered milk and soybean milk for infant affected by dairy cows milk allergy. Nutritional properties of donkey milk have also been compared with sheep milk, one of the best milk used for cheese-making from thousands of years, in order to evaluate the differences between characteristics of protein fractions in three different milks: human, donkey and sheep milk.

INTRODUCTION

Milk is one of the most common causes of food allergies among children under one year age. Studies conducted on the serum of children with hypersensitivity to milk have shown that the proteins mainly responsible of the allergy are α- and β-caseins, however, in some cases, also β-lactoglobulines and α-lactalbumines may be potential allergenic (Businco *et al.* 2000; Iacono *et al.* 1992). When a child is diagnosed with a cow's milk derivatives intolerance and when it is not possible breast feeding, alternative milk sources must be sought. Soy or hydrolysed formulas milks are good substitutes of human milk and also permit the children to grow up well, but they have not a good taste, therefore in some cases are refused by the infants. Furthermore it has been reported some cases of infant with intolerance to hydrolysed cow's milk proteins (Carroccio *et al.*, 2000). In these cases, a valid alternative to soy or hydrolysed milks formula could be equid or goat milks. Several clinical studies have demonstrated that donkey's milk (*Equus asinus*, *Perissodactyla*) could substitute breast feeding in infant affected by severe Ig-E mediated milk allergies. In fact donkey's milk composition in lipid and protein fraction, mineral content and lactose content is comparable to human milk (Salimei *et al.* 2004; Fantuz *et al.*, 2001; Vincenzetti *et al.*, 2005a).

Donkey's milk total protein (13-28 g/l) content is lowest with respect to bovine milk (32-34 g/l) (Salimei *et al.* 2004) but very close to the values for human (9-15 g/l) and mare's milk (Doreau and Boulot, 1989; Malacarne, *et al.* 2002). In goat milk the total protein content is closest to that of bovine milk. The whey proteins in donkey's milk represent the 35-50% of the nitrogen fraction, whereas in cow's milk represent only the 20% (Herrouin *et al.*, 2000). The ratio caseins/whey proteins depends on species and it is high in rapidly growing mammals (lambs and calves). The amount of total caseins in bovine and goat milk is very similar and it is higher with respect to human and equine milk (Table 1). In donkey's milk the total casein content is about 8.7 g/l, very close to the value found in the human milk (Salimei *et al*, 2004).

Bovine milk contains four types of casein (CN) denoted as α_{s1}-, α_{s2}-, β- and κ-casein all of which are produced by specific genes. However whole casein sample separated by polyacrilamide gel electrophoresis (PAGE) results in several bands that represent posttranslational modification of one of the four caseins (different degrees of phosphorylation, glycosilation, proteolysis, formation of disulfide bridges).

Table 1. Composition of the casein fraction of bovine, human, equine and goat milk reported by some authors

	Bovine [a] (g/l)	Human [a] (g/l)	Equine [b] (g/l)	Goat [c] (g/l)
Total caseins	27.2	5.8	14.0	22.0-28.0
α_{s1}-casein	10.0	0.8	2.5	10.0
α_{s1}-casein	3.7	-	0.2	3.0
β-casein	10.0	4.0	11.0	11.0
κ–casein	3.5	1.0	0.25	4.0

a) Martin and Grosclaude, 1993.
b) Miranda *et al.*, 2004.
c) Greppi and Roncada, 2005.

The α_{s1}-casein is the major component of the caseins (up to 40% of the bovine milk caseins); this casein shows an acidic region responsible for the binding of calcium and also three hydrophobic regions (Farrell *et al.*, 2004; Creamer, 2002). The other major component of caseins is β-casein (45% of the casein bovine milk) that shows the largest region of high hydrophobicity whereas at the N-terminal region it shows a large acidic region.

The minor component of caseins is α_{s2}-casein (10% of the CN fraction in bovine milk) and it is the least hydrophobic but the most phosphorylated of the caseins exhibiting varying levels of post-transactional phosphorylation (Farrell *et al.*, 2004; Creamer, 2002).

The k-casein constitutes 10-20% of whole casein in the cow's milk. This family consists of a major carbohydrate-free component and 6 other minimum components with different degrees of phosphorylation and glycosilation (Farrell et al., 2004). In general k-casein consists of two different domains linked together by a peptide positively charged: a N-terminal domain with a net positive charge and with a hydrophobic region and a C-terminal domain with a net negative charge containing several polar aminoacidic residue. k-casein plays an important role for the production of many cheese varieties (Creamer, 2002).

The casein fraction of mare's milk has been studied from several authors (Visser *et al.* 1982; Ochirkhuyag *et al.* 2000; Egito *et al.* 2002; Miranda *et al.* 2004.), leading to the identification and characterization of α_{S1}-, α_{S2}-, β-caseins. The presence of κ-like-casein in mare's milk was controversial and several studies, carried out in order to find this protein (Kotts *et al.* 1976; Ono *et al.* 1989; Ochirkhuyag *et al.* 2000), indicated that a κ-like-casein could be absent or present in small and therefore undetectable quantities. However a more recent work conducted in pony mare's milk (Miranda *et al.* 2004) revealed the presence of the κ casein even if in a small amount (see also table 1). The β-casein of mare's milk can be divided into several sub-fractions that present a high heterogeneity in the degree of phosphorylation, in the similar way of human milk where six β-caseins different in their phosphorus content are present.

The whey proteins are the group of milk proteins that remains soluble in milk serum after precipitation of caseins at pH 4.6 at 15°C. The major component of this fractions are α-lactalbumin (α-LA), β-lactoglobulin (β-LA), serum albumin (SA), immunoglobulins (Ig) and proteose-peptone fraction. Among whey proteins, α-lactoalbumin and β-lactoglobulin are of particular interest because they are considered the principal milk allergens in neonates and children (Carroccio *et al.* 1999). Alfa-lactalbumin has a specific physiological function in the mammary gland, α-LA interacts with the enzyme β-1, 4-galactosyltransferase to form the lactose synthase complex. α-LA modifies the substrate specificity of β-1,4-galactosyltransferase allowing to the formation of lactose from glucose and UDP-glucose (Kuhn *et al.* 1980). All known α-LA contains a tightly bound Ca^{++} that influence the stability and the structure of the protein but it does not affect the lactose synthesis. Donkey's milk α-LA is present in two isoforms A and B with different isoelectric point (Giuffrida *et al.* 1992). Bovine milk contains α-LA at a concentration of about 1.2-1.5 g/l (Curadi *et al.*, 2000; Martin and Grosclaude, 1993) similar to the value in human milk (1.6 g/l) and goat milk (1.2 g/l) whereas in the mare's milk the amount of α-LA is 3.3 mg/ml (Miranda *et al.* 2004; Greppi and Roncada, 2005). In the bovine milk, the concentration of α-LA is higher at the beginning of lactation and decreases near the end of lactation (Butler *et al.*, 1994). This decrease is correlated to the decrease of the lactose concentration at the end of the lactation.

The major whey protein in bovine milk is β-lactoglobulin (β-LG), about 10% of the total milk protein in cow's milk; it exists in several genetic variants but it is generally presents as A or B variant (Creamer and Sawyer, 2002). The structure of β-lactoglobulin is similar to retinol-binding protein (Godovac-Zimmerman *et al.* 1986) and lipocalins suggesting that β-LG may have a role in the transport of fatty acids and vitamin A; β-LG and is considered the principal milk allergen in neonates and children (Carroccio *et al.* 1999). Three genetic variant for donkey β-LG were found: one of them is a β-lactoglobulin A-like with three aminoacid substitutions, the other two are similar to the β-lactoglobulin B-like with two aminoacid exchange (Herrouin *et al.* 2000). In human milk β-LG was not found (Martin and Grosclaude, 1993).

Milk contains several antimicrobial factors such as lactoferrin, lactoperoxidase, lysozyme, N-acetyl-β-D-glucosaminidase that have the ability to kill and inhibit a broad spectrum of bacteria. These factors may function in the digestive tract to reduce the incidence of gastro-intestinal infections or by defending against infections of the mammary gland. Milks of different species contain different amounts of the various antimicrobial factors.

Lactoferrin presents in milk is synthesized by the epithelial cells of the mammary gland, human milk is particularly rich in lactoferrin: the concentration in the colostrum ranges from 1.0 to 16.0 g/l whereas in the mature milk is 1.0-2.0 g/l. In bovine colostrum lactoferrin ranges from 0.2 to 5.0 g/l but decreases in mature milk to 0.1-0.2 g/l. In the mature goat milk the amount of lactoferrin is low ranging from 0.02 to 0.2 g/l (Greppi and Roncada, 2005), in donkey's milk this value is at present unknown. Lactoferrin is an iron-binding protein and acts as antimicrobial factor by depriving micro-organism of the iron necessary for their growth.

Milk peroxidase is known as lactoperoxidase and acts against bacteria by oxidative mechanism; this enzyme plays an important role in the protection of the mammary gland against infections. In bovine milk the concentration is about 0.03 g/l whereas in the human milk the amount is about 20 fold lower than bovine milk. Lactoperoxydase concentration in donkey's milk actually it is not known. Lysozyme exerts its antimicrobial activity by the hydrolysis of glycosidic bonds of mucopolisaccarides in bacterial cell walls. Lysozyme is an enzyme presents in the milk of some species, in human milk the concentration is 0.12 g/l , in bovine and in goat milk is nearly undetectable, in mare's milk is 1.1 g/l (Miranda *et al.*, 2004). About donkey's milk, previous studies revealed a high lysozyme content if compared to mare's milk and bovine milk, that may be responsible for the low bacterial concentration noted (Salimei *et al.* 2004). Other authors reported the presence in donkey's milk of two variants of lysozyme: LYS A and LYS B that differ for three aminoacid substitutions at position 48, 52 and 61 (Herrouin *et al.* 2000).

In table 2 is shown a comparison of the relative amounts of whey proteins and other milk proteins of different milks.

Donkey's milk lactose content is close to that of human milk (70-72 g/l) but it is higher with respect to bovine milk (48 g/l), and goat milk (41-44 g/l), as described by Greppi and Roncada (2005). The high content of lactose is responsible of the good palatability of this milk and of optimisation of intestinal absorption of calcium that is essential on bone mineralization in the infants (Playford *et al.*, 2000). Because of the high contents of lactose but especially of lysozyme donkey's milk revealed to be a good growth medium for probiotic *lactobacilli* strains useful for the production of fermented beverages with therapeutic properties (Chiavari *et al.*, 2005).

Table 2. Comparison of the concentration of non-casein proteins in bovine, human, equine and goat milk reported by some authors

	Bovine [a,b] (g/l)	Human [a,b] (g/l)	Equine [c] (g/l)	Goat [b] (g/l)
Total whey proteins	4.5	6.3	7.4	6.0
α-lactalbumin	1.2	1.6	3.3	1.2
β-lactoglobulin	3.3	-	3.0	3.1
Lysozyme	trace	0.04-0.2	1.1	trace
Lactoferrin	0.1	1.7	n.d	0.02-0.2
Immunoglobulins	1.0	1.1	n.d	1.0
Albumin	0.4	0.4	n.d	0.5

a) Martin and Grosclaude, 1993
b) Greppi and Roncada, 2005
c) Miranda *et al.*, 2004

Finally, donkey's milk showed pharmaceutical and nutritional properties, known since ancient time as shown by Doreau, (1991) and Solaroli *et al.*, (1993).

Donkey Milk Nutritional Properties

Donkey milk is traditionally known as an alternative food for infants, on the other hand scientific references on its composition has been rare for a long time (Oftedal and Jenness, 1988; Schryver *et al.*, 1986; Salimei *et al.*, 2004). Objective difficulties in milking unselected and/or bred in extensive system animals represented one of the main cause of such a gap in both product and nutritional information, that in turn could justify the observed caution in donkey milk utilisation as dietetic food for infancy (MacDonald, 1995). Besides criticism, some advantages were also recognised to the use of donkey milk compared to other elemental formulas containing hydrolysed protein or soy for reasons of palatability, easy of preparation, cost and allergenicity (MacDonald, 1995).

More recent clinical studies confirm donkey milk feeding as a safe and valid treatment of cases of multiple food intolerance (Iacono *et al.*, 1992; Carroccio *et al.*, 2000).

Dairy cow's milk protein intolerance (CMPI) represents the most frequent food intolerance in infancy (European Allergy White Paper, 1997), ranging between 2 and 7.5% of the infant population. Because of the CMPI consequences on intestinal functionality, it is important to consider that patients often show intolerance to other foods used instead of cow's milk, including powdered milk containing soy or hydrolysed protein. The risk of a bad nutrition is considered high in infants with multiple food intolerance (Iacono *et al.*, 1992).

Recent studies (Fantuz *et al.*, 2001; Vincenzetti *et al.*, 2005a) have demonstrated the high similarity of donkey milk compared to human milk. Donkey milk is characterised by a high content of lactose and minerals whose levels, involved in bone mineralisation, do not seem to negatively affect the renal load of solutes of milk itself, important factor involved in renal function (Iacono *et al.*, 1992).

Considering specifically donkey milk protein content, the average of caseins and whey proteins contents of donkey milk were found similar to the respective levels in human milk (Greppi *et al.*, 1996; Vincenzetti *et al.*, 2005b), while a low content of allergenic components of milk and significant levels of antimicrobial factors, such as lactoferrin, lactoperoxidase and lysozyme, have been also evidenced by Fantuz *et al.* (2001).

Most of the scientist believes that the positive results from the paediatric trials are due to the particular composition of donkey milk protein fraction.

Besides the traditional nutrients of the diet, bioactive molecules in functional foods, including milk, are clearly important in providing hormonal, antibacterial, probiotic or pharmacological support (Donnet-Hughes *et al.*, 2000). On this regard, the properties of donkey milk could be partially due not only to its probiotic activity, related to the high lactose content (Coppola *et al.*, 2002), but also to the presence of trophic and protective factors, immunoglobulins and peptides with antibacterial, immunomodulating and hormonal activity. Some of these protective factors, often characterised by interspecific activity, are also present in dairy cows milk, but may not survive the technological processing used during the preparation of infant formulas. This could explain the greater susceptibility to infection and development of allergy reported in formula-fed compared with breast-fed infants (Donnet-Hughes *et al.*, 2000).

It is also important to consider the high lysozyme content of donkey milk in relation with its probable role in preserving the intestinal ecosystem from potentially harmful microbes. The use in dermatology of donkey milk, well known since ancient Rome, seems as well to be related to its lysozyme content, efficacious in soothing skin and scalp inflammations (Cotte, 1991).

Substances with bioactive properties are also in the category of lipids: according to Molkentin (1999), increasing importance is now attributed to the milk content of mono-, di-

and triglycerides, free fatty acids, phospholipids, glycolipids and steroids along with minor constituents, like carotenoids, vitamins and lipoproteins (Dugo *et al.*, 2005).

Considering the fatty acids composition of milk, most interest is actually addressed to polyunsaturated fatty acids (PUFAs) of both the n-3 and n-6 series, essential in immunomodulating activities and growth (Gill and Valivety, 1997) as well as in the development of the neonatal nervous system (DHA). On this regard, results of a study conducted by Doreau *et al.*, (2002) show the interesting free fatty acids profile of donkey milk, with a lower total saturated fatty acids content, an average higher level of total unsaturated fatty acids and a higher PUFA n3/PUFA ratio than that reported for sheep and goats milk. Those experimental observations could support the positive contribution of donkey milk to an efficient cerebral growth and the development of a normal and complete immunitary system in the neonate (Carroccio *et al.*, 2000).

The good palatability of donkey milk and the very close chemical composition compared to human milk seem to encourage the use of this product in infant feeding especially in cases of multiple food hypersensitivity, lactose intolerance cases obviously excluded.

Human milk is obviously considered by pediatric sceintists the best food for infant nutrition during the first months of age; for this reason we decided to compare this milk with donkey milk, the best substitute for human milk, according to the opinion of many scientists (Iacono *et al.*, 1992; Carroccio *et al.*, 2000), and with sheep milk, a product used for cheese-making from thousands of years, known for the very good technological properties of this milk (Chiofalo *et al.*, 2004).

RESULTS

Protein Characterization of Skimmed Milk in Human, Donkey and Sheep

Donkey's milk was obtained by pluriparous asses Ragusana breed 60 days after parturition. The animals were machine milked following the protocol previously described by Salimei *et al.* (2004). Human milk samples were provided by healthy donors 60 days after parturition and also sheep samples were obtained 60 days after parturition by pluriparous Comisana ewes. Skimmed milk was prepared from 20 ml of fresh milk by centrifugation at 3000 g for 30 min. at 15°C. The protein concentration on skimmed milk samples was determined by the method described by Bradford (1976). The total protein content determined on skimmed donkey's milk was 13.7 g/l, whereas in human and sheep skimmed milk was 11.2 and 61.6 respectively.

The skimmed milk samples were analyzed by 15% sodium dodecyl sulphate polyacrilamide gel electrophoresis (SDS-PAGE) done as described by Laemmli (1970), using the Mini Protean III apparatus (Bio-Rad, gel size 7x 8cm x 0.75mm). The markers used were Bio-Rad molecular weight standards, low range (Standard LMW: phosphorylase b, 97.4 kDa; bovine serum albumine, 66.2 kDa; ovalbumin, 45.0 kDa; carbonic anhydrase, 31.0 kDa; soybean trypsin inhibitor, 21.5 kDa; lysozyme, 14.4 kDa). The results of the electrophoretic analysis showed that donkey's skimmed milk has a composition very close to human milk, as expected (Figure 1).

Considering the electrophoretic pattern of donkey's skimmed milk (Figure 1, lane a) the band 1, with a molecular weight of 80.0 kDa, is tentatively assigned to lactoferrin, that seem to be present in donkey's milk even if in low amount. The bands 2 and 3 showing a molecular weight of 66.0 and 60.0 kDa are respectively assigned to seric albumin and to immunoglobulins.

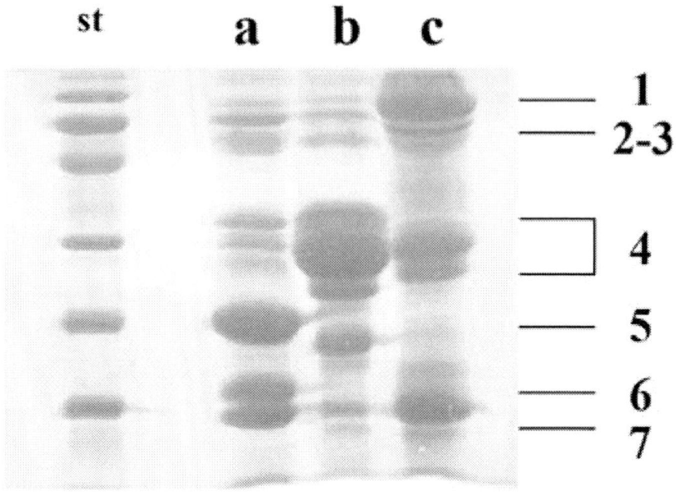

Figure 1. Comparison of the electrophoretic pattern (15% SDS-PAGE) of skimmed milk in donkey (lane a), sheep (lane b). human (lane c) and St: Standard Bio-Rad LMW range (phosphorylase b, 97.4 kDa; bovine serum albumine, 66.2 kDa; ovalbumin, 45.0 kDa; carbonic anhydrase, 31.0 kDa; soybean trypsin inhibitor, 21.5 kDa; lysozyme, 14.4 kDa).

The diffuse band 4, with a molecular weight ranging from 42.0 to 23.0 kDa, corresponds to the casein fraction whereas the large band 5, with a molecular weight of 22.0 kDa, is the β-lactoglobulin, the band 6 corresponds to lysozyme (molecular weight 14.6 kDa) and finally band 7 is the α-lactoglobulin (molecular weight 14.2 kDa). In donkey's milk is considerable the amount of whey proteins with respect to the casein fraction. On the contrary, caseins are the main protein fraction present in sheep skimmed milk (Figure 1, lane b) and lysozyme seems not present. In the human skimmed milk (Figure 1, lane c) is considerable the band corresponding to the lactoferrin but it is not present the β-lactoglobulin.

Casein and Whey Protein Fraction Determination in Human, Donkey and Sheep Milk

Caseins in milk are those phosphoproteins that precipitate from raw skim milk after acidification to pH 4.6 with 10% acetic acid at 15°C, whereas the term whey proteins has been used to identify the group of proteins that remains soluble after precipitation of the casein fraction. The content of total caseins and total whey protein where determined on donkey, sheep and human milk by the method of Bradford (1976). The result is shown in table 3.

Table 3. Determination of total caseins and whey proteins in donkey, sheep and human milk

	Donkey (g/l)	Sheep (g/l)	Human (g/l)
Total whey proteins	7.50	17.0	8.00
Total caseins	6.60	39.0	5.60

Whey Protein Fractions

The whey protein fractions from each milk sample were subjected to reversed- phase chromatography on HPLC (RP-HPLC). Reversed phase column was a C4 Prosphere 300Å (5μm, 4.6 mm I.D., 150 mm. Alltech) connected to an Äkta Purifier (Amersham Biosciences). The column was equilibrated in trifluoroacetic acid (TFA) /H_2O 1:1000 v/v (Buffer A) and the elution was achieved by the following step gradient with buffer B (TFA/H_2O/Acetonitrile 1:100:900 v/v): % D=0, 10 min; % D=20, 10 min.; % D=40, 0.1 min.; % D=60, 40 min. The flow rate was 1 ml/min and fractions of 0.5 ml were collected.

Before the chromatographic course, the samples were clarified by the addition of two volumes of buffer CL (0.1 M bis-tris, pH 8.0 containing 8 M urea, 1.3% trisodium citrate, 0.3 % DTT). 100 μl of clarified samples were directed loaded into RP-HPLC column (Miranda *et al.*, 2004). The results of the chromatographic separation of donkey, sheep and human skimmed milk, are shown in figure 2. Each peak from the chromatographic courses was indicated by a progressive number and the obtained peaks were analyzed by 15% SDS-PAGE.

The electrophoretic and the subsequent N-terminal sequence analysis allowed to the identification of the peaks eluted by the RP-HPLC (figure 3). As shown in figure 2A and 3A, after RP-HPLC of donkey milk whey proteins fraction, four main peaks were recovered, named 4-5, 6, 7, and were identified after 15% SDS-PAGE and N-terminal sequence. N-terminal aminoacid sequence was determined by Edman degradation using an automatic protein sequencer (Protein Sequencer G1000A, Hewlett Packard) on passively transferred electrophoretic bands on PVDF membrane.

In particular the peaks 4-5 with a molecular weight of 14.60 kDa resulted to be lysozyme (sequence KVFSKXELA). The two peaks observed may be two variants of lysozyme: LYS A and LYS B that differ for three aminoacid substitutions at position 48, 52 and 61 as observed by Herrouin *et al.* (2000).

The peak 6 was identified as α-lactoalbumin, molecular weight 14.12 kDA (sequence KQFTKXELSQVLXSM), and the peak 7 with a molecular weight of 22.40 kDa resulted the β-lactoglobulin (sequence TNIPQTMQ). The molecular weights calculated for lysozyme α-lactoalbumin and β-lactoglobulin are in good agreement to those reported for donkey's milk by Herrouin *et al.* (2000). In the sheep milk (figures 2B and 3B) the peaks 8-9-10, with a molecular weights of about 30.0-40.0 kDa, may corresponds to caseins that remained in solution after acidic precipitation.

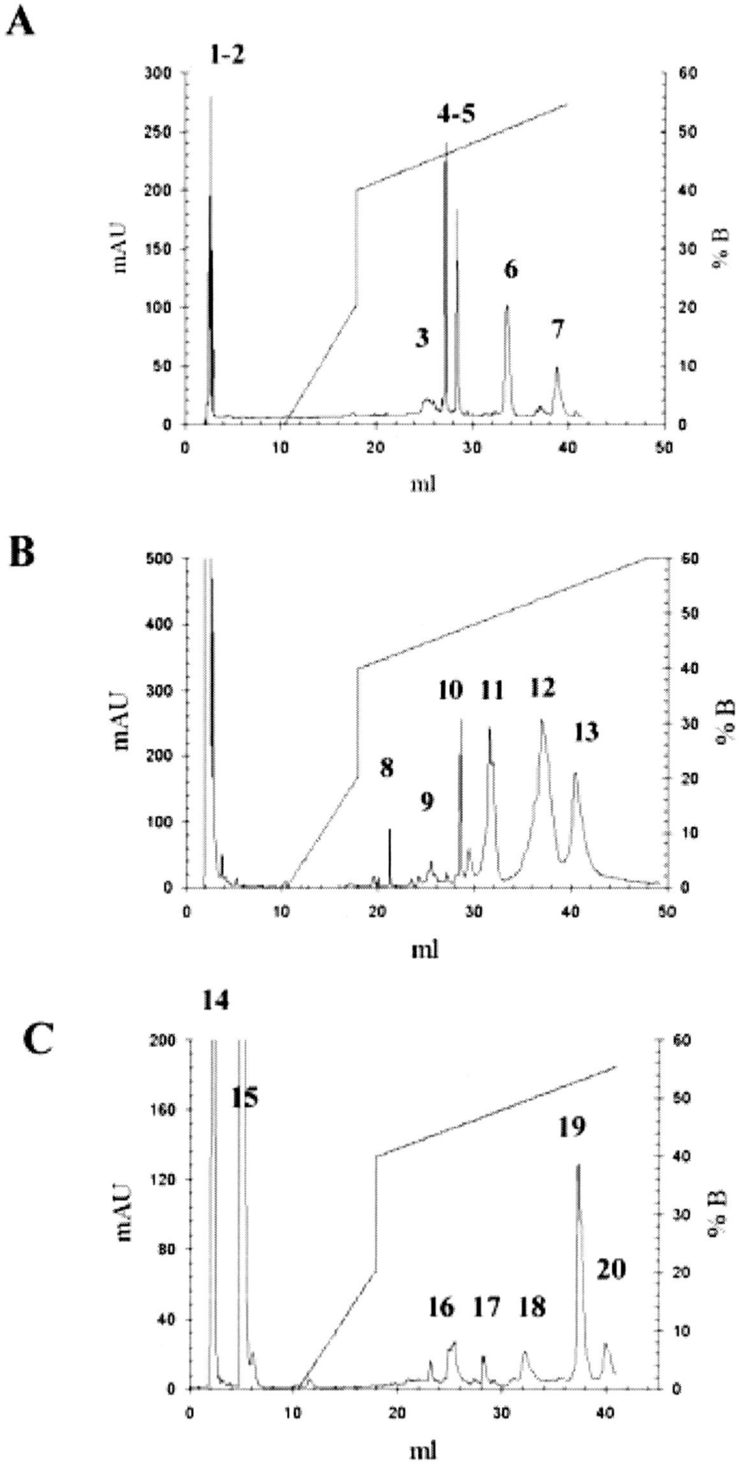

Figure 2. Comparison of a chromatographic separation in RP-HPLC of milk whey proteins from: A)donkey, B) sheep, C) human.

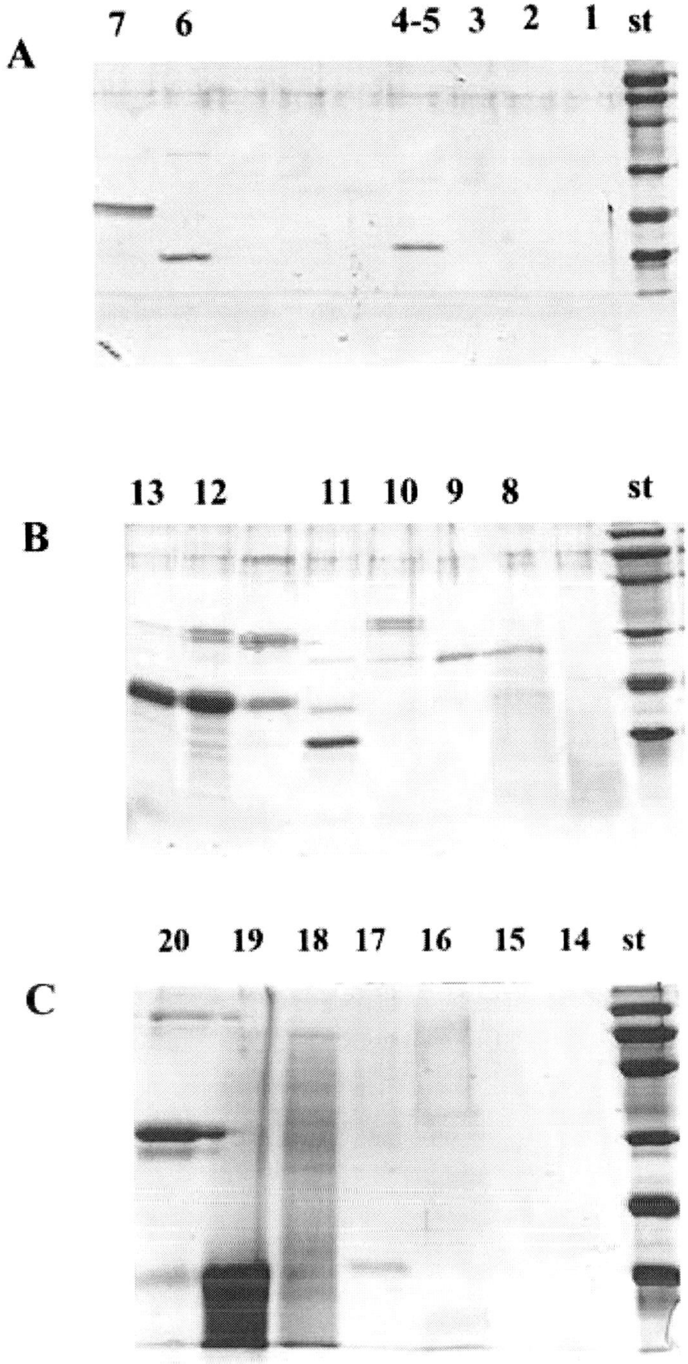

Figure 3. Electrophoretic separation (15% SDS-PAGE) of the peaks eluted from the chromatography on RP-HPLC of milk whey proteins from: A) donkey, B) sheep, C) human. St: Standard Bio-Rad LMW (phosphorylase b, 97.4 kDa; bovine serum albumine, 66.2 kDa; ovalbumin, 45.0 kDa; carbonic anhydrase, 31.0 kDa; soybean trypsin inhibitor, 21.5 kDa; lysozyme, 14.4 kDa).

The peak 11 (13.0 kDa) is identified as α-lactoalbumin, whereas peaks 12 and 13 corresponds to the two genetic variants of the β-lactoglobulin (18.2 kDa). In the bovine milk the main genetic variants A and B occurs at high frequency in most breeds and the presence of one of the two variants may influence the characteristics and the cheese making of the milk (Farrell *et al.*, 2004).

The chromatogram profile on RP-HPLC of human skimmed human milk and the electrophoretic pattern of the peaks eluted (Figure 2C and 3C), are very similar to that observed in donkey's milk except for the β-lactoglobulin that in human milk is not present, as already observed by other authors (Martin and Grosclaude., 1993). In particular, the peak 17 and 19 correspond to the lysozyme and α-lactoalbumin respectively.

A quantitative determination of lysozyme, α-lactoalbumin, β-lactoglobulin was achieved after the RP-HPLC, by measuring the area of the peaks corresponding to the three whey proteins. The standard solutions of egg white lysozyme (2; 1; 0,5; 0,25; 0,15 mg/ml), bovine milk β-lactoglobulin (0.17; 0.30; 0.75; 1.0; 1.5 mg/ml) and bovine milk α-lactalbumin (0.25; 0.35; 0.50; 0.75; 1.0; 1.5 mg/ml) were prepared in buffer CL and 100 μl of each standard solution was separately loaded on the RP-HPLC column. The area of the each standard peak was measured using the walley-to walley integration mode and the quantification was obtained by a calibration curve obtained by relating the concentration in micrograms of each standard loaded in the column versus the area corresponding to each concentration. The quantity of lysozyme, β-lactoglobulin, and α-lactalbumin, given in table 4, was determined by using the calibration curve.

Lysozyme amount in donkey's milk with a value of 1.0 g/l resulted to be higher with respect to sheep milk (trace), whereas is very close to pony mare's milk (1.1 g/l), as observed by Miranda *et al.* (2004). The mean β-lactoglobulin content in donkey's milk (3.75 g/l) was very close to that of bovine milk (3.3 g/l) and pony mare's milk (3.0 g/l), whereas in human milk this protein seems not present (Martin and Grosclaude., 1993). The sheep milk contains the higher amount of β-lactoglobulin with respect to the milks considered in this study. The amount of this protein was calculated for each of the two A and B variants.

Alfa-lactalbumin content shows a value of 1.8 g/l. The data obtained for donkey's milk is in good agreement with α-lactoalbumin content from human (2.2 g/l) and from sheep (1.63 g/l) but it is very low if compared to pony mare's α-lactoalbumin content (3.3 g/l).

Table 4. Quantitative determination of lysozyme, β-lactoglobulin, and α-lactalbumin in donkey, sheep and human milk

	Donkey (g/l)	Sheep (g/l)	Human (g/l)
Lysozyme	1.00	trace	not detectable
β-lactoglobulin	3.75	A 16.7 B 11.0	-
α-lactalbumin	1.80	1.63	2.2

Furthermore the whey protein concentration in donkey milk was calculated at different stages of lactation (60, 90, 120, 160, e 190 after parturition). The lysozyme content was variable during the different stages of lactation with a marked decrease 160 days after parturition, starting from the value of 1.34 g/l 60 days after parturition to the value of 0.76 g/l 190 days after parturition.. β-lactoglubulin content resulted to be stable during the different stages of lactation with a mean value of about 3.75 g/l. Finally donkey's milk α-lactoalbumin content 60 days after parturition was 0.81 g/l and increased two fold 90 days after parturion and remained quite stable with a mean of about 1.8 g/l until the last stage of lactation.

Casein Fractions

Whole caseins from each milk sample, precipitated after acidification to pH 4.6 with 10% acetic acid at 15°C were resuspended in buffer CL and 100 μl directly loaded into RP-HPLC column. The results, obtained from each precipitated casein samples is given in figure 4. Also in this case the peaks eluted from the RP-HPLC were submitted to 12% SDS-PAGE and (only for the donkey's milk samples) to the N-terminal sequence analysis. In donkey's milk eight main peaks were recovered (figures 4A) each of them submitted to 12% SDS-PAGE (figure 5A). The protein amount of peak 3 (one diffuse band of 35.80 kDa) and peak 4 (main band of 30.10 kDa) was not sufficient for their identification by N-terminal sequence; the peak 5 showed a more intense band with molecular weight of 33.30 kDa, identified after sequencing as α_{s1}- casein (RPKLPHRQPE) and a weaker band of 35.10 not identified. Both peak 7 and peak 8 with a molecular weight of 37.50 kDa, were identified as β-caseins. Furthermore, the β-casein sequence of the peak 8 (REKEALNV) showed a substitution E→A in the fifth aminoacid.

Therefore by RP-HPLC performed on donkey's milk precipitated caseins, it was possible to separate and identify α_{S1}-caseins having N-terminal sequence RPKLPHRXPE and variable molecular weights: 30.0, 31.30, 33.0, kDa, and β-caseins, having N-terminal sequence REKEELNVS and molecular weights varying from 33.0 to 37.5 kDa.

The observed heterogeneity within each class of caseins may be due to the existence of genetic variants or to the different degree of post-translational modifications (Stelwagen, 2003). Actually, it was not possible to determine in donkey's milk the presence of other type of caseins like αs_2-, γ- and κ- that were found, although in a low amount, in mare's milk (Doreau, 1991). It may be possible that these variants of caseins in donkey's milk are present in minute and therefore undetectable quantities making difficult their characterization. However, the presence of α_{s2} casein in human milk has not been demonstrated (Miranda *et al.*, 2004).

The sheep milk, as expected, showed the highest amount of caseins (Figures 4B and 5B) with respect to the other two milks analyzed.

The RP-HPLC chromatographic profile of the human milk casein fraction resembles that of donkey's milk (Figure 4C).

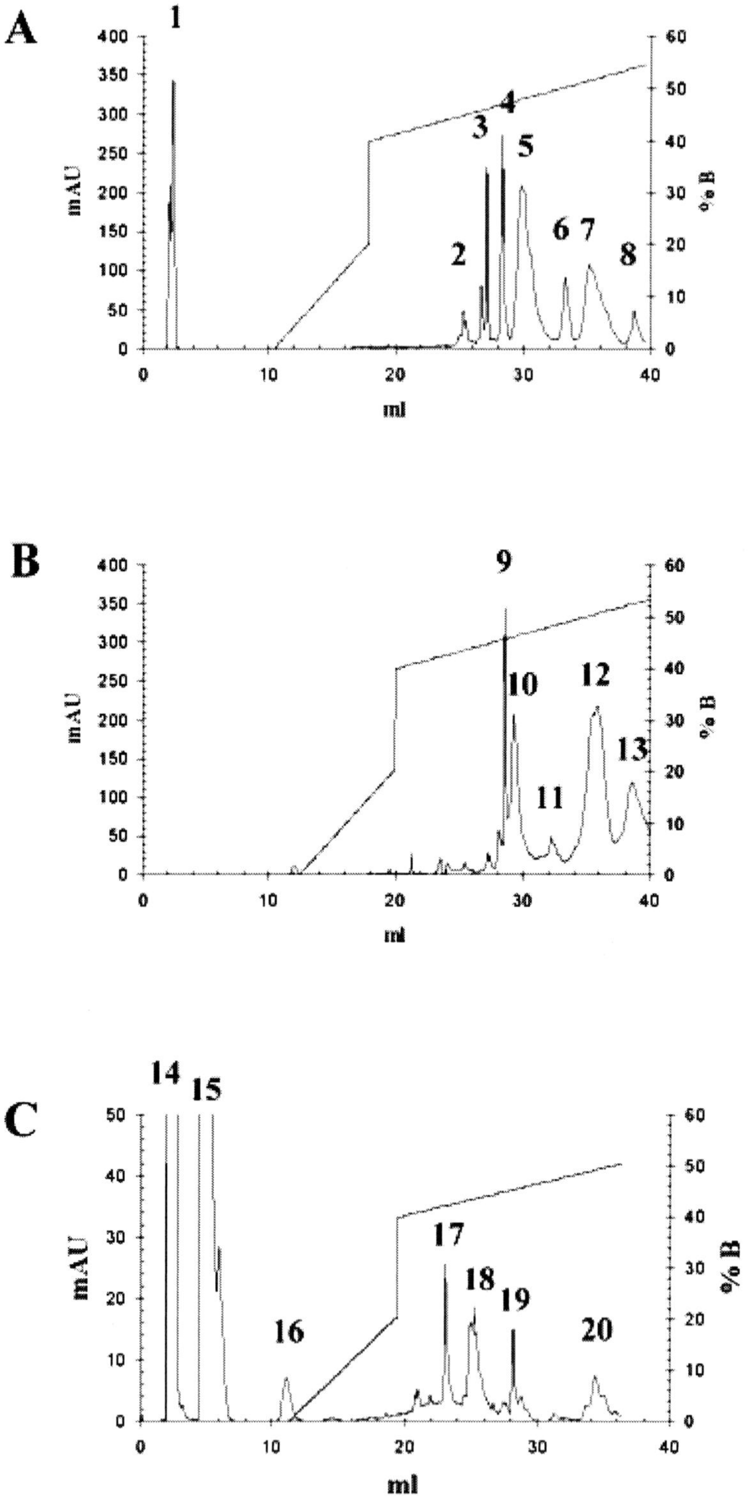

Figure 4. Comparison of a chromatographic separation in RP-HPLC of milk casein fractions from: A)donkey, B) sheep, C) human.

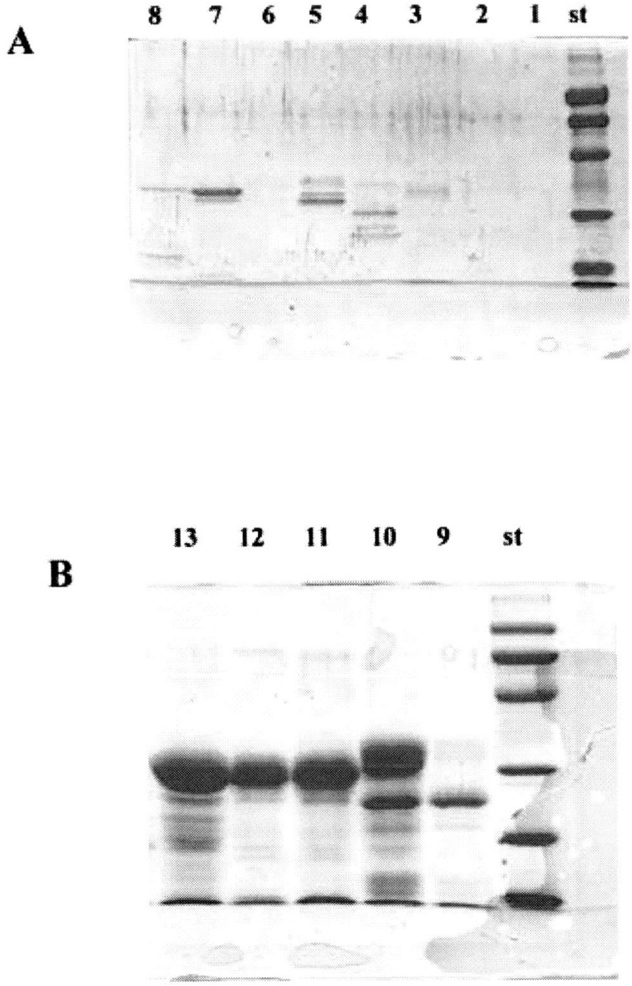

Figure 5. Electrophoretic separation (12% SDS-PAGE) of the peaks eluted from the chromatography on RP-HPLC of milk casein fractions from: A) donkey, B) sheep. St: Standard Bio-Rad LMW (phosphorylase b, 97.4 kDa; bovine serum albumine, 66.2 kDa; ovalbumin, 45.0 kDa; carbonic anhydrase, 31.0 kDa; soybean trypsin inhibitor, 21.5 kDa; lysozyme, 14.4 kDa).

CONCLUSION

When a mother cannot breastfeed, or chooses not to breastfeed, the use of a milk substitute must provide the best option to meet the nutritional and health needs of the infant. Donkey milk has been widely used to replace human milk in the past, because chemical composition and protein content are close to that of human milk, and also because the allergenicity of donkey milk is low. Logically, donkey milk is today still used mainly in those countries where traditionally breeding of donkeys is maintained, specifically in Asia, Africa and in some countries of Eastern Europe. Recently, use of donkey milk as a possible substitute for human milk has appeared in some countries of Western Europe. The recent studies of the paediatric scientists have demonstrated that infant formulae, which are based on dairy cows milk, are less adapted than donkey milk. In fact, donkey milk digestibility is higher than cows' milk and similar to human milk, because of the high whey proteins content and the few casein content.

In Europe donkey milk is also used for cosmetology, since the age of Queen Cleopatra. Some cosmetic products made with donkey milk, such as soaps and/or bath creams, are commercialized and appreciated by the consumers because of the good image of donkey milk. Obviously, these products remain marginal because the similar products made with cows' milk are less expensive.

In most of the world, sheep's milk is used essentially for cheese making, rather than in some countries part of this milk is transformed into yoghurt or whey cheeses. Many cheeses made with sheep's milk have an Appellation of Origin, so that production is confined to specific regions, only a minority are made on an industrial basis. In fact sheep's milk coagulates well and is very suitable for making high-quality cheese: coagulation and curd formation are faster than for cows' milk, because of the high content of casein in sheep milk. The chemical composition of sheep's milk, especially considering protein fraction, describes well the manufacturing properties of this milk and helps in understanding why sheep's milk has never been used for human nutrition.

Human milk is obviously the best food for children. The chemical composition of human milk has been used as a guide for the formulation of infant formulae and/or human milk substitutes. It is very important to remember that, unlike infant formulae, the composition of human milk is not uniform. Significant changes in the composition of breast milk occur not only between individual women, but also throughout the day, throughout the lactation period, with changing maternal diet and as a result of other external factors such as exercise or metabolic illness. Today there is still much to learn about the biology of human milk, and consequently there is much to study about the best composition of infant formula: donkey milk can be considered the closest natural milk to human milk, and the results obtained by paediatric scientists seem to confirm the nutritional value of this milk, known from ancient times in many countries of the world.

REFERENCES

Bradford, M.M. (1976) A rapid and sensitive method for the quantitation of microgram quantities of protein utilizing the principle of protein-dye binding. *Anal. Biochem*, *72*, 249-254.

Businco, L., Gianpietro, P.G., Lucenti, P.J., Lucaroni, F., Pini, C., Di Felice, G., Iacovacci, P., Curadi, C. and Orlandi, M. (2000) Allergenicity of mare's milk in children with cow's milk allergy. *J. Allergy Clin. Immunol.*, *105*, 1031-1034.

Butler, J.E., Navarro, P. and Heyermann, H. (1994). Heterogeneity of bovine IgG2 VI. Comparative specificity of monoclonal and polyclonal capture antibodies foe IgG2a (A9). *Vet. Immunol. Immunopathol.*, *40*, 119-133.

Carroccio, A., Cavataio, F. and Iacono, G. (1999) Cross-reactivity between milk proteins of different animals. *Clin. Exp. Allergy, 29*, 1014-1016.

Carroccio, A., Cavataio, F., Montalto, G., D'Amico, D. and Alabrese, L. (2000) Intolerance to hydrolysated cow's milk proteins in infants: characteristics and dietary treatment. *Clin. Exp. Allergy, 18*, 1597-1603.

Chiavari, C., Coloretti, F., Nanni, M., Sorrentino, E. and Grazia, L. (2005) Use of donkey's milk for a fermented beverage with lactobacilli. *Lait, 85*, 481-490.

Chiofalo, B., Liotta, L., Zumbo, A. and Chiofalo, V. (2004) Administration of olive cake for ewe feeling: effect on milk yield and composition. *Small Rum. Res., 55*, 169-176.

Coppola, R., Salimei, E., Sorrentino, E., Nanni, M., Ranieri, P., Belli Blanes, R. and Grazia, L. (2002) Behaviour of *Lactobacillus rhamnosus* strains in ass's milk. *Ann. Microbiol., 52*, 55-60.

Cotte, J. (1991) Future of milk components in cosmetology. *Lait, 71*, 213-224.

Creamer, L.K. (2002) Milk Proteins / Casein Nomenclature, structure and association properties. In: H. Roginski, J.W. Fuquay, P.F. Fox (Eds.), *Enciclopedia of Dairy Science*, (pp. 1895-1902). London, UK: Academic Press, Elsevier Science.

Creamer, L.K., and Sawyer, L. (2002) Milk Proteins / Beta-Lactoglobulin. In: H. Roginski, J.W. Fuquay, P.F. Fox (Eds.), *Enciclopedia of Dairy Science*, (pp. 1932-1939). London, UK: Academic Press, Elsevier Science.

Curadi, M.C., Orlandi, M., Greppi, G.F., Toppino, P.M., Barzaghi, S. and Cattaneo, T.M.P. (2000) Identification of protein fractions in mares' colostrum and milk. *Milchwiss., 55*, 446-449.

Donnet-Hughes, A., Duc, N., Serrant, P., Vidal, K. and Schiffrin, E.J. (2000) Bioactive molecules in milk and their role in health and disease : The role of transforming growth-factor beta. *Immun. Cell Biol., 78*, 74-79.

Doreau, M. (1991) Le lait de jument. *Prod. Anim., 4*, 138-141.

Doreau, M. and Boulot, S. (1989) Recent knowledge on mare production : a review. *Livest. Prod. Sci., 22*, 213-235.

Doreau, M., Gaillard, L.L., Chobert, J.M., Leonil, J., Egito, A.S. and Haertlè, T. (2002) Composition of mare and donkey milk fatty acids and protein and consequences on milk utilization. *Proc. 4° Congr. Ital. Society Ippologia* (SIDI), Campobasso, Italy, pp. 51-71.

Dugo, P., Kumm, T., Lo Presti, M., Chiofalo, B., Salimei, E., Fazio, A., Cotroneo, A. and Mondello, L. (2005) Determination of triacylglycerols in donkey milk by using high performance liquid chromatography coupled with atmospheric pressure chemical ionization mass spectrometry *J. Separation Sci., 28* (9-10): 1023-1030.

Egito, A.S., Miclo, L., Lopez, C., Adam, A., Girardet, J.M. and Gaillard, J.L. (2002) Separation and characterization of mare's milk alpha(s1)-, beta-. Kappa-caseins, gamma-caein-like, and proteoso peptone component 5-like peptides. *J. Daity Sci., 85*, 697-706.

European Allergy White Paper (1997) The UCB Institute of Allergy, pp.29.

Fantuz, F., Vincenzetti, S., Polidori, P., Vita, A., Polidori, F. and Salimei, E. (2001) Study on the protein fractions of donkey milk. *Proc. XIV Congr. Ital. Soc. Anim. Prod.* (ASPA), Firenze, pp. 635-637.

Farrell Jr., H.M., Jemenez-Flores, R., Bleck, G.T., Brown, E.M., Butler, J.E., Creamer, L.K., Hicks, C.L., Hollr, C.M., Ng-Kwai-Hang, K.F.and Swaisgood, H.E. (2004) Nomenclature of the proteins of cow's milk-sixth revision. *J. Dairy Sci., 87*, 1641-1674.

Gill, I. and Valivety, R. (1997) Polyunsaturated fatty acids. 1. Occurrence, biological activities and applications. *Trends Biotechnol., 15*, 401-409

Giuffrida, M.G., Cantisani, A., Napoletano, L., Conti, A. and Godovac-Zimmerman, J. (1992) The amino-acid sequence of two isoforms of a-lactalbumin from donkey (*Equus asinus*) milk is identical. *Biol. Chem. Hoppe Seyler, 373*, 931-935.

Godovac-Zimmerman, J., Conti, A., Liberatori, L. and Braunizer, G. (1986) the amino acid sequence of betalactoglobulin I from horse colostrum: betalactoglobulins are retinol-binding proteins. *Biol Chem Hoppe Seyler, 366*, 601-608.

Greppi, G.F., Iametti, S., Bonomi, F., Sessa, F., Feligini, M. and Enne, G. (1996) Minor whey components in horse and donkey milk. In G. Enne and G.F. Greppi (Eds.), *Food and Health - Role of animal products*. (pp. 149-154). Parsi, France, Elsevier Biofutur.

Greppi, G.F., and Roncada, P. (2005) La componente proteica del latte caprino. In G. Pulina (ed.) *L'alimentazione della capra da latte* (pp. 71-99). Bologna, Italy., Avenue Media publisher.

Herrouin, M., Mollé, D., Fauquant, J., Balestra, F., Maubois, J.L. and Léonil, J. (2000) New genetic variants identified in donkey's milk whey proteins. *J Protein Chem, 19*, 105-115.

Iacono, G., Carroccio, A., Cavataio, F., Montalto, G., Soresi, M. and Balsamo, V. (1992) Use of ass's milk in multiple food allergy. *J Pediatric Gastroent Nutr., 14*, 177-181.

Kotts, C. and Jenness, R. (1976) Isolation of k-casein-like proteins from milks of various species. *J. Dairy Sci.*, 59, *816-822.*

Kuhn, N.J., Carrick, D.T. and Wilde, C.J. (1980) Lactose synthesis: The possibility of regulation. *J. Dairy Sci., 63*, 328-336.

Laemmli, U.K. (1970) Cleavage of structural proteins during the assembly of the head of bacterophage T4. *Nature, 227*, 680-685.

MacDonald, A. (1995) Which formula in cows milk protein intolerance. The dietitians dilemma. *European J. Clin.Nutr., 49 (suppl.1),* s56-s63.

Malacarne, M., Martuzzi, F., Summer, A. and Mariani, P. (2002). Protein and fat composition of mare's milk: some nutritional remarks with reference to human and cow's milk. *Int. Dairy J., 12*, 869-877.

Martin, P. and Grosclaude, F. (1993) Improvement of milk protein-quality by gene technology. *Livest. Prod. Sci., 35*, 95-115.

Miranda, G., Mahé, M.F., Leroux, C. and Martin, P. (2004) Proteomic tools to characterize the protein fractions of Equidae milk. *Proteomics, 4*, 2496-2509.

Molkentin, J. (1999) Bioactive lipids naturally occurring in bovine milk. *Nahrung, 43*,185-189.

Ochirkhuyag, B., Chobert, J.M., Dalgalarrondo, M. and Haertlé, T. (2000) Characterization of mare's caseins. Identification of α_{S1}- and α_{S2} caseins. *Lait,* 80, *223-235.*

Oftedal, O.T. and Jenness, R. (1988) Interspecies variation in milk composition among horses, zebras and asses (Perissodactyla: Equidae) *J. Dairy Res., 55*, 57-66.

Ono, T., Kohno, H., Odagiri, S. and Takagi T (1989) Subunit components of casein micelles from bovine, ovine, goat and equine milks. *J Dairy Res., 56*, 61-68.

Playford, R.J., MacDonald, C.E., Johnson, W.S. (2000) Colostrum and milk-derived peptide growth factors for the treatment of gastrointestinal disorders. *Am. J. Clin. Nutr., 72*, 5-14.

Salimei, E., Fantuz, F., Coppola, R., Chiofalo, B., Polidori, P. and Varisco, G. (2004) Composition and characteristics of ass's milk. *Anim. Res.*, *53*, 67-78.

Schryver, H.F., Oftedal, O.T., Williams, J., Cymbaluk, N.F., Antczak, D. and Hintz, H.F. (1986) A comparison of the mineral composition of milk of domestic and captive wild equids (*Equus przewalski, E. zebra, E. burchelli, E. caballus, E. asinus). Comp. Biochem.Physiol.*, *85A*, 233-235.

Solaroli, G., Pagliarini, E., and Peri, C. (1993) Composition and nutritional quality of mare's milk. *Ital. J. Food Sci.*, *1*, 3-10.

Stelwagen, K. (2003) Milk Protein. In: H. Roginski, J.W. Fuquay, P.F. Fox (Eds.), *Enciclopedia of Dairy Science*, (pp. 1835-1842). London, UK: Academic Press, Elsevier Science.

Vincenzetti, S., Polidori, P., Fantuz, F., Mariani, P.L., Cammertoni, N., Vita, A. and Polidori, F. (2005a) Donkey's milk caseins characterization. *Ital. J. Anim. Sci., , 4 (Suppl. 2),* 427-429.

Vincenzetti, S., Polidori, P., Salimei, E., Mariani, P.L., Allegri, S., Cammertoni, N., Quadrini, B. and Vita A. (2005b) Purification and identification of αs1 and β-caseins from asses milk. *Vet. Res. Comm.*, *29 (Suppl. 2),* 211-213.

Visser, S., Jennes, R. and Mullin, R. (1982) Isolation and characterization of β- and γ- caseins from horse milk. *Biochem. J.*, *203*, 131-139.

In: Protein Research Progress
Editor: A.B. Boscoe and C. R. Listow, pp. 189-206
ISBN 978-1-60456-663-3

Chapter 6

NEW TRENDS IN PROTEIN FEEDING AND FLOCK MANAGEMENT FOR IMPROVING SHEEP MILK AND CHEESE FEATURES

Marzia Albenzio**, *Antonella Santillo and Agostino Sevi
Dipartimento PRIME and Istituto per la Ricerca e le Applicazioni Biotecnologiche per la Sicurezza e la Valorizzazione dei Prodotti Tipici e di Qualità (BIOAGROMED)
Università di Foggia, Via Napoli 25, 71100 Foggia (Italy)

ABSTRACT

The increase of intensive production systems in dairy sheep flocks necessitates specifications on feeding strategies and housing structures to optimize milk production and animal welfare, and to control the impact of animal husbandry on the environment. The nutritional and processing properties, and thereby the market value of milk, largely depends on its protein content. The efficiency of dietary nitrogen (N) utilization for milk protein synthesis is rather low in dairy animals so farmers are driven to use high protein level diets for sustaining milk production. High protein level diet can result in a number of deleterious events for animals, farmers, and environment, such as: i) an increase in the milk production cost, ii) an overloading of renal draining activity, iii) an increase of N output to the environment, iv) high levels of ammonia pollution in animal houses, which originates from animals' urine and feces and can be injurious to both livestock and stock persons' health. In particular, an increased N excretion in urine and feces, as a consequence of poor dietary N absorption, can result in high microorganism and ammonia concentrations in animal houses. This requires high ventilation rate for sustaining the performance and health of farmed livestock. In addition, it has been postulated that poor welfare conditions can modify the digestive tract physiology, producing a modification in mobility and nutrient utilization and, as a result, higher amount of incompletely digested material. Based on these considerations, increasing the protein supply to the ewes' mammary gland by reducing total protein degradability in the

* Corresponding author: Marzia Albenzio (m.albenzio@unifg.it); Tel. +39-0881-589327 Fax ++39-0881-589331

rumen seems to be a more suitable strategy than increasing the protein level of the diet, because it improves the efficiency of dietary protein utilization by the lactating ewe, especially when specific limiting amino acids are encapsulated to prevent bacterial deamination in the rumen to make them available for absorption in the small intestine. A number of recent studies have addressed the control of micro-climate and of air pollution in animal houses as well as the specification of good management practices aimed at optimizing milk production. Little is known about the influence of protein diet on cheese quality and proteolysis during ripening as well as on the effects of sheep housing conditions on ovine cheese manufactured from raw milk, which is a focus aspect to link flock management and the quality of milk to be processed into dairy products. In this review new achievements on the effects of both feeding treatment and flock management on ovine milk and cheese features are reported.

Keywords: sheep housing, protein feeding, management practices, milk quality, proteolysis in cheese.

Dietary Protein and Amino Acids Requirement

As monogastric animals, ruminants require a nutrient source of essential amino acids. However, because rumen microbes are the primary source of protein they modify the amino acid pattern of the diet. Consequently the supply of amino acids from the rumen to the small intestine for absorption assumes great importance [1].

A large part of dietary protein and non-protein nitrogen is degraded to ammonia in the rumen and, depending on fermentation conditions, a variable amount of dietary ammonia together with ammonia from endogenous sources is incorporated into microbial protein. Degradation of dietary protein depends on how readily it is fermented and on its retention time in the rumen.

The newborn ruminant is very similar to a monogastric animal. While reticulo-rumen and omasum obviously are present, they only begin to function when solid food is first eaten, which occurs at 2-3 weeks of age in the lamb. Even then, ewes' milk remains the lamb's main food for several more weeks and provides a valuable supplement to the microbial protein for the rumen for months thereafter. The milk escapes the reticulo-rumen by way of the oesophageal groove, so passing directly to the abomasum and the intestine [2].

In diet formulation for lactating ewe, protein content and quality are of critical importance because they must be suitable for lactation requirements. Quality of dietary protein is evaluated in terms of its ruminal degradability and of its amino acid composition. The protein requirement corresponds to the amount of amino acid daily supplemented and available to intestinal absorption. For ruminants several factors must also be taken into account: protein synthesis by rumen micro flora, its digestibility and its amino acid composition. Rumen microflora act processes of deamination and decarboxylation on amino acids supplied with diet (Table 1).

The concept of ideal protein has been used for the protein that provides absorbed amino acids in the proportion that gives maximum efficiency of utilization.

Table 1. A comparison of the essential amino acid profiles of body lean tissue and milk with that of ruminal bacterial and protozoa and some common feeds (from Shwab, 2003)

Item	Arg	His	Ile	Leu	Lys	Met	Phe	Thr	Trp	Val	EAA
Animal products											
Lean tissue[a]	16.8	6.3	7.1	17.0	16.3	5.1	8.9	9.9	2.5	10.1	-
Milk [b]	7.2	5.5	11.4	19.5	16.0	5.5	10.0	8.9	3.0	13.0	-
Rumen microbes											
Bacteria c	10.4	4.1	11.5	15.9	16.5	5.1	10.1	11.3	2.7	12.4	-
Protozoa d	9.3	3.6	12.7	15.8	20.6	4.2	10.7	10.5	2.8	9.7	-
Forages [e,f]											
Alfaalfa hay	12.5	4.7	10.3	17.9	12.4	3.8	11.6	10.6	3.6	12.7	41.2
Alfaalfa silage	10.9	4.7	11.1	17.9	12.1	3.8	11.7	10.7	2.7	14.1	35.6
Maize silage	6.2	5.7	10.6	27.2	7.9	4.8	12.1	10.1	1.4	14.1	31.6
Grass hay	11.7	4.9	10.0	18.8	10.5	3.9	11.8	10.9	3.7	13.6	33.1
Grass silage	9.4	5.1	10.9	18.8	10.1	3.7	13.4	10.2	3.3	15.0	32.6
Grains [e]											
Barley	13.4	6.1	9.2	18.5	9.6	4.5	13.5	9.1	3.1	13.0	37.7
Maize	11.5	7.8	8.2	27.9	7.1	5.3	11.5	8.8	1.8	10.0	40.1
Oats	16.6	5.9	9.1	17.7	10.1	4.2	12.5	8.4	2.9	12.6	41.2
Sorgum	9.4	5.7	9.3	31.9	5.4	4.2	12.3	7.8	2.5	11.6	42.8
Wheat	13.6	7.1	9.6	19.3	8.1	4.6	13.3	8.4	3.5	12.3	34.4
Plant proteins [e]											
Brewers' grains	14.7	5.1	9.8	20.0	10.4	4.3	11.7	9.1	2.5	12.1	39.2
Canola meal	16.5	6.6	9	15.9	13.2	4.4	9.5	10.4	3.4	11.1	42.6
Maize DDG w/sol	10.7	6.6	9.8	25.4	5.9	4.8	12.9	9.1	2.3	12.4	37.8
Maize gluten meal	7.1	4.7	9.1	37.2	3.7	5.2	14.1	7.5	1.2	10.3	45.2
Cottonseed meal	26.0	6.6	7.3	13.8	9.7	3.7	12.5	7.6	2.8	10.0	42.6
Linseed meal	20.9	4.8	11.0	14.5	8.7	4.2	11.1	8.9	3.7	12.3	42.2
Peanut meal	27.6	6.0	8.1	15.9	8.3	2.9	12.1	6.7	2.4	9.8	40.1
Soia bean meal	16.2	6.1	10.1	17.2	13.9	3.2	11.6	8.7	2.8	10.2	45.3
Synflower meal	20.8	6.2	9.9	15.2	8.0	5.6	11.0	8.7	2.9	11.7	42.2
Animal proteins [e]											
Blood meal	7.8	1.3	2.2	22.7	18.9	2.1	12.1	7.7	2.8	15.4	56.4
Feather meal	16.2	2.7	11.4	19.9	6.0	1.8	11.6	11.1	1.7	17.6	42.7
Fish meal	13.1	6.4	9.2	16.2	17.2	6.3	9.0	9.4	2.4	10.8	44.5
Meat and bone mea	19.5	5.3	7.7	17.2	14.5	3.9	9.4	9.1	1.6	11.8	35.7
Whey, dry	5.0	4.5	12.1	21.2	17.6	3.3	7.0	14.1	3.5	11.7	42.2

[a] Average values of empty, whole body carcasses as reported in three studies. [b] Average values as reported in three studies. [c] The mean of average values for over one hundred dietary treatments. [d] Average values from 15 literatures reports. [e] Calculated from values presented in National Research Council (2001).
[f] Legume and grass hays and silages are mid-maturity. EAA, essential amino acids; DDG, distillers' dried grains.

It is worth to notice that the optimal composition varies according to physiological status and it is specific to a particular type of production. Identification of the optimal amino acid composition for a particular productive process can be achieved indirectly by identifying the limiting amino acids in a reference protein whose amino acid composition is known, for example, a microbial protein [3, 4].

Microbial protein is by far the largest source of protein passing to the small intestine. It is part of microbial biomass from bacteria, protozoa and fungi. Since only a small part of the energy of the diet is available for microbial growth under the anaerobic conditions prevailing in the rumen, there is a general relationship between the supply of fermentable energy (Non Structural Carbohydrates) and the yield of microbial biomass. When microbial protein production is limited or under high amino acid needs, microbial protein may not meet the amino acid requirements for maintenance and production [1]. As a consequence, some amino acids appear to be of critical importance and limiting in term of milk production [5].

Several groups [6, 7, 8] have assessed the possibility of increasing the protein supply to the ewes' mammary gland by reducing dietary protein degradability. This is a suitable strategy for improving the efficiency of dietary protein utilisation by the lactating ewe, especially when specific limiting amino acids are encapsulated to prevent bacterial deamination in the rumen to make them available for absorption in the small intestine. Rumen-protected amino acids (RPAA) are free amino acids that have been coated or encapsulated with materials that allow for most of the amino acids to escape ruminal degradation but be released in the abomasum or small intestine [9]. Protected amino acids are important in dairy cattle nutrition to maintain amino acid balance, support milk production and minimize wasting of nitrogenous compounds.

Methionine, lysine and histidine have been identified as the most limiting for milk protein synthesis in dairy cattle [10]. Methionine has been identified as first limiting and lysine as second limiting in microbial protein for nitrogen retention of both growing cattle and lambs. Histidine has been identified as possible third limiting for sheep, while methionine and lysine are co-limiting amino acids for milk protein synthesis when cows are fed maize silage-based diets with little or no protein supplementation [9]. Ewes fed a diet with a theoretically better balanced amino acid composition (Table 2) showed a significant increase in milk yield, in milk protein and fat content, and a slight improvement of renneting parameters. No differences appeared, instead, for milk amino acid composition [4]. The authors suggest that the supplementation of a feed ratio with Rumen-Protected (RP) methionine and lysine may improve the protein metabolism of lactating ewes, thus making it possible to reduce crude protein content of diet without affecting milk yield and composition.

It is well know that the efficiency of dietary N utilization for milk protein synthesis in dairy animals is quite low (15 to 35%) [11, 12] so farmers are driven to use high protein level diets for sustaining milk production in lactating animals. This can result in a number of deleterious events for animals, farmers and the environment, such as i) an increase in the milk production cost; ii) an overloading of animal renal draining activity, iii) an increase of N output to the environment, iv) high levels of noxious gases in animal houses, which originates from animals' urine and feces and can be injurious to both livestock and stock persons' health, v) increased N excretion in urine and feces can result in high microorganisms in animal houses.

Table 2. Theoretical amino acid composition of the digestible protein from diets[a] with a balanced (Diet A) and unbalanced (Diet B) amino acid profile (from Sevi et al., 1997)

Amino acids	Diet A		Diet B	
	%	Index	%	Index
Lysine	7.5	100	7	100
Hystidine	2.2	30	2.1	30
Arginine	5.0	67	4.6	66
Threonine	4.3	57	5	71
Methionine	2.3	31	1.7	24
Valine	6.1	81	5.7	81
Phenilalanine	5.3	71	6.1	87
Isoleucine	4.8	64	5.5	78
Leucine	7.9	105	7.4	106
Tryptophan	1.4	19	1.7	25
Serine	4.3	58	4.5	64
Aspartic acid	10.4	139	9.8	139
Glutamic acid	14.8	197	13.1	186
Glycine	5.6	74	6.6	95
Alanine	6.5	87	6.2	89
Tyrosine	4.5	60	3.7	53
Proline	5.0	67	6	86
Cystine	2.1	28	3.3	47

[a]Diets were composed of alfalfa hay and wheat straw for both groupd and, for each group, by one of two concentrates, obtaine by mixing the same fed components, but different as to their amino acidic compositio1 n.

Previous experiments have demonstrated that an increase in the crude protein (CP) level of the ration from 13 to 16% [7] and from 13% to 19% [13] has a beneficial effect on ewe milk yield and protein content, but it also results in higher blood urea concentrations [7] and increased N excretion in urine [13]. Conversely, Cowan et al. [14] did not find any difference in the yield and protein content of the milk from ewes receiving CP concentrations of 11.6 and 14.3 % of dry matter. Sevi et al. [8, 15] have suggested that the efficiency of utilization of dietary N in the lactating ewe increases with decreasing protein levels in the diet from 16 to 13%, especially when limiting amino acids (lysine and methionine) in the ration are encapsulated to prevent bacterial deamination in the rumen.

Apart from the contradictory results on the relationship between CP level of diet and milk yield and composition, the concentration of urea in the milk is always regarded as a good indicator of efficient use of dietary protein by the animals, and it is currently used to evaluate and to adjust the diet [16]. In our study, bulk milk produced by the ewes receiving the low crude protein level (13% vs. 16% of DM) diet displayed a higher casein content and a lower urea concentration. Therefore, it can be argued that the low crude protein level improve nitrogen utilization in the ewes without detrimental effect on the efficiency of milk production for cheese-making.

MANAGEMENT PRACTICES AIMED AT IMPROVING MILK QUALITY

Milk must have a desirable chemical composition and must be of satisfactory hygienic quality. This is essential in relation to public health, the suitability of milk for processing, and the quality of milk products. Hygienic, nutritional and renneting parameters of milk may be distinguished.

The health of the udder can have a profound effect on the quality and processing characteristics of milk. The most widely used indicator of udder health is somatic cell count (SCC), a measure of the number of white blood cells, known as leucocytes in milk [17]. An elevated SCC usually indicates the presence of mastitis. Mastitis is caused by pathogenic bacteria entering the mammary gland via the teat canal and multiplying within the udder sinuses or epithelia, or in the teat duct. Mammary inflammation during mastitis causes a range of physical, microbiological and chemical changes in milk.

The microorganisms found in raw milk may come from several sources: organisms from the udder, the environment (e.g. water, soil), milking equipment, bulk milk storage tank, tank in processing plant [18]. The psychrotrophs can be very detrimental to milk quality, especially as related to heat-stable enzymes which can degrade milk proteins during raw milk storage.

The climatic extremes and the limitation of natural pastures as a food source together with the increased demand for sheep cheese was the trigger for the intensification of dairy sheep farming in the Mediterranean Basin, in the second half of the twentieth century. This intensification was featured by promoting the adoption of controlled technical parameters in sheep houses and implementing feeding procedures. Protein supplementation in grazing dairy ewes is a popular strategy for upgrading sheep nutrition in order to reduce the variability and improve ewe milk yield and composition.

Feeding strategies together with several other management practices may improve the overall quality of milk for cheese making. A proper evaluation of technical parameters in sheep housing requires knowledge on the effects of these factors on animal welfare, environment, milk yield and composition, and renneting ability, thus being of great practical interest for farmers in dairy sheep systems.

Stocking density is a critical factor when housing lactating ewes, because space allocation is known to affect both the performance and welfare of livestock. Sevi et al. [19] report that ewes housed with a space allocation of 2 m^2 per animal may have a beneficial effect on their health and performance, with milk yield and casein content increased of around 10%, as well as the coagulation properties of milk (+8%) (Table 3). This can be regarded as a direct effect on the hygienic quality of milk of the lower levels of aerial microorganisms recorded in the room where the low stocking density group (2m^2 per ewe) was housed compared to high stocking density group (1 m^2 per ewe). Higher levels of microorganisms also affected the mammary system of the animals with higher microbial loads and somatic cell counts being recorded in milk from ewes housed under high stocking density conditions. High milk SCC is an index of decreased secreting epithelium function leading to a reduction in milk casein content [20].

Table 1. Effects of some management practices on milk yield and composition

Parameter	Stocking density	Bentonite	Stocking density x Bentonite	Renewal of litter	Renewal x Bentonite	Paraformaldehyde	Air space allocation	Ventilation rate in winter	Ventilation rate in summer
	($2m^2$/ewe)	0.5 kg/m^2	($2m^2$/ewe) x 0.5 kg/m^2	every 4 weeks	every 4 weeks x 0.5 kg/m^2	5g/kg body weight	$7m^3$/ewe	$50m^3$/h/ewe	$70m^3$/h/ewe
Milk yield, g/d	**	**	**	**	NS	NS	***	*	*
Protein, %	NS	***	-	***	***	*	NS	NS	NS
Fat, %	NS	***	*	***	***	*	NS	NS	***
Casein, %	**	-	**	-	-	-	**	*	NS
Clotting time, min	*	NS	***	NS	NS	**	NS	NS	***
SCC, log_{10} cells/ml	*	**	*	**	NS	*	***	***	*

NS, not significant, * $P<0.05$, ** $P < 0.01$, *** $P < 0.001$. The effects of each treatment refer to a negative control.

When space allocation per ewe is not adequate to the number of stocked animals, improving the hygiene of the bedding may be a suitable strategy to sustain animal health and milk production. A litter treatment with bentonite [21], a material with a high water-absorbing capacity, can partially alleviate adverse effects of high stocking density by reducing enzyme and microbial activity in the bedding [8, 21]. Milk produced by high densely stocked ewes, which were housed on a straw litter treated with bentonite, displayed lower concentrations of bacteria in their milk together with lower airborne microorganism loads. Furthermore, an increase of milk yield (3%), of milk fat (10%) and casein content (5%), and of milk clotting time (8%) was recorded in this group. When no litter treatment was used higher SCC (+40%) was found in milk due to greater bacterial colonization of the udder (Table 3). A high somatic cell count is generally accompanied by an enhanced influx of serum components from the extra-cellular fluid into the milk [22]. These components include a range of hydrolytic enzymes which further modify the milk composition via the breakdown of casein and fat [23].

Paraformaldehyde is a polyoxymethylene containing 90 to 98% formaldehyde which has a recognised bacteriostatic effect on the microorganism naturally present in droppings. Litter treatment with paraformaldehyde was shown to influence milk hygienic quality reducing somatic cell count and microbial cell load of 10, and 15% respectively, when compared with a control group (Table 3). In particular, the milk produced by ewes subjected to litter treatment with paraformaldehyde had a mesophile microorganisms content which was always < 500,000 cfu/ml. This is a threshold indicated by the EU directive 92/46 for ewes milk intended for cheese production without heat treatment. As well as some authors [20, 19] indicate 500,000 SCC/ml of milk as the limit of acceptability of ewe milk to be put on the market. The use of paraformaldehyde for litter disinfection markedly increased protein (10%) and fat (20%) content in milk yielded by the ewes in the test group compared to the control group (Table 3). A quick and complete coagulation of sheep milk requires high contents of casein and fat as well as an adequate concentration of calcium and phosphorus, and a pH of 6.5 to 6.7. In the mentioned experiment, the improvement of clotting time (20%) in the test group was ascribed not only to improved milk composition but also to the low level of endogenous (mainly contained in leukocytes) and microbial enzymes, which act a post-secretory proteolytic and lipolytic activity. At least in ewes, addition of paraformaldehyde to the bedding is a procedure that may be applied at intervals suitably far apart to be not economically prohibitive. Its use, however, would remain limited to circumstances in which there are difficulties in processing or marketing milk of low hygienic quality which could not be easily resolved using other cheaper methods [24]. In a subsequent study [25] the effects of litter renewal intervals on the yield and quality of ewe milk were evaluated as an alternative strategy to litter treatment with anti-microbial products. Renewal of litter at four weeks intervals can sustain health status of the mammary gland and improve ewe performance at the same levels as litter treatment with bentonite.

An adequate airspace allocation (7 m^3/animal) results in a lower airborne microorganism concentration, higher milk yield (20%) and milk casein content (10%), and improved renneting ability and hygienic quality of milk in terms of somatic cell count (-10%) (Table 3). Inadequate airspace may be a limitation to high efficiency of production and good health in farmed livestock, especially when animals cannot benefit from efficient ventilation systems, as generally occurs in sheep husbandry [26].

Ventilation plays a main role in sustaining the welfare and performance of farmed livestock, by affecting thermal exchanges between the animal's body surface and the environment and by removing aerial pollutants, which originate from animals and their excreta. In particular, low ventilation rates can fail in removing efficiently the moisture and gases, which originate from the respiratory activity of animals and the decomposition and fermentation of manure and result in increased relative humidity and higher air concentrations of ammonia and carbon dioxide. Very high ventilation rates, instead, can result in higher air dust concentrations, probably due to reduced humidity levels and to turbulent air currents maintaining dust particles suspended in the air for a longer time.

Lactating cows exposed to heat stress conditions can experience lowered protein concentration in milk. In five of ten studies examining heat stress effects on animal performance, milk protein concentration was significantly reduced by an average of 0.18 units. Decrease in milk protein percentage under heat stress may be ascribed to lowered microbial protein synthesis in the rumen and lowered protein intake due to lowered intake of dry matter. Other effects of thermal stress include reduction in potassium and sodium, increase in growth hormone, prolactin, and SCC. In cow the immune system is probably under greater strain during heat stress conditions and therefore is less able to manage subclinical mammary infections [27].

Abdalla et al. [28] report that, also in sheep, feed intake decreases under heat stress, especially when sheep are fed on low quality feed [29], due to both the effort of reducing heat production and the slower feed transit through the digestive tract. Our research group reported that feed administration in late afternoon was beneficial to ewes in minimizing the impact of thermal stress on their immune function and udder health [30]. With a pending high heat load situation, a moderate ventilation rate (66 m^3/h per ewe) improves the well-being of the lactating ewe compared to a low ventilation rate (35 m^3/h per ewe), as suggested by behavioural, endocrine and immune indicators, but has no effects of the yield and quality of milk. Under such conditions, a fan ventilation system, programmed to operate over upper critical air temperature (30 °C) and relative humidity (70%), has been proved to be economically unattractive, because it involves about a threefold greater energy cost and does not lead to remarkable improvements of ewe welfare and productivity compared to a moderate ventilation regimen (66m^3/h per ewe) [31]. Under moderate heat stress, instead, the deleterious effects of a low ventilation rate (35 m^3/h per ewe) on milk yield and on clotting properties of milk are more marked (-13% and -10%, respectively) when compared to a moderate ventilation regimen (70 m^3/h per ewe) [32] (Table 3).

The importance of ventilation during the winter season, instead, is often underestimated and this may result in a serious limitation to high efficiencies of production and good health in intensive dairy sheep systems. Indeed, during the winter season, when ewes are in early and mid lactation, and displayed higher productive levels, the adverse effects of inadequate ventilation regimen on the yield and quality of ewe milk are even more marked. Data from our experiments [33] demonstrate that exposure of ewes to low (23 m^3/h per ewe) and very high ventilation rates (73 m^3/h per ewe) results in increased noxious gases, dust and airborne microorganism concentrations compared to a moderate ventilation rate (47 m^3/h per ewe).

Throughout the trials, the low ventilation regimen led to a reduced yield and a deteriorated coagulating behavior of milk compared to high ventilation rates (Table 3). When the bulk milk was collected, after 6 weeks of exposure to the three ventilation regimens, both the low and the very high ventilation rates resulted in higher levels of somatic cell and of mesophilic bacteria counts compared to the moderate ventilation rate. The greater bacterial growth led to increased SCC, as leukocyte infiltration in the alveoli is one of the main defence mechanisms of the animal against invading bacteria [34]. Furthermore, enzymes produced by the bacterial flora may act as plasminogen activators [35]. Higher plasmin (PL) activity is often related to higher milk SCC, as somatic cells release a number of plasminogen activators or proteolytic enzymes. In bulk milk from the ewes exposed to low and high ventilation rates the greater plasmin acivity and a higher plasminogen (PG) to plasmin ratio could be ascribed to the impaired milk hygienic quality. It is well known that plasmin is the principal indigenous proteinase in milk and hydrolyzed casein to peptides reducing milk ability for cheesemaking [23, 36].

In a recent research, Sevi et al. [15] investigated combined effects of dietary protein level (13% and 16% of dry matter) and ventilation regimen (23 m^3/h and 47 m^3/h) on nitrogen utilization and milk quality. Bulk milk produced by the ewes receiving the low crude protein diet (13%) displayed higher casein (10%) and lower urea contents (-20%) (Table 4). Therefore the choice of a proper dietary crude protein level plays a main role in sustaining protein synthesis in the mammary gland [15]. Milk produced by the ewes subjected to the low ventilation rate (23 m^3/h/ewe) displayed a lower hygienic quality in terms of SCC and mesophilic cell load, respect to those exposed to the moderate ventilation regime (47 m^3/h/ewe). In fact, the 16% CP diet resulted in a higher feces and water excretion and, in the ewes exposed to a mean ventilation rate of 23 m^3/h, in a higher fecal nitrogen emission. The lower efficiency in removing airborne microorganisms may be claimed to explain the impaired hygienic quality of the milk obtained from ewes subjected to the low ventilation rate.

Table 4. Chemical composition of the bulk milk from ewes fed on a low (13%) and moderate (16%) crude protein level diet and exposed to a low (23 m^3/h per ewe) (LPLV and MPLV) and moderate (47 m^3/h per ewe) ventilation rate (LPMV and MPMV). Values are means ± SE (from Albenzio et al., 2007)

	LV		MV			Effects, P		
Item	LP	MP	LP	MP	SEM	Protein level	Ventilation rate	Protein level x Ventilation rate
Milk Fat, %	5.93	5.62	5.76	5.84	0.26	NS	NS	NS
Milk Protein,%	5.69	5.77	5.69	5.66	0.19	NS	NS	NS
Milk Casein, %	4.51	4.24	4.58	4.13	0.12	**	NS	NS
Milk Urea, mg/dl	37.98	45.16	38.48	45.99	0.91	***	NS	NS

NS, not significant, **$P < 0.01$, ***$P < 0.001$. In the same row, means followed by different letters are significantly different at $P<0.005$.

In addition, the ewes subjected to low ventilation rate displayed a higher plasminogen activity in milk, whereas no differences emerged among treatments in the conversion of zymogen to plasmin.

EFFECTS OF MANAGEMENT PRACTICES ON PROTEOLYSIS IN OVINE CHEESE

Sheep milk is used almost only for cheesemaking. Therefore, knowledge regarding the effects of feeding and flock management on milk features should be extended to cheese. Few papers have been published to investigate the relationship between management strategies (i.e. dietary protein level and management practices in sheep housing) and nutritional and hygienic quality of milk to be processed into cheese. These remarks are of great importance for typical cheeses which are manufactured from raw milk to preserve nutritional features, and a desirable microflora during cheese ripening [37]. Cheeses made with raw milk have a very different microflora at the genus level, and also in term of species within the same genus [38, 39]; such bio-diversity is generally not found in pasteurised milk cheese.

In the context of international trade and regulations, where hygienic aspects are predominant, it appears important to study the consequences and implications of milk pasteurisation on the ripening process and ultimately on the sensory characteristics of cheese. Elimination of milk microorganisms, activation/inactivation of indigenous milk pro-enzymes and enzymes, denaturation of serum proteins, modification of the rennet ability of milk and activity of starter acid bacteria are the properties of milk which may be modified by mild heat treatments (e.g. 72–75°C for 15–30 s). Reducing the indigenous milk microflora may have considerable consequences on the biochemical changes which occur in cheese during ripening. It has been generally found that raw milk cheeses have more intense flavour and sometimes they are also described as too strong, atypical and less uniform than pasteurised milk cheeses. However, it can be expected that raw milk cheeses reach their optimum sensory quality earlier than pasteurised milk cheeses and that the microbial diversity is sought by both the producers and the consumers, because it is considered as a special feature of traditional cheeses

Canestrato pugliese is a famous Italian PDO (Protected Designation of Origin) ovine cheese; it may be produced according to a traditional protocol, which includes heating of the curds in hot whey (80°C for 30 s), from pasteurized or raw milk [37]. Canestrato pugliese has been subjected to studies regarding proteolysis, lipolysis, microbiology, technology, volatile profile and sensory analysis [37] [40] [41].

As a consequence of the higher levels of protein and casein in the milk of the ewes fed a low dietary protein level, higher concentrations of the protein (+20%) and casein content (+30%) were found in fresh cheese curd [42]. Casein is the substrate of proteolytic enzymes (i.e. plasmin, coagulants, microbial) which act hydrolyzing casein matrix into peptides with high molecular weight, at the beginning of cheese ripening. Proteolysis is the major biochemical process in cheese [43]; it depends on the rennet enzymes, microbial proteinases, moisture and pH of curd, salt content, and temperature, humidity and time of ripening [44]. Proteolysis is primarily responsible for changes in cheese texture and gives a significant contribution to flavor. Interestingly proteolytic process behaved differently in cheeses obtained from ewes fed on a diet with low protein level in the latest stage of ripening (90 days). At the end of ripening, a higher concentration of proteolytic products was observed in the cheeses obtained from ewes subjected to a moderate ventilation rate (47 m^3/h per ewe), as confirmed by nitrogen fraction concentration (Figure 1). It can be argued that, under the same dietary protein level (13% CP of dry matter), moderate ventilation rate (50 m^3/h/ewe) exerts a long-term effect on the proteolytic features of cheese. It is well known that, among the agents contributing to proteolysis, non-starter bacteria and their enzymes are of major importance in cheese ripening [43]. The use of a moderate ventilation regimen in sheep housing results in a reduced adventitious microflora in raw milk and a prevalence of a suitable dairy microflora. In a previous study on Canestrato pugliese cheese [37] the use of raw milk for cheemaking led to a large number of cheese-related bacteria associated with increased concentration of water – soluble and pH 4.6 –soluble N (Figure 2A, 2B).

When collected for Canestrato pugliese cheesemaking during summer, the bulk milk from the less ventilated ewes had higher microbial load and somatic cell count compared to that from the more ventilated animals so it requires a heat treatment prior to milk processing. The low ventilation regimen (35m^3/h per ewe) resulted in a weaker caseous matrix of the curd releasing higher concentrations of fat and protein in the whey compared to the moderate ventilation regimen (70 m^3/h per ewe). A higher casein content and lower non-casein nitrogen and water soluble proportions were found in the moderate ventilation rate as well as a fewer number of bands derived from CNs hydrolysis at 45 d of ripening (Figure 3). Cheese from moderate ventilation treatment also exhibited a higher PL activity levels at 45 days of ripening, suggesting the existence of a higher proteolytic potential, leading to a more intense proteolysis during further cheese ageing [45].

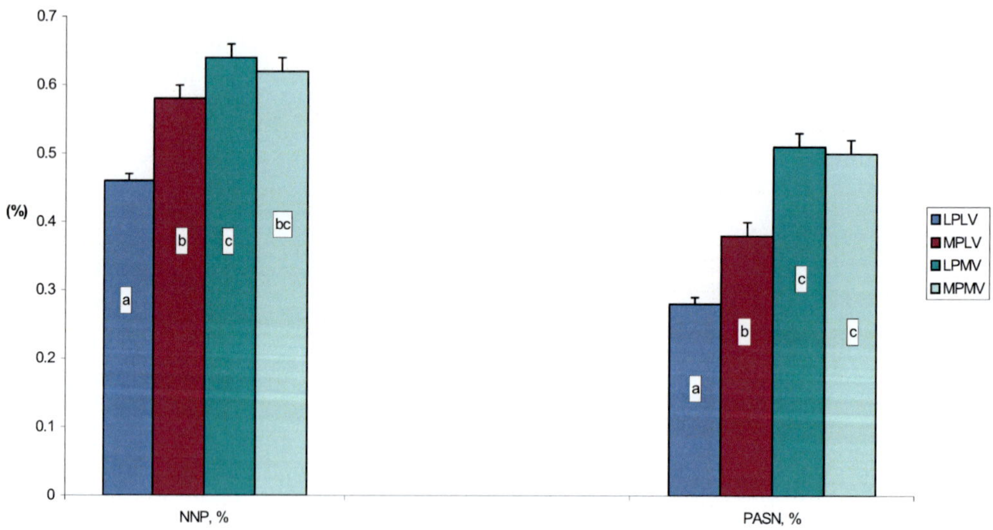

Figure 1. Nitrogen fractions of Canestrato pugliese cheese manufactured from the milk of ewes fedon a low (13%) and moderate (16%) dietary protein level and exposed to a low (23 m3/h/ewe) (LPLV and MPLV) and moderate (47 m3/h/ewe) ventilation rate.

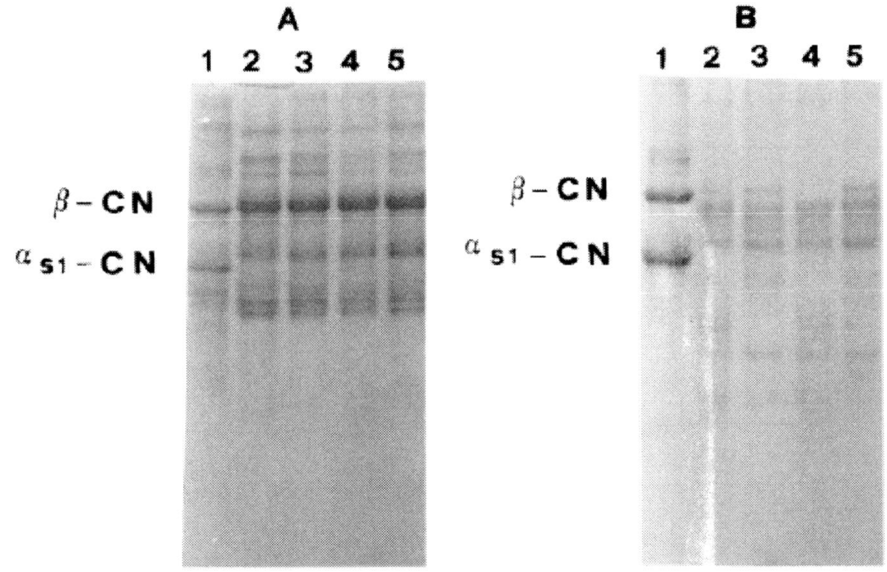

Figure 2.Urea-PAGE of water-insoluble (A) and water-soluble (B) fractions of Canestrato Pugliese cheeses. Lanes: 1, ovine Na-caseinate (CN) standard; 2,3, Cheeses made with raw milk; 4, cheese made with pasteurized milk; 5, cheese made with raw milk and heating the curd in hot whey (Albenzio et al., 2001).

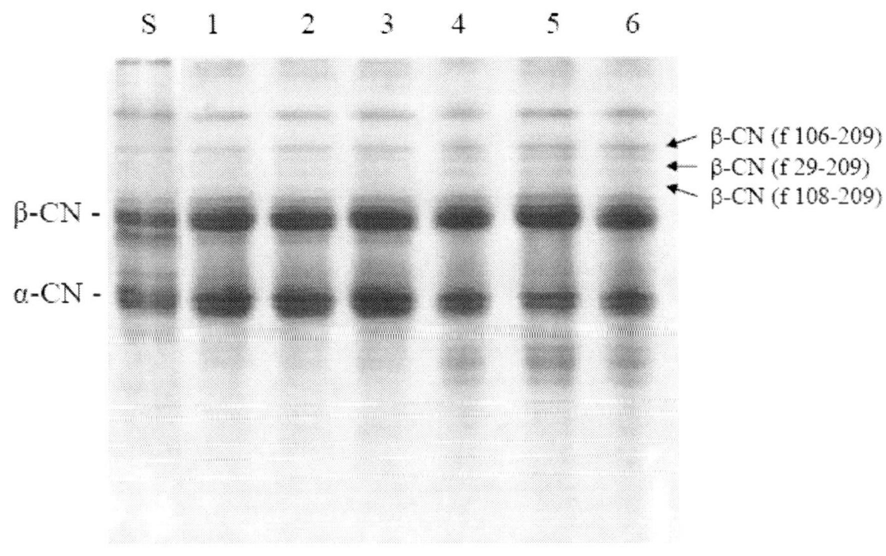

Figure 3. Urea-PAGE of pH 4.6-insoluble fractions of Canestrato Pugliese cheese manufactured from ewes subjected to different ventilation regimens. Lane S (Standard): Ovine Na Caseinate purified, lanes 1, 4 cheeses obtained from milk of ewes subjected to low ventilation regimen (23 m^3/h/ewe) at 1 and 45 days of ripening; lanes 2, 5 cheeses obtained from milk of ewes subjected to moderate ventilation regimen (50 m^3/h/ewe) at 1 and 45 days of ripening: lanes 3, 6 cheeses obtained from milk of ewes subjected to programmed ventilation regimen (70 m^3/h/ewe) at 1 and 45 days of ripening (Albenzio et al., 2005b).

The contribution of plasmin to proteolysis is relevant to cheese quality through hydrolysis of casein [46]. The increase in the plasmin activity in cheeses, as a result of either plasminogen activation or exogenous plasmin addition, ameliorates the flavour and the overall quality of cheese [47].

During winter season, cheese produced from a moderate ventilation regimen (50 m^3/h per ewe) resulted in the highest activity of PL in cheese, which resulted in an accelerated proteolysis, as confirmed by the highest levels of water soluble nitrogen (WSN) being recorded in those cheeses. This fraction contains numerous small- and medium-sized peptides, free amino acids and their degradation products, organic acids and their salts, which are widely used as reliable indexes of cheese ripening [48, 49]. The more intense proteolysis in the cheeses obtained from the moderate ventilation regimen is also displayed by the higher expression of degradation products, with high and low electrophoretic mobility, in the urea-PAGE of the pH 4.6-insoluble N fractions. It is worth to notice that, during ripening, the decrease of PL and PG activities coincided with an increase in non protein nitrogen (NPN), which suggests that NPN molecules derived from CN-hydrolysis may inhibit the plasmin-plasminogen enzyme system [50].

Conclusion

Most sheep milk is processed into dairy products so its content in total protein and casein plays a major role in giving it a good market value. In addition, a lot of typical ovine cheeses are manufactured from raw milk so hygienic quality of milk has a relevant part in preventing indigenous microflora from competition of spoilage microorganisms and from heat treatments. Indeed, reduction in indigenous milk microflora may have considerable consequences on the biochemical changes which occur in cheese during ripening.

In the light of these considerations, both feeding and management strategies are of great importance to the quality of sheep milk cheese. Proper feeding strategies give the substrate from which the mammary epithelium can draw nitrogen for milk protein synthesis, and proper flock and housing management avoid reduced utilization of dietary protein, reduced synthesis of casein in the mammary gland, altered post-secretory changes of casein during milk storage and cheese ripening. Evidence exists that physiological disturbance from inadequate housing conditions can impair the efficiency of dietary nitrogen utilization as well as high somatic cell count levels and bacteria counts in milk can lower the synthesis of casein and enhance its degradation before and during milk processing into dairy products. These events and severe heat treatments cumulate their effects in impoverishing sheep milk and cheese features.

References

[1] Merchen, N.R. and Titgemeyer, E. C. (1992). Manipulation of amino acid supply to the growing ruminant. *Journal of Animal Science* 70, 3238-3247.

[2] Chen, X. B. and Ørskov, E.R. (1994). Amino acid nutrition in sheep. Ch. 13, pgg.307-328. In *Amino acids in farm animal nutrition*. Edited by J.P.F. D'Mello, CAB International, Wallingford, Oxon, UK.

[3] Sevi, A., Taibi, L., Muscio, A., Dell'Aquila, S. (1996). Effects of graded levels of dietary rumen-protected lysine on Comisana ewes' milk yield and composition. *Scienza e tecnica lattiero-casearia*, 47, 182-192.

[4] Sevi, A., Muscio, A., Scapicchio, P., Dell'Aquila, S., Taibi, L. (1997). Dietary amino acid composition: effects on Comisana ewes' milk yield and composition. *Scienza e Tecnica Lattiero-casearia*, 38, 229-240.

[5] Fraser, D.L., Ørskov, E.R., Whitelaw, F.G., Franklin, M.F. (1991). Limiting amino acids in dairy cows given casein as the sole source of protein. *Livestock Production Science*, 28, 235-252.

[6] Cannas, A., Pes, A., Mancuso, R., Vodret, B., Nudda, A. (1998). Effect of dietary energy and protein concentration of milk urea nitrigen in dairy ewes. *Journal of Dairy Science*, 81, 499-508.

[7] Jaime, C., Purroy, A. (1995). Level and quality of protein in rations for lactating ewes. *Ann. Zootech*., 4, 135-142.

[8] Sevi, A., Muscio, A., Cordola, L., Dantone, D. (1998). Milk yield of Comisana ewes fed RP methionine and lysine at two levels of dietary protein content. *Italian Journal of Food Science*, 10, 137-146.

[9] Schwab, C. G. (2003). Ruminally protected amino acids. In *Encyclopedia of Dairy Sciences*. Vol. II, pgg. 991-997. Edited by Roginski, H., Fuquay, J.W., Fox, P.F. by MPG Books Lt, Bodmin, Cornwall, UK.

[10] Lynch, G.P., Elsasser, T.H., Jackson, C., Rumsey, T.S., Camp, M.J. (1990). Nitrogen metabolism of lactating ewes fed rumen-protected methionine and lysine. *Journal of Dairy Science*, 74, 2268-2276.

[11] NRC (1988). Nutrient Requirements of Dairy Cattle. Update 1989. National Academy of Science, Washington, DC.

[12] Tamminga, S. (1992). Nitrogen management of dairy cows as a contribution to pollution control. *Journal of Dairy Science*, 75, 345-357.

[13] Gonzalez, J.S., Robinson, J.J., McHattle, I. (1984). The effect of level of feeding on the response of lactating ewes to dietary supplements of fish meal. *Animal Production*, 40, 39-45.

[14] Cowan, R.T., Robinson, J.J:, McHattie, I., Pennie, K. (1981). Effects of protein concentration in the diet on milk yield, change in body composition and the efficiency of utilization of body tissue for milk production in ewes. *Animal Production*, 33, 111-120.

[15] Sevi A., Albenzio M., Annicchiarico G., Caroprese M., Marino R., Santillo A. (2006) Effects of dietary protein level on ewe milk yield and nitrogen utilization, and on air quality under different ventilation rates. *Journal of Dairy Research* 73 197-206.

[16] Cannas A. (2002). Feeding of lactating ewes. *In Dairy Sheep Feeding and Nutrition,* pp 123-166. (Ed. G. Pulina). Bologna: Avenue media.

[17] Auldist, M. (2003). Effects of processing characteristics. Vol. II, pgg. 2002-2007. In *Encyclopedia of Dairy Sciences*. Edited by Roginski, H., Fuquay, J.W., Fox, P.F. by MPG Books Lt, Bodmin, Cornwall, UK.

[18] Slaghuis, B., Wolters, G., Reinemann, D. (2003). Milking hygiene. Vol.III, pgg.2016-2021. In *Encyclopedia of Dairy Sciences*. Edited by Roginski, H., Fuquay, J.W., Fox, P.F. by MPG Books Lt, Bodmin, Cornwall, UK.

[19] Sevi, A., Massa, S., Annicchiarico, G., Dell'Aquila, S., Muscio, A. (1999). Effect of stocking density on ewes milk yield, udder health and micro-environment. *Journal of. Dairy Research*, 66, 489-499.

[20] [20] Duranti, E., Casoli, C. (1991). Variations in the nitrogen composition and renneting characteristics of ewes' milk in relation to somatic cell content. *Zootecnica e Nutrizione Animale*, 17, 99-105.

[21] Sevi A., Taibi L., Muscio A., Albenzio M., Dantone D., Dell'Aquila S. (2001) Quality of ewe milk as affected by stocking density and litter treatment with bentonite. *Italian Journal of Food Science, 13, 77-86.*

[22] Sheldrake, R.F., Hoare, R.J.T., McGregor, G.D. (1983). Lactation stage, parity, and infection affecting somatic cells, electrical conductivity, and serum albumin in milk. *Journal of Dairy Science*, 66, 542-547.

[23] Albenzio M., Caroprese M., Marino R., Santillo A., Taibi L., Sevi A. (2004). Effects of somatic cell count and stage of lactation on the plasmin activity and cheese-making properties of ewe milk. *Journal of Dairy Science,* 87, 533-542.

[24] Sevi A., Muscio A., Casamassima D., Albenzio M. (2000). Effects of litter treatment with paraformaldehyde on bacteriological, nutritional and renneting characteristics of ewes milk. *Advances in Food Science*, *22*, 14-21.

[25] Sevi A., Albenzio M., Muscio A., Casamassima D., Centoducati P. (2003). Effects of litter management on airborne particulate in sheep houses and on the yield and quality of ewe milk. *Livestock Production Science,* 81, 1-9.

[26] Sevi, A., Taibi, L., Albenzio, M., Annicchiarico, G., Muscio, A. (2001). Airspace effects on the yield and quality of ewe milk. *Journal of Dairy Science*, 84, 2632-2640.

[27] Staples, C.R., Thatcher, W.W. (2003). Stress, heat, in dairy cattle: effects on milk production and composition. Vol. IV, pgg. 2592- 2597. In *Encyclopedia of Dairy Sciences*. Edited by Roginski, H., Fuquay, J.W., Fox, P.F. by MPG Books Lt, Bodmin, Cornwall, UK.

[28] Abdalla, E.B., Kotby, E.A., Johnson, H.D. (1993). Physiological responses to heat induced hyperthermia of pregnant and lactating ewes. *Small Ruminant Research*, 11, 125-134.

[29] Costa, M.J.R.P., Siva, R.G., Souza, R.C. (1992). Effect of air temperature and humidity on ingestive behaviour of sheep. *International Journal of Biometeorology*, 36, 218-222.

[30] Sevi, A., Annicchiarico, G., Albenzio, M., Taibi, L., Muscio, A., Dell'Aquila, S. (2001) Exposure of solar radiation and feeding time and production of lactating ewes under high ambient temperature. *Journal of Dairy Science,* 84, 629-640.

[31] Sevi A.; Albenzio M., Annicchiarico G., Caroprese M., Marino R., Taibi L. (2002). Effects of ventilation regimen on the welfare and performance of lactating ewes in summer. *Journal of Animal Science, 80, 2362-2372.*

[32] Sevi, L. Taibi, M. Albenzio, G. Annicchiarico, R. Marino and M. Caroprese (2003). Influence of ventilation regimen on micro-environment and on ewe welfare and milk yield in summer. *Italian Journal of Animal Science, 2*, 197-202.

[33] Sevi, A., Taibi, L., Albenzio, M., Caroprese, M., Marino, R., Muscio, A. (2003). Ventilation effects on air quality and on the yield and quality of ewe milk in winter. *Journal of Dairy Science*, 86, 3881-3890.

[34] Burvenich, C., Detilleux, J., Paape, M.J., Massart-leen, A.M. (2000). Physiological and genetic factors that influence the cows resistance to mastitis, especially during early lactation. In: Zacconi, A. (Ed.), *Proceedings of the International Symposium on Immunology of Ruminant Mammary Gland*, Stresa, Italy, June 11-14, 2000, pp.9-20.

[35] Fajardo-Lira, C.E., and Nielsen, S.S. (1998). Effect of psychrotrophic microorganisms on the plasmin system in milk. *Journal of Dairy Science* 81 901-908.

[36] Albenzio M., Caroprese M., Santillo A., Marino R., Muscio A., Sevi A. (2005). Proteolytic patterns and plasmin activity in ewe milk as affected by somatic cell count and stage of lactation. *Journal of Dairy Research, 72*, 86-92.

[37] Albenzio M, Corbo MR, Shakell-Ur-Rehaman, Fox PF, De Angelis M, Corsetti A, Sevi A, and Gobbetti M. (2001) Microbiological and biochemical characteristics of Canestrato pugliese cheese made from raw milk, pasteurised milk or by heating the curd in hot whey. *International Journal of Food Microbiology*, 67, 35–48.

[38] Fontecha, J., Pelaez, C., Juarez, M., Requena, T., Gomez, C. (1990). Biochemical and microbiological characteristics of artisanal hard goat's cheese. *Journal of Dairy Science* 73, 1150-1157.

[39] Macedo, A. C., Malcata, F. X., Hogg, T. A. (1995). Microbiological profile in Serra ewe's cheese during ripening. *Journal of Applied Bacteriology*, 79, 1-11.

[40] Corbo, M.R., Albenzio M., De Angelis, M., Sevi, A. and Gobbetti M. (2001). Microbiological and biochemical properties of Canestrato pugliese hard cheese supplemented with bifidobacteria. *Journal of Dairy Science, 84, 551-561.*

[41] Coda, R., Brechany, E., De Angelis, M., De Candia, S., Di Cagno, R., Gobbetti, M. (2006). Comparison of the compositional, microbiological, biochemical, and volatile profile characteristics of nine Italian ewes' milk cheeses. *Journal of Dairy Science*, 89, 4126-4143.

[42] Albenzio M., Santillo A., Caroprese M., Marino, Annicchiarico G., Sevi A. (2007). Effects of ventilation rate and of dietary protein level in intensive dairy sheep system on the features of Canestrato pugliese cheese. *Journal of Dairy Research*, 74, 26-33.

[43] Fox, P.F. (2003). Significance of indigenous enzymes in milk and dairy products. Pages 255-277 in *Handbook of Food Enzymology*. Eds J.R. Whitaker, A.G.J. Voragen and D.W.S. Wong. New York: Marcel Dekker, Inc.

[44] Park, Y. W. (2001). Proteolysis and lipolysis of goat milk cheese. *Journal of Dairy Science* 84, E84-E92.

[45] Albenzio M., Santillo A., Caroprese M., Marino R., Centoducati P., Sevi A. (2005). Effect of different ventilation regimens on ewe milk and Canestrato Pugliese cheese quality in summer. *Journal of Dairy Research,* 72, 447-455.

[46] Kelly, A. L. AND McSweeney, P. L. H. (2002). Indigenous proteinases in milk *Advanced Dairy Chemistry,* 1, 494-519.

[47] Farkye, N.Y. (2003). Indigenous enzymes in milk; other enzymes. In P.F. Fox, P.L.H. McSweeney (Eds.), *Advanced Dairy Chemistry*. Vol. I, Proteins (3rd ed., pp. 571-603). New York, USA: Kluwer Academic-Plenum Publishers.

[48] McSweeny, P.L.H. AND Fox, P.F. (1997). Chemical methods for the characterization of proteolysis in cheese during ripening. *Le Lait* 77, 41-76.

[49] Fox, P.F. (1989) Proteolysis during cheese manufacture and ripening *Journal of Dairy Science*, 72, 1379-1400.

[50] Albenzio M., Marino R., Caroprese M., Santillo A., Annicchiarico G., Sevi A. (2004). Quality of milk and Canestrato Pugliese cheese from ewes exposed to different ventilation regimens. *Journal of Dairy Research,* 7, 434-443.

In: Protein Research Progress
Editor: A.B. Boscoe and C. R. Listow, pp. 207-231
ISBN 978-1-60456-663-3

Chapter 7

NOVEL APPROACH FOR PROTEIN RESEARCH

Peiqiang Yu*
College of Agriculture and Bioresources,
University of Saskatchewan
51 Campus Drive, Saskatoon,
Canada, S7N 5A8

ABSTRACT

An understanding of the structure and matrix of the whole protein is often vital to understanding its digestive behavior, utilization and availability in both humans and animals. Traditional approach to study dietary protein is to focus on total protein composition and amino acid profile using "wet" chemical analysis as suggested by NRC, called "a chemical approach". However, using the NRC chemical approach to determine energy and nutritive values is not accurate in comparison with biological approach. The fatal shortcoming of this approach is that the NRC chemical approach does not consider food/seed/feed intrinsic structure, matrix and molecular chemical make-up. It only considers total chemical composition. Traditional "wet" chemical analysis can determine total protein composition and amino acid profile, but fails to detect protein inherent structure, molecular chemical make-up and protein matrix due to destruction of feed protein tissues during the processing for chemical analysis. In this chapter, a novel approach is introduced to reveal protein structure, molecular chemical make-up, and protein matrix within intact tissue at cellular and subcellular level with advanced synchrotron-based analytical technology (SFTIRM). This technology provides new concept and methodology for dietary protein research.

Keywords: Protein Imaging; Molecular Chemical Make-up; Protein Matrix; Synchrotron; Protein Utilization and Availability, Nutrition.

---

*Corresponding author: SAF Research Chair, College of Agriculture and Bioresources, University of Saskatchewan, 6D10 Agriculture Building, 51 Campus Drive, Saskatoon, Canada, S7N 5A8, Tel: +1 306 966 4132, Fax: +1 306 966 4150, E-mail:peiqiang.yu@usask.ca

1. INTRODUCTION

1.1. Protein Intrinsic Structure, Molecular Chemical Make-Up and Protein Matrix: Importance

Protein value, digestive behavior, and availability are closely related to not only total protein content and protein amino acid profile, but also protein intrinsic structure, molecular chemical make-up, and protein matrix, such as protein to non-structural carbohydrate matrix, protein to structural carbohydrate matrix, and protein to lipid matrix (For example, a recent study clearly shows that protein to starch granule matrix in barley tissue may play an important role in determining protein digestive behavior, utilization and availability (Yu et al., 2004a).

1.2. Limitation and Shortcoming of Chemical Approach

As we know, traditional approach to study plant/seed/feed protein value is to focus on total protein composition and amino acid profile [using "wet" chemical analysis as suggested by such as NRC (1996, 2001)] without considering protein structure, molecular chemical make-up and protein matrix, This approach is called "a chemical approach". However, the numerical research show that the NRC chemical approach can not accurately predict plant/seed/feed energy and nutritive values in comparison with biological approach, mainly due to the fact that the NRC chemical approach does not consider plant/seed/feed intrinsic structure, molecular chemical make-up, and protein matrix within an intact tissue. This chemical approach only considers total chemical composition.

1.3. "Wet" Chemical Analysis Fails to Detect Protein Intrinsic Structures and Matrix

In other words, not because we do not want to consider the protein inherent structure, molecular chemical make-up, and protein matrix in relation to protein nutritive value, mainly because we did not have an advanced research tool to detect them and reveal them. Conventional "wet" chemical analysis (such as AOAC, 1990) which we use for a long time has limitation and shortcoming. It can only determine total protein composition, but fails to detect the protein intrinsic structures and matrix. This is because the conventional "wet" chemical analysis looks for a specific component (such as protein) through homogenization of the samples and separation of the component of interest from the complex matrix. As a result, the object of the analysis is destroyed and the information about the spatial origin and distribution of the component of interest is lost (Budevska, 2002).

1.4. Advanced Synchrotron-Based Analytical Technology and Objectives

The objective of this article was to introduce a novel approach to detect protein structure, molecular chemical make-up, and protein matrix within intact tissues at cellular and subcellular level with advanced synchrotron-based analytical technology. This advanced technology will provide a new concept and methodology for protein research and provide us with a greater understanding the protein and animal or human interface.

2. Advanced Synchrotron-Based Analysis Technology Available for Protein Structure and Matrix Research

2.1. What Is a Synchrotron?

A synchrotron is a giant particle accelerator that turns electrons into light. A synchrotron facility usually consists of six major components. These include the Electron Gun, Linear Accelerator (also called LINAC), Booster Ring, Storage Ring, Beamlines, and End Experimental Stations (Wetzel et al., 1998; Martin, 2002). In a short definition, the synchrotron is a facility that produces extremely bright light, million times brighter than sunlight. The extremely bright synchrotron light makes it possible to detect molecular chemical structure within intact biological tissues at ultraspatial resolution within cellular dimensions (Miller, 2000 and 2002; Martin, 2002; Marinkovic et al., 2002; Dumas, 2003).

2.2. Synchrotron Light-Based Fourier Transform Infrared Microspectroscopy

Advanced synchrotron light-based Fourier transform infrared microspectroscopy (SFTIRM), is one of the advanced synchrotron-based technology. This synchrotron based bioanalytical technique has been developed as a non-destructive and non-invasive technique (Wetzel et al., 1998; Dem et al., 1999; Holman et al., 2002; Jackson and Mantsch, 2000; Marinkovic et al., 2002; Miller, 2002). This technique, taking advantages of extremely bright synchrotron light (which is usually 100~1000 times brighter than conventional Globar source and has small effective source size), is capable of exploring the molecular chemistry within microstructures of biological samples with high signal to noise ratio at ultraspatial resolutions (Miller et al. 1998; Wetzel et al. 1998; Raab and Martin, 2001; Wetzel, 2001; Marinkovic et al. 2002; Miller et al. 2000 and 2002; Dumas, 2003; Yu et al. 2003a and 2004b). This technique is able to link chemical information (qualitative and relatively quantitative analytical results) to structural information of biological samples (Budevska, 2002). This technique can provide us four kinds of information simultaneously: tissue composition, tissue structure, tissue chemistry and tissue environment. The technique can be used to increase the fundamental understanding of plant/feed/seed structures at the cellular level and bring a new level of understanding of analytical information (Budevska, 2002; Yu et al. 2004c and 2005a; Yu, 2005a).

2.3. Application of Synchrotron Infrared Microspectroscopy in Protein Research

To date there has been very little application of this novel technique to the study of plant/seed/feed protein in relation to protein nutritive value (Yu, 2004). This novel technique shows the potentials which can be used for protein inherent structure, protein molecular make-up and protein matrix study.

3. Image Protein Distribution and Intensity within Intact Tissue at Cellular and Subcellular Level

3.1. Imaging Molecular Chemistry of Protein Tissues

With advanced synchrotron-based analytical technology -SFTIRM, now it is achievable to image or map protein distribution and intensity or concentration within intact plant/feed/seed tissues at cellular and subcellular level, which make it possible to rapidly characterize plant/seed/feed inherent protein structure in the inherent tissue. The chemical-microstructures of plant/seed/feeds have been of interest in our laboratory in regard to nutritive value and nutrient utilization (Yu et al., 2004a). The photomicrographs of the plant/seed/feed tissues, even including electronic micrograph, are not very informative with respect to molecular chemical make-up and structural-chemical features of the inherent structures at ultraspatial resolution. With the synchrotron-based technology, the structural-chemical information of various plant/feed/seeds can be revealed by imaging or mapping at a cellular and subcellular level (Yu et al., 2003a; 2004b, 2004c).

Figure 1 show chemical imaging micro-structure of four different seed examples: corn, barley, Winterfat, flaxseed and wheat (Yu et al., 2003a, 2004c, and 2005b). Each seed has very unique inherent structure. For example, Figure 1 shows barley tissue chemical image from pericarp, seed coat, aleurone layer to part of endosperm within intact tissue. It represents maps of biological component of the seed tissue with single pixel spectra measuring an area as small as 10×10 μm of the sample. It shows visible and protein images in representation of protein concentration or intensity and distribution.

Using the advanced synchrotron-based technology (SFTIRM), the distribution and relative concentration of the biological components associated with the seed inherent structure were also revealed. The protein images were taken from the region of the visible image outlined by the rectangle area. For example, from Figure 1, we know that protein do not present in pericarp region and seed coat regions (which are not known before), but protein present much more in endosperm region and was unequally distributed in the tissue (usually we use false colors to represent chemical intensity and distribution, blue stand for high, red stand for low, white stand for no chemicals).

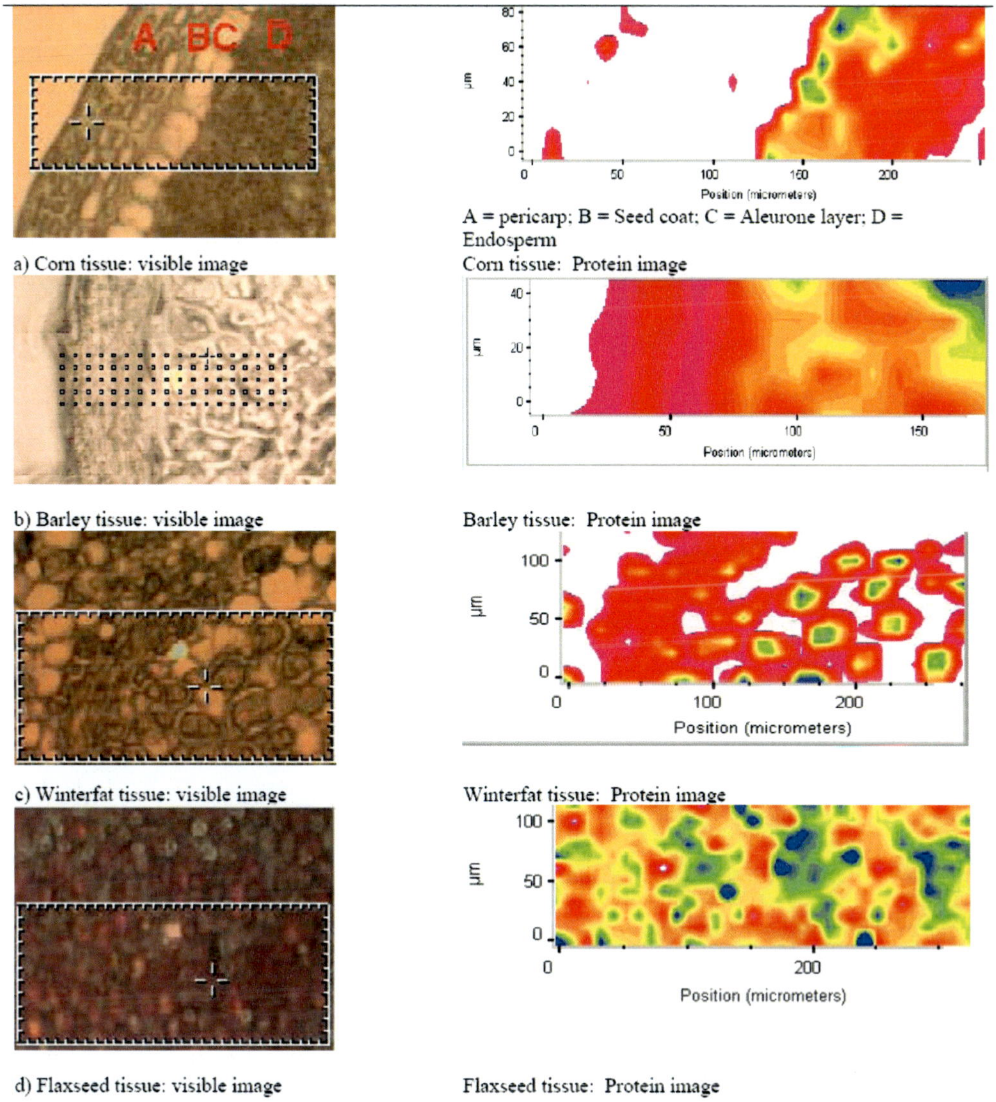

Figure 1. Protein microstructural chemical images (visible and chemical image) of the plant seed tissues at cellular and subcellular level (a) protein image of corn; (b) protein image of barley; (c) protein image of Winterfat; (d) protein image of flax seed tissue, and (e) protein image of wheat tissue to show protein intensity and distribution.

We can also using ultra-structural mapping technique to show other biological component imaging such as lipid, lignin, cellulose, hemicellulose imaging (Figure 2), as reported in Yu et al (2004c). The results show that lignin only present in outside regions of corn which is in pericarp region.

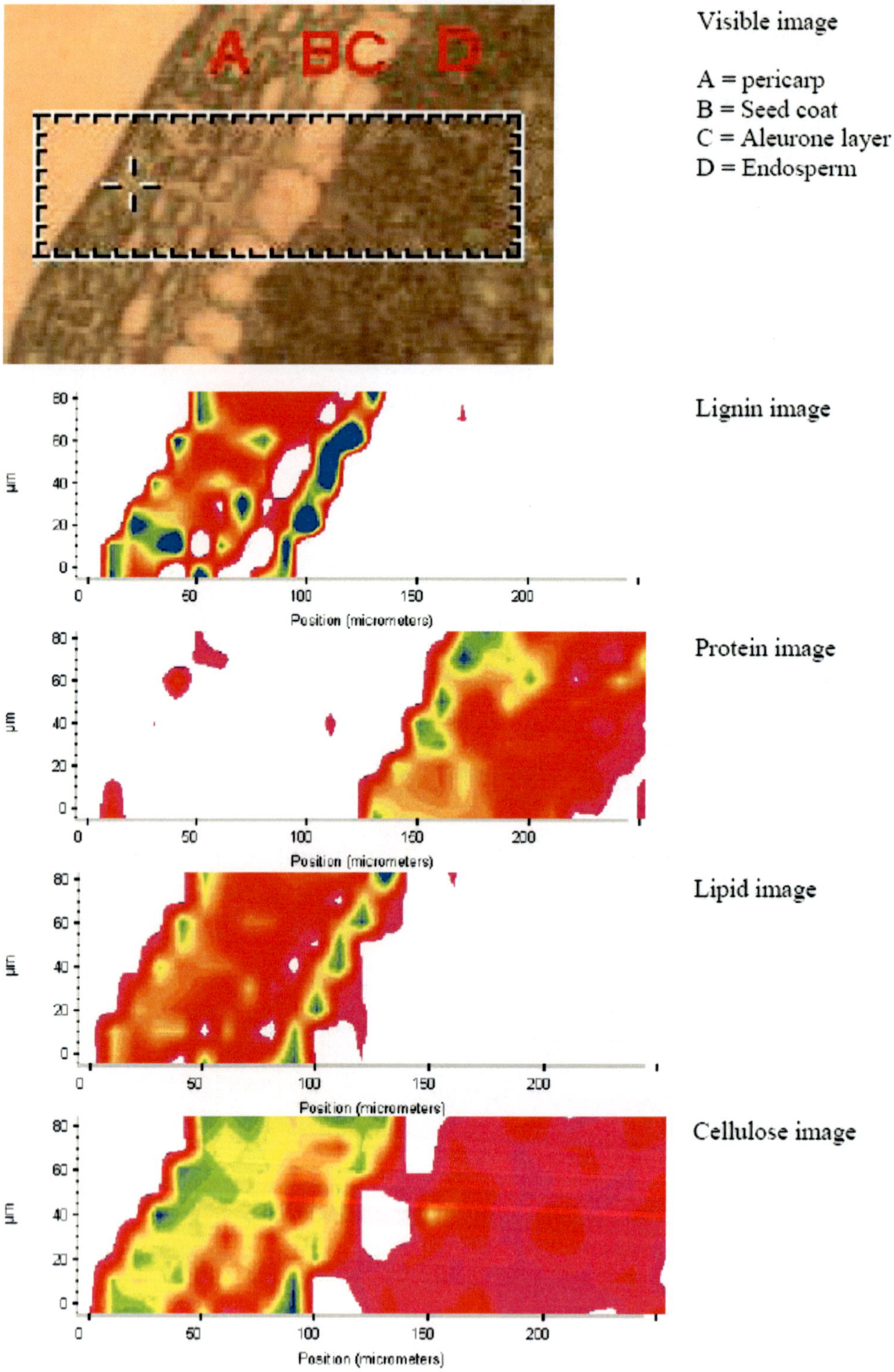

Figure 2. Biological components images of corn (lignin, protein. Lipid and carbohydrate).

3.2. Agglomerative Hierarchical Cluster Analysis (CLA) for Molecular Chemistry Imaging

If a chemical function group band from plant/seed/feed tissues is not very unique and has more or less inference with other functional group bands, we can use agglomerative hierarchical cluster analysis (CLA) for chemical imaging as published in Yu (2005b). The CLA analysis is one of multivariate analysis, of which function performs an (agglomerative hierarchical) cluster analysis of an infrared spectra data set and displays the results of CLA as images and as dendrograms. First, it calculate distance matrix, which contains information on the similarity of chemical functional group spectra. Then, in hierarchical clustering, the algorithm searches within the distance matrix for the two most similar infrared spectra (minimal distance). These spectra are combined into a new object (called a "cluster" or called "hierarchical group"). The spectral distances between all remaining spectra and the new cluster are re-calculated (Cytospec, 2004). For cluster analysis, Ward's algorithm was used in our studies (Yu, 2005b) for clustering. Spectral distances were computed as D-values. In a dendrogram, spectral distances were given on the Y-axis. Horizontal lines illustrated the merging process, eg the combination of two clusters to a new cluster of spectra. It is a technique which clusters IR spectra based on similarity with other spectra (Jackson and Mantsch, 2000; Miller, 2002). This method can give results with a better discrimination of different seed structures using seed chemical image (Yu, 2005b) and seed group spectrum (Yu, 2005c). Figure 3 is an example to show that the clusters in the pericarp region, aleurone layer and endosperm region are different between seeds and between seed structures within the seed (Yu, 2005c).

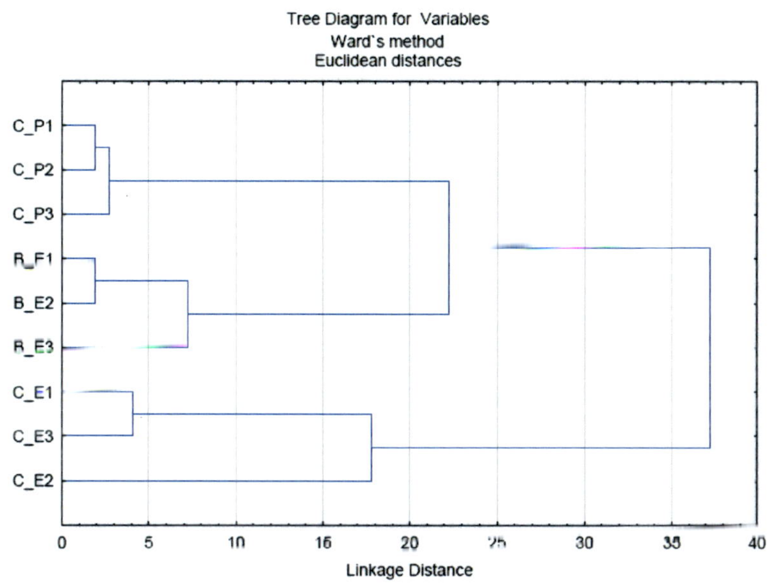

Figure 3. The CLA cluster of plant seed intrinsic structures showing that the clusters in the pericarp region (C_P IR group), aleurone layer (C_E IR group) and endosperm region (B_E IR group) are different between feeds and between feed structure within the feed [CLA analysis: (1) Select spectral region: 4000 to 800 cm-1; (2) Distance method: Euclidean; (3) Cluster method: Ward's algorithm].

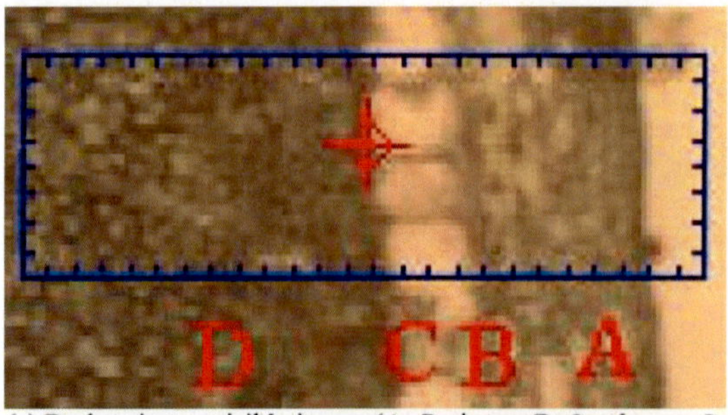

(a) Barley tissue: visible image (A: Pericarp; B: Seed coat; C: Aleurone; D: Endosperm)

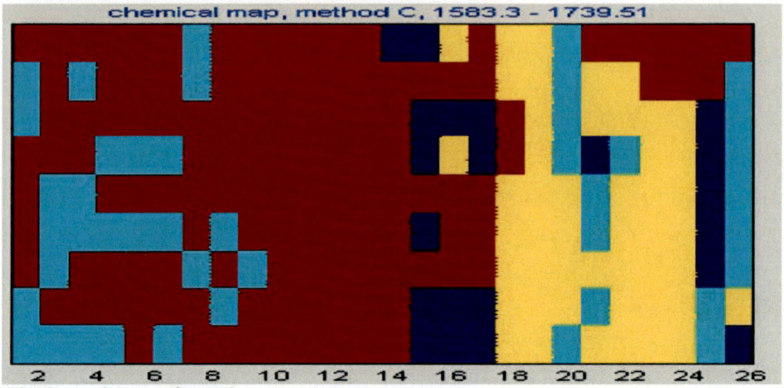

(b) Cluster image of protein

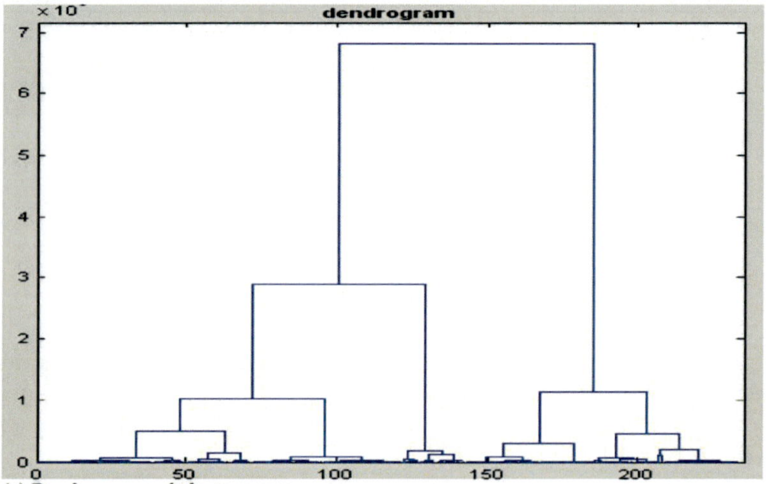

(c) Dendrogram and cluster trees

Figure 4. The CLA cluster chemical image of protein clearly showing that the clusters in the pericarp region and endosperm region are different [CLA analysis: (1) Select spectral region: 1739.51 to 1583.30 cm-1; (2) Distance method: D-Values; (3) Cluster method: Ward's algorithm; (4) Number of Cluster: 4 in the cluster image].

It displays the results in the form of a dendrogram. From this diagram, three classes can be distinguished below a linkage distance less than 23, with C-E group forming a separate group. Depending on the aggregation level (horizontal axis) different explanations can be inferred. The C_P group and B_E group form the two distinct groups just below an aggregation (linkage) distance 8. They form a single group at an aggregation distance of about 22. The C_E group can be grouped together with the C_P and B_E groups for a linkage distance equal to about 38. Figure 3 clearly shows significantly different spectral clusters in the different seed structures: pericarp region, aleurone layer and endosperm region using whole spectral region from 4000 to 800 cm^{-1} for CLA group spectrum analysis. In other words, different seed structures have different spectra clusters and different seeds have different spectra clusters due to different molecular structure (Yu, 2005c). Figure 4 show CLA protein imaging that gave satisfactory results and was conclusive in showing that it has a great potential to discriminate and classify functional groups differences existing in different seed structure regions. The CLA analysis approach places synchrotron-based technology at the forefront of those potential techniques that can be used in the rapid characterization of plant/seed/feed micro-chemical structure.

3.4. Implications of Molecular Chemistry Imaging

From above biological component image studies, it is concluded that with the advanced synchrotron technology, now we can "see" inside of tissues in a chemical sense, at a cellular and subcellular level, not only visible image but also chemical images. With the synchrotron-based technology, we can image protein distribution and intensity across the tissue at ultra-spatial resolution within cellular dimensions. Very important implications from these studies are: we can chemically define the plant/feed/seed intrinsic structures and analyze individual parts with intact tissue (for example, we can analyze chemical differences in seed coat region without destruction of the seed tissues between two different varieties of barley seeds); we can compare plant/feed/seeds (species and varieties) according to their spectroscopic characteristics/biological components spatial distribution and intensity at a cellular level; we can relate plant/feed/seeds intrinsic structures using ultra-structural imaging/mapping to their quality, nutritive value, animal digestive behaviors and nutrient utilization.

4. Reveral Protein Structural Matrix

4.1. Protein Inherent Structural Matrix Associated with Nutritive Value and Digestive Behavior

As we know, protein matrix is highly associated with nutritive value and digestive behavior, such as protein to structural carbohydrate (cellulose, hemicellulose) matrix, protein to non-structural carbohydrate matrix. In barleys, protein to starch granule matrix in endosperm region may play an important role in determine fermentation characteristics and degradation behaviors as well as nutrient availability (Yu et al., 2004a). Harrington is a malting-type barley variety. As with most barley varieties, its extent and rate of biodegradation are high (Table 1). However, if ruminants consume rapidly degradable barley, the incidence of digestive disorders such as bloat and acidosis can increase. This reduces animal performance and can lead to death in acute cases (Nocek and Tamminga, 1991; Givens et al., 1993).

Table 1. Effects of variety and processing method on protein degradation characteristics, degraded protein balance and nutrient availability

Items	Barley variety				SEM
	Harrington		Valier		
	Roller (0.533 mm)	Hammer mill (2 mm)	Roller (0.533 mm)	Hammer mill (2 mm)	
Protein degradation characteristics (by the Øskov model)					
Protein Soluble fraction (S, g/kg)	57 c	157 b	24 d	211 a	6.7
Protein potential degraded fraction (D, g/kg)	843 a	650 b	694 b	733 ab	29.3
Undegraded protein fraction (U, g/kg)	99 b	193 ab	282 a	56 b	31.3
Degradation rate (Kd, /h)	0.120 b	0.302 a	0.112 b	0.147 b	0.0249
Undegraded protein fraction (RUP, g/kg)	383 b	301 b	541 a	271 b	28.0
Undegraded protein fraction (RUP, g/kg DM)	55 b	44 b	81 a	41 b	4.08
Effective degraded fraction (EDCP, g/kg)	617 a	699 a	459 b	729 a	28.0
Effective degraded fraction (EDCP, g/kg DM)	81 a	91 a	62 b	99 a	3.7
Predicted protein supply and availability					
Degraded protein balance (OEB, g/kg DM)	-34 b	-23 ab	-51 c	-15 a	3.7
Absorbable microbial protein (AMCP, g/kg DM)	44 a	50 a	34 b	54 a	2.0
Absorbable rumen bypass feed protein (ARUP, g/kg DM)	47 b	37 b	69 a	35 b	3.5
Total absorbable intestinal Protein (MP, g/kg DM)	95 b	91 b	107 a	93 b	1.5

Valier barley has been recently developed as specialized barley for ruminants (Gibson et al., 1994). Its extent and rate of rumen degradation are low, which should be beneficial to animal performance and health when animals are fed this variety (Yu et al. 2003b). The hypothesis in the study (Yu et al. 2004a) was that the different degradation characteristics between the two barley varieties may be partly due to differences in the protein to starch granule matrix of the endosperm tissue. As we know, traditional "wet" chemical analysis cannot detect such biological differences because the inherent seed structures that may inhibit digestion are destroyed during processing for analysis. The "wet" chemical analysis (AOAC, 1990) also showed that the total chemical composition were similar between the two barleys (Yu et al., 2003b). The purpose of this study (Yu et al., 2004a) was to use the new synchrotron-based new technology (SFTIRM) to identify differences in the structural matrix makeup of Harrington and Valier barley endosperm tissues, so as to obtain a better understanding of the intrinsic structure of barley endosperm tissue in relation to rumen degradation characteristics. The photomicrograph (Figure 5) of cross-sections of the endosperm tissue of Harrington and Valier barley showed that Harrington has larger starch granules that are less densely packed in the protein matrix, relative to Valier.

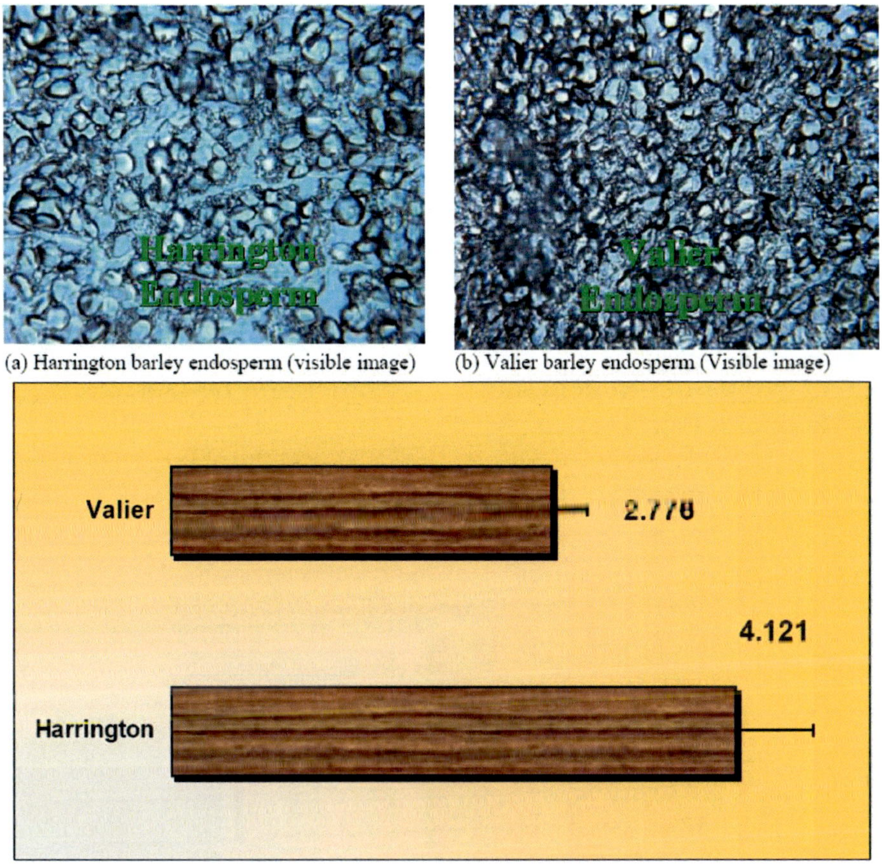

Figure 5. Protein to starch matrix: ratios of starch to protein synchrotron-based FTIR absorption unit in the barley endosperm tissues: comparison of Harrington and Valier.

As with traditional analytical methods, it is difficult to pinpoint exact differences. No specific chemical information on barley micro-endosperm tissue is obtained through such visible images. Using the synchrotron-based analytical technique, the new information cam out that the results showed that Harrington had a wider range of starch to protein IR absorbance intensity ratio (1.406 to 10.119 vs. 1.419 to 4.274), suggesting that it is more heterogeneous than Valier in endosperm chemical makeup. Valier had a lower ratio of starch to protein IR absorbance intensity than Harrington (4.121 ± 0.576 vs. 2.776 ± 0.270), which implies that the starch granules in Valier are more closely associated with the protein matrix. This closer association may prevent the starch granules from being rapidly degraded in the rumen (Yu et al., 2004a). The lower starch to protein ratio matrix in the micro-endosperm tissue of Valier barley may provide a clue to reduced effective degradability and the somewhat slower rate of rumen nutrient degradation observed in the in situ study (Yu et al. 2003b). The greater association of the protein matrix with the starch granules in the endosperm tissue of Valier barley may limit access of ruminal microorganisms to the starch granules and thus reduce the rate and extent of degradation relative to that of Harrington barley. This study (Yu et al., 2004a) indicated that Harrington and Valier barley differ in chemical/structural endosperm matrix makeup. The lower ratio of starch to protein absorbance intensity for Valier indicated a tighter association between starch and protein in the endosperm. This may explain why Valier exhibits a lower extent and rate of rumen degradation than Harrington barley. More systematic research is being carried out to verify the theory that has been developed in the study (Yu et al., 2004a) and to further investigate the potential to use the synchrotron (SFTIRM) as a research tool to evaluate rumen fermentation behavior of cereal grains (A. Walker and J.J. McKinnon, Personal contact). The above research was to use an animal model to illustrate protein inherent structural matrix associated with nutritive value and digestive behavior. It is also true in human nutrition.

4.2. Protein with Other Biological Components Ratio Imaging

With the assistance of the advanced synchrotron-based technology, another way to study protein matrix is to look at biological component ratio feature in the plant/feed/seed tissues by using ultra-structural ratio imaging such as protein to starch ratio images, protein to total carbohydrate ratio image (Figure 6). Different biological component ratio images in the tissue (e.g. protein-to-total-starch-ratio image; hemicellulose-to-total-carbohydrate ratio image) show structural chemical features, representing biological component ratio distribution and intensity in the inherent structure. Such biological component ratio imaging has advantage to indicate relative biological component contents, which could be used to determine plant/feed/seed quality, characteristics and nutritive values. It can be used to compare different plant/feed/seed varieties and species according to the inherent structural spectral characteristics, chemical functional group ratio intensity and distribution (Yu et al., 2004c).

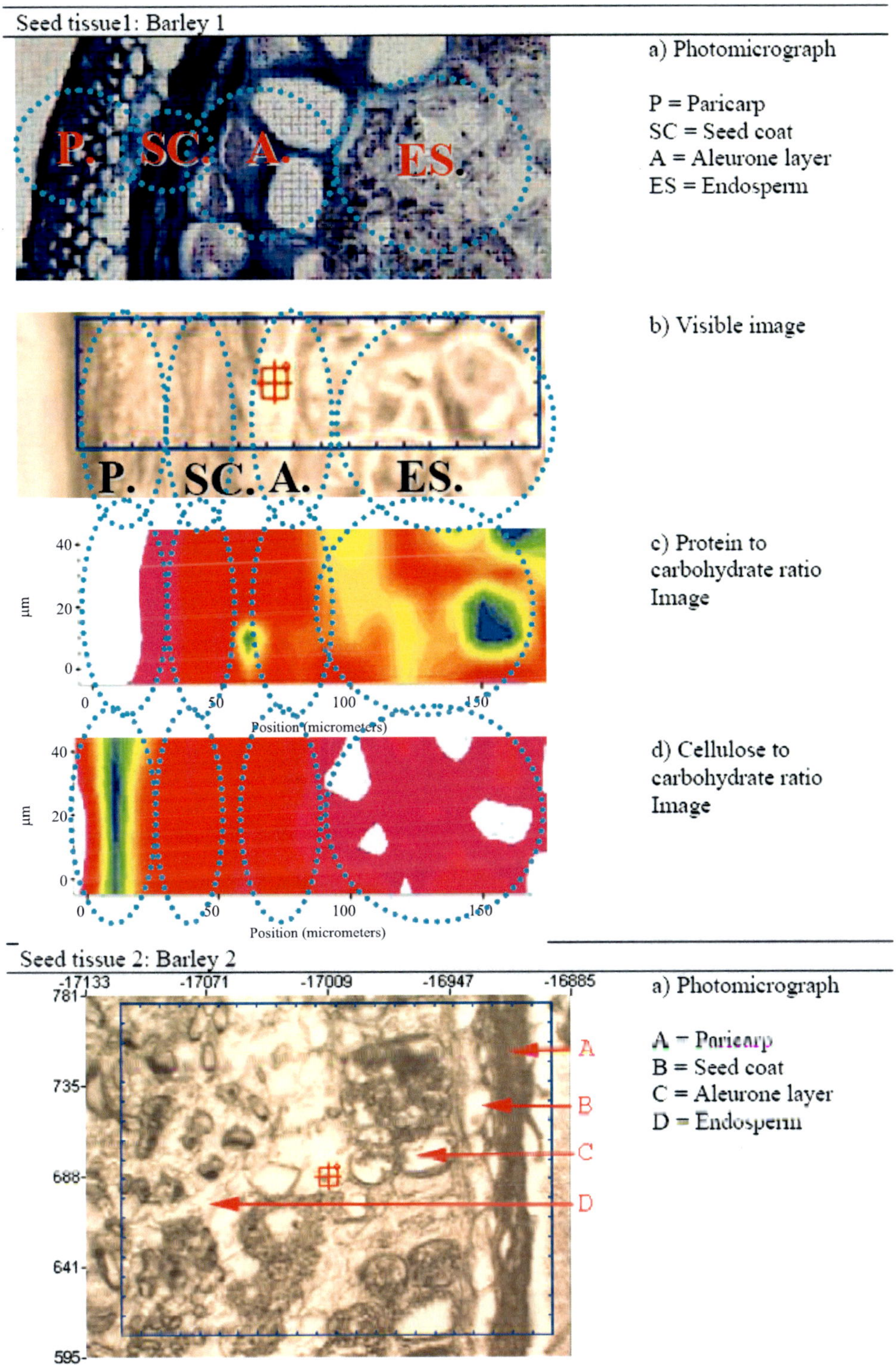

Figure 6. Biological component ratio images of Seed tissues (barleys and flaxseed) show the distribution and intensity of protein to carbohydrate ratio and cellulose to carbohydrate ratio within intact tissues, using the synchrotron-based analytical technique (SFTIRM).

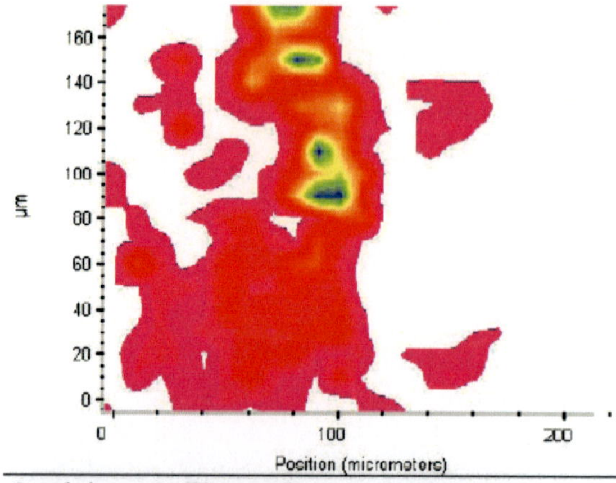

Protein-to-total-starch-ratio image

Seed tissue 3: Flaxseed

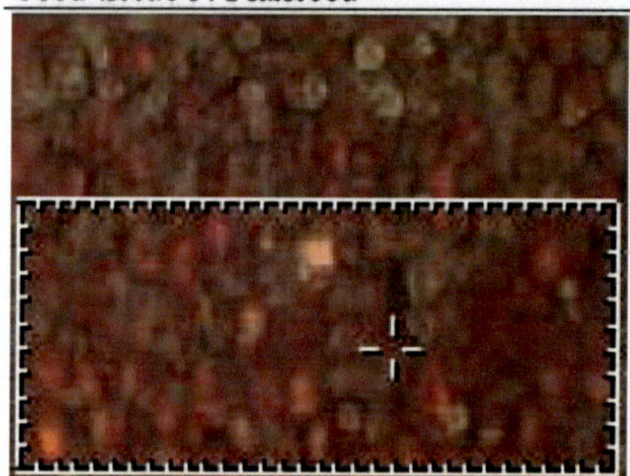

Flaxseed tissue (F1 type)

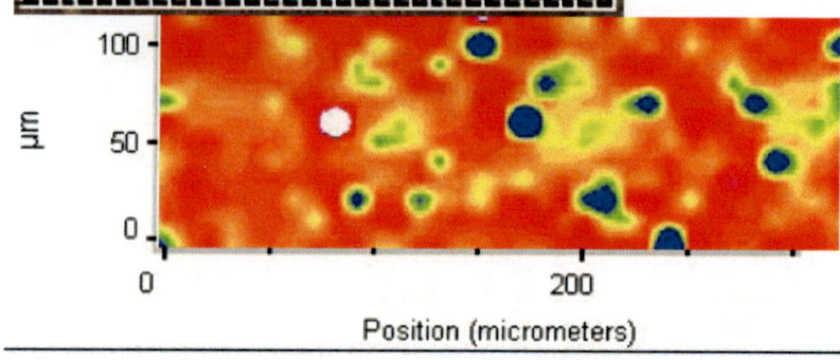

protein to carbohydrate ratio image

Figure 6. (Continued).

5. EXPLORE PROTEIN MOLECULAR CHEMICAL STRUCTURES

5.1. Molecular Chemical Features of Protein Internal Structures in Relation to Protein Value

With the advanced synchrotron-based technology, now we are also able to locate and find the relatively "pure" protein tissue within intact tissue at cellular and subcellular level (Wetzel et al., 2003; Yu et al., 2005d). When we analyze protein molecular structure and molecular composition, we have to use the protein tissue as pure as possible to reduce other biological components impact on protein molecular structure analysis, such as carbohydrate scattering effect on protein amide I and II spectrum, if we want to know accurate protein structural chemical make-up. With the advanced synchrotron-based technology, we can explore protein molecular structure chemical composition, such as protein secondary structure (α-Helix and β-Sheet and their ratios) of plant/feed/seed tissue as well as processing effects and treatment effects, and relation this information to nutrient utilization and availability (Yu et al., 2004d; 2005c).

Protein secondary structures include main α-helix and β-sheet and small amount of β-turn and random coil (Carey 1996; Martin, 2000; Dumas, 2003). The percentage and ratio of the protein secondary structures may influence protein quality, nutrient utilization, availability and digestive behavior (Yu et al. 2004d, 2005c). High percentage of β-sheet structure may partly cause low access to gastrointestinal digestive enzymes, which may result in a low protein value and low protein availability. Recent research (Seguchi et al., 2004) shows that processing affects protein secondary structure profiles and food quality.

5.2. Effect of Heat Processing on Protein Molecular Chemical Features and Relation to Protein Value

In the study, Yu et al. (2005c) used the synchrotron-based technology to reveal ultra-structural chemical features of protein secondary structures of flaxseed tissues affected by varieties (brown and golden) and heating processing (raw and roasting) within intact tissue, in relation to protein nutritive value, utilization and availability and quantify protein secondary structures using multi-component peak modeling methods Gaussian and Lorentzian. The detailed comparison of these two multi-peak modeling methods in detecting plant/seed/feed tissue inherent structure has been reported in Yu (2005d). The study (Yu et al., 2005c) showed that with the synchrotron-based technology, the protein molecular chemical makeup of the flaxseed tissues could be revealed at ultra-spatial resolution within cellular dimensions. The protein secondary structure differed between the golden and brown flaxseed tissues in terms of percentage and ratio of model-fitted α- helix and β-sheet. By using multi-component peaks modeling (Yu, 2005d) at protein amide I region, the results show that the golden contained relatively higher percentage of model-fitted α-helix (47 vs. 37%), lower percentage of model-fitted β-sheet (37 vs. 46%) and higher ratio of model-fitted α-helix to β-sheet (1.3 vs. 0.8), indicating potential high protein value, high nutrient utilization and availability in the golden flaxseeds (Table 2). The roasting reduced percentage of model-fitted α-helix (from 47 to 36%), increased percentage of model fitted β-sheet (from 37 to 50%) and reduced α-helix to β-sheet ratio (1.3 to 0.7) of the golden flaxseed tissues. However, the roasting did not affect percentage and ratio of model-fitted α-helix and β-sheet in the brown flaxseed tissue. These results indicated different sensitivities of protein secondary structure to the heat processing between the flaxseed varieties and that the roasting may affect protein value, nutrient utilization and availability in the golden seeds but not in the brown.

In this study, multivariate statistical analyses were also used to discriminate and classify inherent protein structures of the raw and roasted brown and golden flaxseed protein tissues (Figure 7).

Although the cluster analysis showed that the four flaxseed (F1 to F4) cannot be fully distinguished, however, when combined brown and golden flaxseed together and emphasized on the raw and roasting effects, the PCA analysis could fully distinguish between the raw and roasted flaxseeds in protein amide I FSD spectrum (Theory about FSD see Kauppinen et al., 1981) and the raw and roasted can be grouped in separate ellipses (Figure 7). The first three PCs explain 36.7, 19.4, and 16.5% of the variation in the protein amide FSD spectrum data set.

Table 2. Relative percentage of model-fitted α-helix and β-sheet and their ratios in protein secondary structures of flax seed tissues: comparison of raw and roasted golden and brown flax seeds[z, y, x,]

Feeds	Sample number	% of model-fitted α-helix	% of model-fitted β-sheet	Ratio of model-fitted α-helix to β-sheet
Raw Golden Flax Seed (F1)	20	47 a	37 c	1.28 a
Roasted Golden Flax Seed (F2)	20	36 b	50 a	0.73 b
Raw Brown Flax Seed (F3)	20	37 b	46 b	0.81 b
Roasted Brown Flax Seed (F4)	20	34 b	48 ab	0.77 b
SEM		1.3	1.2	0.051

[z] The data analyzed with model: Y = mean + feeds + fitting method + feeds× fitting method error, (where, Feeds = F1, F2, F3, F4; Fitting methods = Gaussian, Lorentzian) showing no significant differences of fitting method and interaction.

[y] Means with the same letter in the same column are not significantly different ($P>0.05$);

[x] SEM=standard error of mean.

5.3. Effect of Seed Variety on Protein Molecular Chemical Features and Relation to Protein Value

In another study of protein molecular chemistry of barley varieties (CDC Dolly, LP955, Harrington, Valier, AC Metcalfe, and Sisler) (Table 3), barley protein structural chemical make-up also existed differences among the varieties in terms of protein secondary structure profile. The results (Table 3) show barley protein contained 18 to 34% of model-fitted α-helix, 14 to 25% of model-fitted β-sheet, 44 to 69% others.

AC Metcalfe, Sisler, and LP955 contained highest percentage of α-helix (30 to 34%). Dolly and Valier contained lowest percentage of model-fitted α-helix (18 to 23%). Harrington was in between which was 25%. For protein β-sheet, AC Metcalfe, and LP955 contained highest percentage (22 to 25%). Dolly and Valier contained lowest percentage (13 to 17%). Barley varieties contained different α-helix to β-sheet ratios from 1.36 to 1.95, but not to a significant extent. The α-helix to others ratio (0.30 to 0.99 and β-sheet to others ratio (0.22 to 0.76) were significantly affected by barley varieties. The PCA analysis results (Figure 8) clearly showed that the protein molecular chemical make-up and inherent structure were largely different between Valier and Harrington, CDC Dolly and Valier, and CDC Dolly and AC Metcalfe. However, protein inherent structure between Dolly and Harrington, Dolly and LP955, Dolly and Sisler are not fully distinguished, indicating they are closely related each other.

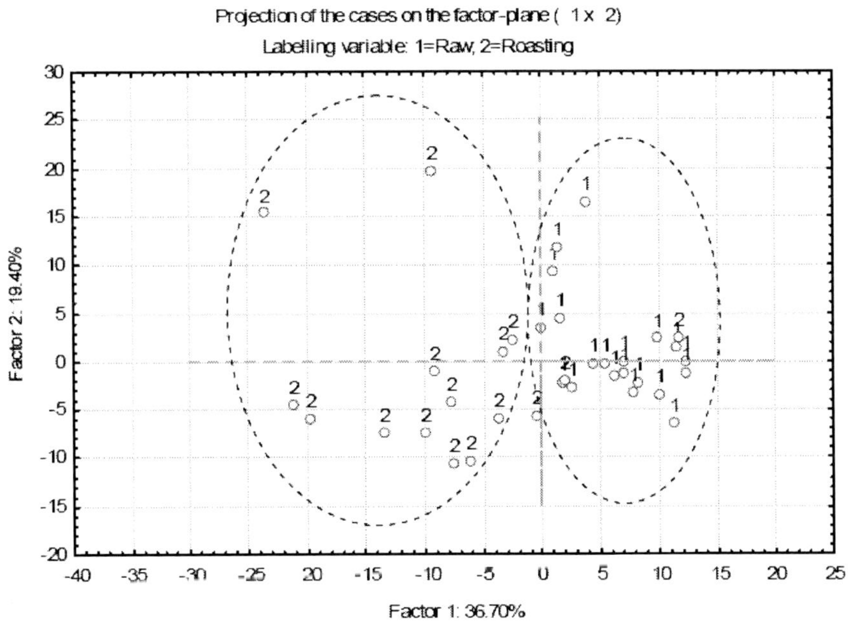

(a) PCA Component 1 vs. PCA Component 2

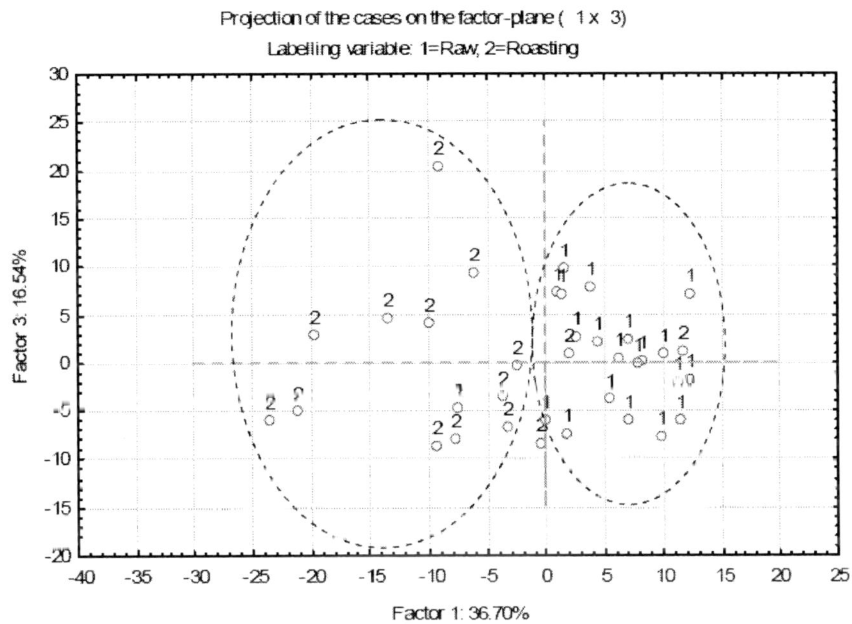

(b) PCA Component 1 vs. PCA Component 3

Figure 7. PCA analysis of synchrotron-based protein FSD amide I spectrum (1700-1620 cm-1) obtained from raw and roasted flaxseed tissues at a cellular level [1=Raw Flax Seed, 2=Roasted Flax Seed]: a) Scatter plot of the 1st principal component versus the 2nd principal component; b) Scatter plot of the 1st principal component versus the 3rd principal component (The 1st principal component explains 36.7% of the variance while the 2nd principal component explains 19.4% of the variance; the 3rd principal component explains 16.5% of the variance).

Table 3. Effect of barley varieties on protein secondary structure profile: Percentage and ratios of model-fitted α-helix and β-sheet in protein secondary structures of barley seed tissues at a cellular level: Comparison of six different varieties of barley (total samples for each variety: 22 to 30)

Items	Sample number	Protein secondary structure profile					
		% of model-fitted α-helix (±SD)	% of model-fitted β-sheet (±SD)	% of others (±SD)	Ratio of model-fitted β-sheet to α-helix (±SD)	Ratio of model-fitted α-helix to others (±SD)	Ratio of model-fitted β-sheet to others (±SD)
Barley Varieties							
Dolly	30	23 (±6) c	17 (±4) bc	60 (±9) b	0.80 (±0.24) ab	0.40 (±0.15) cd	0.30 (±0.11) c
Valier	22	18 (±11) c	14 (±6) c	69 (±13) a	0.99 (±0.47) a	0.30 (±0.25) d	0.22 (±0.14) c
Harrington	24	25 (± 10) b	17 (±4) bc	58 (±10) bc	0.82 (±0.48) ab	0.47 (±0.26) cd	0.31 (±0.10) c
LP955	30	31 (±15) a	25 (±13) a	44 (±21) d	0.95 (±0.57) a	0.99 (±0.91) a	0.76 (±0.49) a
AC Metcalfe	26	34 (±8) a	22 (±5) a	44 (±11) d	0.67 (±0.15) b	0.86 (±0.38) ab	0.55 (±0.20) b
Sistler	22	31 (±7) a	18 (±6) b	51 (±7) cd	0.64 (±0.32) b	0.64 (±0.24) bc	0.37 (±0.16) c
SEM		2.0	1.4	2.6	0.081	0.094	0.051

SEM= Standard error of mean.

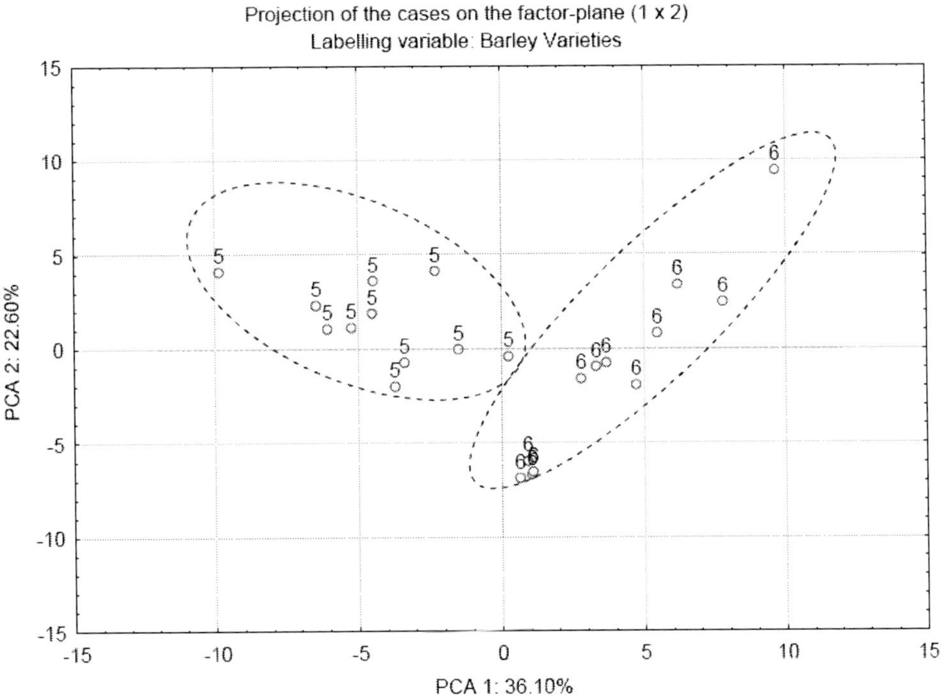

(a) Harrington (=6) vs. Varlier (=5): PCA Component 1 vs. PCA Component 2 (The 1st, 2nd and 3rd principal component explains 36.1, 22.6 and 16.0% of the variance).

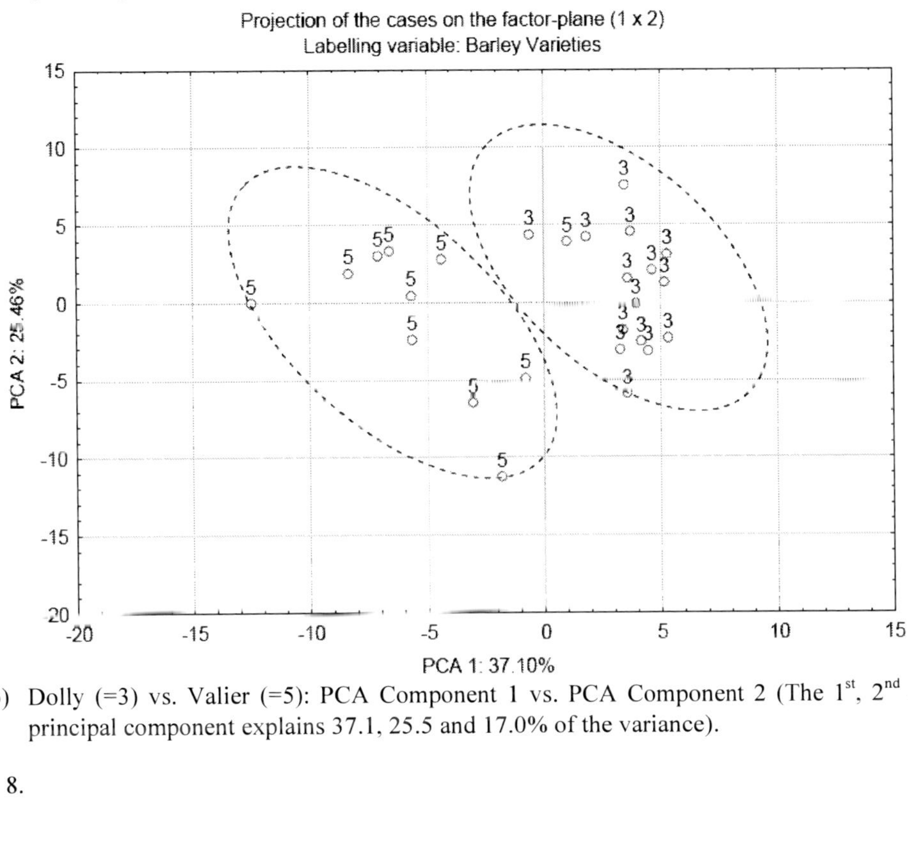

(b) Dolly (=3) vs. Valier (=5): PCA Component 1 vs. PCA Component 2 (The 1st, 2nd and 3rd principal component explains 37.1, 25.5 and 17.0% of the variance).

Figure 8.

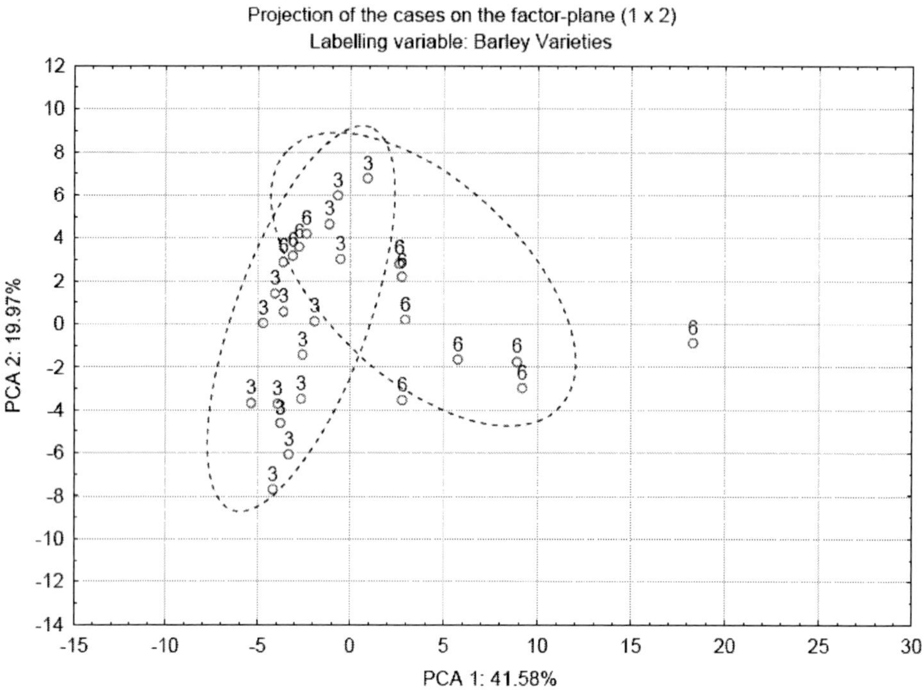

(c) Dolly (=3) vs. Harrington (=6): PCA Component 1 vs. PCA Component 2 (The 1st, 2nd and 3rd principal component explains 41.6, 20.0 and 13.9% of the variance).

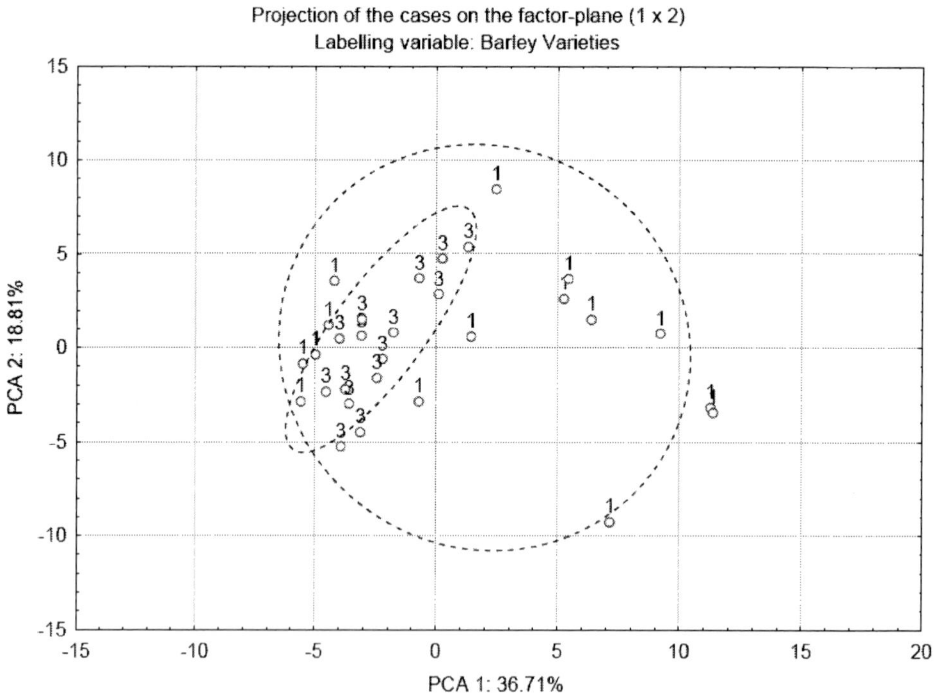

(d) Dolly (=3) vs. LP955 (=1): PCA Component 1 vs. PCA Component 2 (The 1st, 2nd and 3rd principal component explains 36.7, 18.8 and 11.6% of the variance).

Figure 8. (Continued).

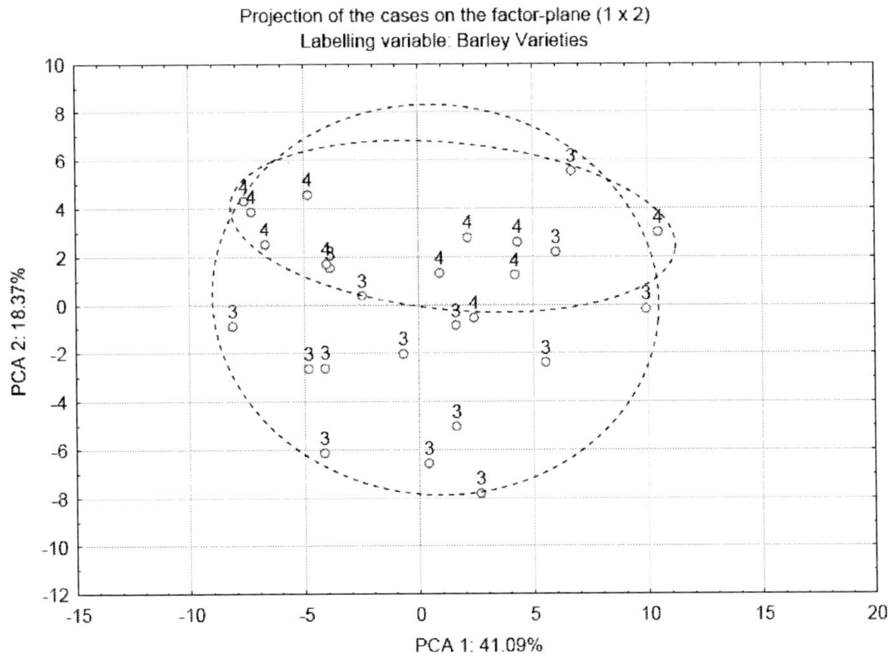

(e) Dolly (=3) vs. Sisler (=4): PCA Component 1 vs. PCA Component 2 (The 1st, 2nd and 3rd principal component explains 41.1, 18.4 and 16.4% of the variance).

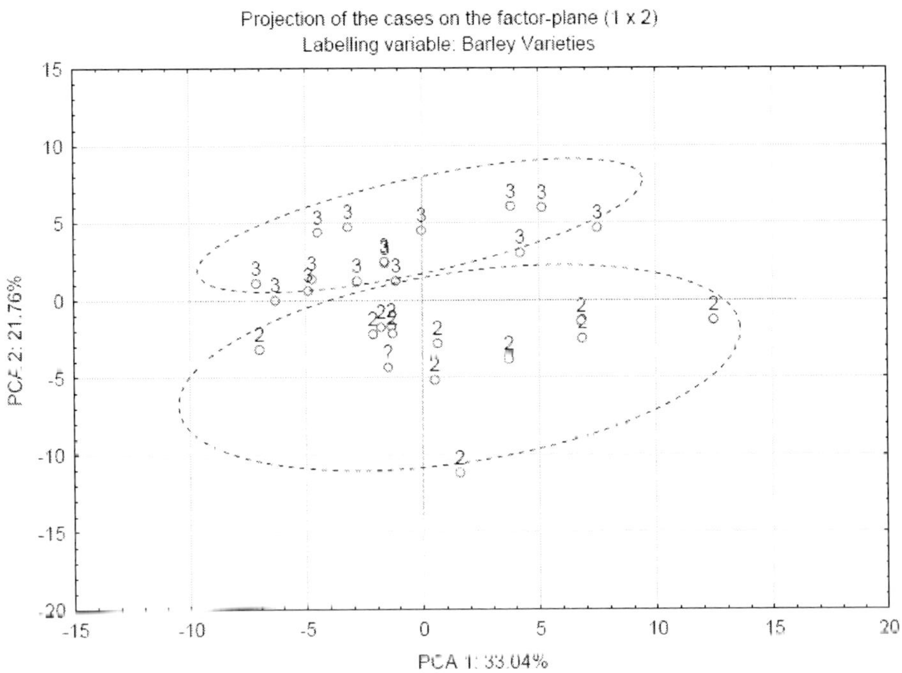

(f) Dolly (=3) vs. AC Metcalfe (=2): PCA Component 1 vs. PCA Component 2 (The 1st, 2nd and 3rd principal component explains 33.0, 21.8 and 15.0% of the variance).

Figure 8. Scatter plot of the 1st principal component vs. the 2nd principal component of PCA analysis of synchrotron-based protein FSD protein amide I spectrum (1710 to 1576 cm-1) obtained from various barley seed tissues at a cellular level (pixel size 10 μm×10 μm) [1= LP955; 2=AC Metcalfe; 3=Dolly;

4=Sisler; 5=Valier; 6=Harrington]: (a) Harrington (=6) vs. Varlier (=5) (The 1st, 2nd and 3rd principal component explains 36.1, 22.6 and 16.0% of the total variance); (b) Dolly (=3) vs. Varlier (=5) (The 1st, 2nd and 3rd principal component explains 37.1, 25.5 and 17.0% of the total variance); (c) Dolly (=3) vs. Harrington (=6) (The 1st, 2nd and 3rd principal component explains 41.6, 20.0 and 13.9% of the total variance); (d) Dolly (=3) vs. LP955 (=1) (The 1st, 2nd and 3rd principal component explains 36.7, 18.8 and 11.6% of the total variance); (e) Dolly (=3) vs. Sisler (=4) (The 1st, 2nd and 3rd principal component explains 41.1, 18.4 and 16.4% of the variance) and (f) Dolly (=3) vs. AC Metcalfe (=2) (The 1st, 2nd and 3rd principal component explains 33.0, 21.8 and 15.0% of the total variance).

All results demonstrate the potential of the ultraspatially resolved synchrotron–based technology (SFTIRM) to localize relatively "pure" protein in the seed tissues and reveal protein molecular structure in terms of protein secondary structure profile at a cellular level without destruction of the inherent structure of seed tissue.

Conclusion and Future Research Direction

Traditional wet chemical analysis can determine total plant/feed/seed protein composition and amino acid profile, but fails to detect protein inherent structure, molecular chemical make-up, and protein matrix due to destruction of plant/feed/seed tissues during the processing for chemical analysis. With the advanced synchrotron-base analytical technology (SFTIRM), we are able to "see" the plant/feed/seed tissue in a chemical sense. This analytical technology can be used to reveal protein inherent structure, protein distribution and intensity, protein molecular chemical make-up, and protein matrix within intact plant/feed/seed tissue at cellular and subcellular level. This technology show high potential which can be used a new research tool for dietary protein study. The future research will be focus on quantification of the relationship between protein inherent structures, molecular chemical make-up and protein matrix and protein availability and digestive behavior. It is believed that by using the advanced synchrotron technology (SFTIRM), it will make a significant step and an important contribution to plant, feed and seed molecular structural-chemical research.

Acknowledgments

National Synchrotron Light Source in Brookhaven National Laboratory (NSLS-BNL, New York, USA) is supported by the U.S. Department of Energy contract DE-AC02-98CH10886. Synchrotron Radiation Center (SRC, University of Wisconsin, Madison) is supported by the NSF under Award No. DMR-0084402. The author is grateful to Nebojsa Marinkovic, Lisa Miller, Wang Qi, Alexander Ignatov, and Jennifer Bohon at U10B and U2B (NSLS-BNL, New York) and Robert Julian at Port 031 (SRC, University of Wisconsin, Madison) for helpful data collection, John McKinnon, Colleen Christensen (CLS) and David Christensen for support and help discussion during the study, and Brian Rossnagel (CDC, University of Saskatchewan) and Vern Racz, and Philip Thacker (University of Saskatchewan) for supplying plant and seed samples. This research has been financially supported from Natural Sciences and Engineering Research Council of Canada (NSERC, Individual Discovery Grant) and Saskatchewan Agricultural Development Fund (ADF).

REFERENCES

AOAC, 1990. *Official Methods of Analysis* (15th Ed.). Association of Official Analytical Chemists, Arlington, VA.

Budevska, B.O. (2002). In *Handbook of Vibrational Spectroscopy*, vol. 5. Applications of Vibrational Spectroscopy in Life, Pharmaceutical and Natural Sciences, pp. 3720–3732. [Chalmers, JM. and Griffiths P.R., editors]. New York: John Wiley and Sons, Inc.

Carey, F. A. 1996. Organic Chemistry. 3rd ed. McGraw-Hill Companies, Inc.

Cytospec, 2004. *Software for infrared spectral imaging*. V. 1.1.01

Diem, M., Boydston-White, S., and Chiriboga, L. 1999. Infrared spectroscopy of cells and tissues: Shining light onto a novel subject. *Applied Spectroscopy*. 53: 148A-161A.

Dumas, P. 2003. Synchrotron IR Microspectroscopy: A Multidisciplinary Analytical Technique. The 6th Annual Synchrotron CLS Users' Meeting and Associated Synchrotron Workshops – WinXAS and Infrared. University of Saskatchewan, Canada. Nov 13-15, 2003.

Gibson, L.A., Bowman, J.G.P., Oberthur, L.E., Blake, T.K., 1994. Determination of genetic markers associated with ruminant digestion of barley. *Proc. Western Section*, American Society of Animal Science, V45.

Givens, D.I., Clark, P., Jacklin, D., Moss, A.R., Savery, C.R., 1993. Nutritional aspect of cereal, cereal grain by-products and cereal straw for ruminants. *HGCA Research Review* 24, 1-180. Home-grown Cereal Authority, Hamlyn House, Highgate Hill, London, England.

Holman Hoi-Ying N, Bjornstad KA, McNamara MP, Martin MC, McKinney WR and Blakely EA. 2002 Synchrotron infrared spectromicroscopy as a novel bioanalytical microprobe for individual living cells: cytotoxicity considerations. *J. Biomedical Optics*. 7, 1–10.

Jackson, M., and Mantsch, H. H. 2000. Infrared spectroscopy Ex Vivo Tissue Analysis. Pages 1-25 in R.A. Meyers, ed. *Encyclopedia of Analytical Chemistry*. John Wiley and Sons Ltd. Chichester, UK

Kauppinen, J. K. Moffatt, D. J. Mantsch, H. H., and Cameron, D. G. 1981. Fourier selfdeconvolution: a method for resolving intrinsically overlapped bands. *Applied Spectroscopy*. 35: 271-276.

Marinkovic, N.S., Huang, R., Bromberg, P., Sullivan, M., Toomey, J., Miller, L.M., Sperber, E., Moshe, S. Jones, K.W., Chouparova, E., Lappi, S., Franzen S., and Chance, M.R. 2002. Center for Synchrotron Biosciences' U2B beamline: an international resource for biological infrared spectroscopy. *J. Synchrotron Rad*. 9, 189-197.

Martin, M.C. 2002. Fourier-Tranform Infrared Spectroscopy. Available: http://infrared.als.lbl.gov/. Accessed Oct, 2002.

Miller LM, Carlson CS, Carr GL and Chance MR. 1998. A Method for Examining the Chemical Basis for Bone Disease: Synchrotron Infrared Microspectroscopy. *Cell Mol. Bio.* 44, 117-127.

Miller L. M. 2000. The impact of infrared synchrotron radiation on biology: past, present, and future. *Synchrotron Radiation News* 13, 31-37.

Miller, L. M. 2002. Infrared Microsperctroscopy and Imaging. Retrieved in October from http://nslsweb.nsls.bnl.gov/nsls/pubs/nslspubs/imaging0502/irxrayworkshopintroduction.ht

Nocek, J.E., Tamminga, S., 1991. Site of digestion of starch in the gastrointestinal tract of dairy cows and its effect on milk yield and composition. *J. Dairy Sci.*, 74, 3598-3629.

NRC. 1996. *Nutrient Requirement of Beef cattle* (7th Ed). National Research Council, Washington, D.C., National Academy Press.

NRC. 2001. *Nutrient requirement of dairy cattle* (7th Ed). National Research Council, Washington, D.C., National Academy Press.

Raab, T.K; Martin, M.C. Visualizing rhizosphere chemistry of legumes with mid-infrared synchrotron radiation. *Planta* 213, 881-887.

Seguchi, M., Takemoto, M., Mizutani, U., Ozawa, M., Nakamura C., and Matsumura, Y. 2004. Effects of secondary structures of heated egg white protein on the binding between prime starch and tailings fractions in fresh wheat flour. *Cereal Chem.* 81: 633-636.

Wetzel, D. L., Eilert, A. J., Pietrzak, L. N., Miller, S. S., and Sweat, J. A. 1998. Ultraspatially resolved synchrotron infrared microspectroscopy of plant tissue in situ. *Cell. Mol. Bio.* 44, 145-167.

Wetzel, D.L. 2001. When molecular Causes of wheat quality are known, molecular methods will supercede traditional methods. *Proc. Int'l Wheat Quality Conf.* II, Manhattan, Kansas, USA, May 2001, pp 1-20.

Wetzel, D.L., Srivarin, P., and Finney, J.R. 2003. Revealing protein infrared specrtral detail in a heterogeneous matrix dominated by starch. *Vibrational spectroscopy*. 31, 109-114.

Yu, P., McKinnon, J.J., Christensen, C.R., Christensen, D.A., Marinkovic, N.S. and Miller, L.M., 2003a. Chemical imaging of micro-structures of plant tissues within cellular dimension using synchrotron infrared microspectroscopy. *J. Agric. Food Chem.* 51, 6062-6067.

Yu P., Meier, J., Christensen, D.A. Rossnagel, B. and McKinnon, J.J. 2003b. Using the NRC-2001 model and the DVE/OEB system to evaluate nutritive values of Harrington (malting-type) and Valier (feed-type) barley for ruminants. *Anim. Feed Sci. Technol.* 107, 45-60.

Yu, P. 2004. Application of advanced synchrotron-based Fourier transform infrared microspectroscopy (SR-FTIR) to animal nutrition and feed science: a novel approach. *Bri. J. Nutri.* 92: 869-885.

Yu, P., Christensen, D.A., Christensen, C.R., Drew, M.D., Rossnagel, B.G. and McKinnon, J.J. 2004a. Use of synchrotron FTIR microspectroscopy to identify chemical differences in barley endosperm tissue in relation to rumen degradation characteristics. *Can. J. Anim. Sci.* 84, 523-527.

Yu, P., McKinnon, J.J., Christensen, C.R. and Christensen, D.A. 2004b. Using synchrotron transmission FTIR microspectroscopy as a rapid, direct and non-destructive analytical technique to reveal molecular microstructural-chemical features within tissue in grain barley. *J. Agric. Food Chem.* 52: 1484-1494.

Yu, P., McKinnon, J.J., Christensen, C.R. and Christensen, D.A. 2004c. Imaging molecular chemistry of pioneer corn. *J. Agric. Food Chem.* 52, 7345-7352.

Yu, P., McKinnon, J. J., Christensen, C. R., Christensen, D. A. 2004d. Using synchrotron-based FTIR microspectroscopy to reveal chemical features of feather protein secondary structure: comparison with other feed protein sources. *J. Agric. Food Chem.* 52, 7353-7361.

Yu, P. 2005a. Molecular chemistry imaging to reveal structural features of various plant feed tissues. *J. Structural Biol.* 150, 81-89.

Yu, P. 2005b. Application of cluster analysis (CLA) in feed chemical imaging to accurately reveal structural-chemical features of feeds within cellular dimension. *J. Agric. Food Chem.* 53, 2872-2880.

Yu, P. 2005c. Applications of cluster analysis (CLA) and principal component analysis (PCA) in feed structure and feed molecular chemistry research using synchrotron-based FTIR microspectroscopy. *J. Agric. Food Chem.* 53: 7115-7127.

Yu, P. 2005d. Multi-Component Peak Modeling of Protein Secondary Structures: Comparison of Gaussian with Lorentzian Analytical Method for Plant Feed and Seed Molecular Biology and Chemistry Research. *Applied Spectroscopy*. 59:1372-1380.

Yu, P., C.R. Christensen, D.A. Christensen, J.J. McKinnon. 2005a. Ultrastructural-chemical makeup of yellow- (*Brassica Rapa*) and brown-Seeded (*Brassica Napus*) canola within cellular dimensions, explored with synchrotron reflection FTIR microspectroscopy. *Can. J. Plant Sci.* 85, 533-541.

Yu, P, R. Wang and Y. Bai. 2005b. Reveal protein molecular structural-chemical differences between two types of winterfat (forage) seeds with physiological differences in low temperature tolerance using synchrotron-based FTIR microspectroscopy. *J. Agric. Food Chem.* 53: 9297-303.

Yu, P., McKinnon, J.J., Soita, H.W., Christensen, C.R. and Christensen, D.A. 2005c. Use of synchrotron-based FTIR microspectroscopy to determine protein secondary structures of raw and heat treated brown and golden flaxseeds: a novel approach. *Can. J. Anim. Sci.* 85, 437-448.

INDEX

B

C

D

H

J

K

L

M

N

O

T

PROTEIN RESEARCH PROGRESS